Studies in Infrastructure and Control

The book series aims to publish top-quality state-of-the-art textbooks, research monographs, edited volumes and selected conference proceedings related to infrastructure, innovation, control, and related fields. Additionally, established and emerging applications related to applied areas like smart cities, internet of things, machine learning, artificial intelligence, etc., are developed and utilized in an effort to demonstrate recent innovations in infrastructure and the possible implications of control theory therein. The study also includes areas like transportation infrastructure, building infrastructure management and seismic vibration control, and also spans a gamut of areas from renewable energy infrastructure like solar parks, wind farms, biomass power plants and related technologies, to the associated policies and related innovations and control methodologies involved.

Mubashir Hussain Wani · Faizal Hafiz ·
Akshya Kumar Swain

Digital Twins for Building Optimization: Enhancing Energy and Comfort

Springer

Mubashir Hussain Wani
Department of Electrical Engineering
Jacobs Engineering Inc.
Dubai, United Arab Emirates

Faizal Hafiz
SKEMA Business School
Université Côte d'Azur
Nice, France

Akshya Kumar Swain
Department of Electrical, Computer
and Software Engineering
The University of Auckland
Auckland, New Zealand

ISSN 2730-6453 ISSN 2730-6461 (electronic)
Studies in Infrastructure and Control
ISBN 978-3-032-23273-1 ISBN 978-3-032-23274-8 (eBook)
https://doi.org/10.1007/978-3-032-23274-8

This Springer imprint is published by the registered company Springer Nature Switzerland AG
The registered company address is: Gewerbestrasse 11, 6330 Cham, Switzerland

To my mother and father, for their unwavering faith, sacrifices, and quiet strength.

To my wife, Shahin, and son, Mahir, whose presence makes every moment cherishable.

Mubashir Hussain Wani

Preface

In an era where buildings are expected to operate as efficient, flexible, and intelligent components of modern energy systems, the role of Building Energy Management Systems (BEMS) has become more critical than ever. This book emerges from the convergence of academic research and professional engineering practice, with the aim of presenting a disciplined and application-driven perspective on intelligent control, optimization, and Digital Twin (DT) frameworks for Building Energy Systems (BES).

The foundations of this book are rooted in the author's doctoral (Ph.D.) research at the University of Auckland, which investigated the application of Computational Intelligence (CI) and control methodologies to Building Energy Management (BEM). That research explored how physically informed modeling, optimization, and learning-based techniques can be combined to improve energy efficiency and occupant comfort under realistic operational constraints. Rather than pursuing algorithmic novelty in isolation, the work emphasized system-level understanding, objective formulation, and the practical limitations inherent to building energy dynamics.

In parallel with academic research, the author's professional role as an Electrical Design Engineer in the Building Services sector has provided sustained exposure to real-world challenges associated with BEMS design, integration, and operation. Industrial experience spanning Building Management Systems, energy modeling, and the development of Building Digital Twins (BDTs) has reinforced a central insight reflected throughout this book: effective intelligence in buildings arises not from standalone algorithms, but from the careful integration of control engineering, physical modeling, and data-driven decision-making within deployable architectures.

This book therefore seeks to bridge the gap that often exists between theoretical research and practical implementation. Computational Intelligence (CI) is treated as an enabling layer—complementary to first-principles understanding and engineering judgment—rather than as a black-box solution. Emphasis is placed on interpretability, robustness, and scalability, with attention given to the realities of building operation, uncertainty, and multi-objective performance trade-offs. To maintain practical relevance and support reproducibility, many of the modeling, simulation, and optimization workflows presented in this book are demonstrated through Python™-based implementations. When scaled (adapted, modified or extended) suitably, these implementations can function as a flexible and transparent interface connecting building energy simulation engines, data-driven models, and control algorithms.

The material is organized into three coherent parts. Part I establishes the background and system context, connecting renewable energy integration, Demand-Side Management (DSM), control engineering, and Digital Twin (DT) concepts. Part II focuses on Computational Intelligence (CI) techniques for building energy optimization and control, emphasizing application-driven problem formulation and performance evaluation. Part III advances toward Digital Twin (DT)–enabled intelligent BEMS, where simulation models, real-time data, optimization, and control are integrated within closed-loop decision frameworks.

This book is intended for senior undergraduate students, graduate students, doctoral researchers, and practicing engineers working in electrical engineering, energy systems, smart buildings, and clean energy domains. It is not intended to provide definitive answers, but rather to equip the reader with the conceptual tools, engineering intuition, and methodological foundations required to engage critically with a rapidly evolving field.

Dubai, United Arab Emirates — Mubashir Hussain Wani
Nice, France — Faizal Hafiz
Auckland, New Zealand — Akshya Kumar Swain

Acknowledgements The completion of this book has been made possible through the support, guidance, and encouragement of many individuals and institutions, to whom I am sincerely grateful.

I would first like to acknowledge my academic mentors and collaborators, whose guidance and intellectual engagement have shaped both my doctoral (Ph.D.) research and the broader ideas developed in this book. In particular, I am grateful to my doctoral supervisor, Assoc. Prof. Akshya Swain, for his consistent support, insightful guidance, and encouragement throughout my research journey. I also thank Assoc. Prof. Abhishek Ukil for his valuable advice, constructive feedback, and continued support during my doctoral studies. I am further thankful to Dr. Faizal Hafiz for many enriching technical discussions and thoughtful comments which helped refine several research directions reflected in this work.

I gratefully acknowledge the support of Callaghan Innovation, New Zealand, for funding my doctoral research through the R&D Fellowship Grant. This support enabled the research activities that form a substantial foundation of the material presented in this book, including industry collaboration with Fisher and Paykel Technologies Ltd., New Zealand. I would like to express my appreciation to the team at Fisher and Paykel Technologies, particularly Maximiliano Ploder and Rolf Koole, for their mentorship, technical insight, and constructive engagement during the collaborative research.

I am thankful to the University of Auckland for providing an excellent research environment and access to critical infrastructure. In particular, I acknowledge the support of the NeSI High Performance Computing (HPC) facilities and the Nectar Research Cloud, which were instrumental for large-scale simulations and co-simulation studies involving EnergyPlus™ and MATLAB™, during my doctoral (Ph.D.) studies. I also extend my thanks to the relevant eServices teams for facilitating access to these resources.

Special thanks are due to Rachel Coxhead for her assistance in enabling access to the University of Auckland IT Services, Libraries, and Learning Services, which supported various stages of research and manuscript preparation. I am also grateful to the departmental administrative staff for their continued assistance and support.

Beyond academic and institutional support, I am deeply grateful to my parents for their unwavering encouragement, patience, and belief in my pursuits. I also wish to thank my wife for her constant support, understanding, and strength throughout this journey. Their support has been a continual source of motivation and balance.

Finally, I acknowledge with gratitude the professional environment and industry exposure gained through my work in Building Services Engineering (BSE), which has strongly influenced the practical orientation and applied perspective of this book.

January 2026 Dr. Mubashir Hussain Wani

Competing Interests The authors have no competing interests to declare that are relevant to the content of this manuscript.

Contents

About the Authors

Mubashir Hussain Wani is a Chartered (Electrical) Engineer and researcher focused on intelligent energy systems and cyber-physical infrastructures for buildings. He obtained his Ph.D. in Electrical and Computer Engineering from The University of Auckland, New Zealand, in 2024, where he was awarded the MBIE R&D Fellowship—NZ (through Callaghan Innovation) for industry-driven research on advanced control techniques for Building Energy Management Systems (BEMS).

His research centers on the application of Computational Intelligence (CI) in smart buildings and smart grids, integrating Digital Twins (DTs), Multi-Objective Optimization (MOO), and Machine Learning (ML) to enhance energy efficiency, system resilience, and occupant comfort. He earned an M.Sc. in Energy and Sustainability with Electrical Power Engineering from the University of Southampton in 2016, and a B.Eng. (Hons.) in Electronics and Communication Engineering in 2015.

His professional experience spans diverse R&D and engineering roles, including work as an R&D Fellow at Fisher and Paykel Technologies, serving as a Lab Supervisor for Embedded Systems at The University of Auckland, holding integration and automation engineering positions in India, and serving as a Sales Applications Engineer in power management and protection systems in New Zealand. He has also contributed to electrical design for ELV and ICT systems on large-scale smart and cognitive city developments in Saudi Arabia and the UAE.

Dr. Wani has published in leading IEEE conferences and peer-reviewed journals and is committed to connecting academic research with practical engineering solutions across infrastructure, energy, and intelligent system design.

Faizal Hafiz received the Ph.D. degree in Electrical Engineering from The University of Auckland, Auckland, New Zealand, in 2020.

He is currently a Postdoctoral Research Fellow with the Artificial Intelligence (AI) Institute, SKEMA Business School, Université Côte d'Azur, Sophia Antipolis, France. He was a New Zealand International Doctoral Scholar in Computational Intelligence (CI) and Control from 2016 to 2020. Previously, he served as a Research Assistant Professor at King Saud University, Riyadh, Saudi Arabia, from 2010 to

2016, and as a Utility Assistant Manager with Reliance Energy Ltd., Mumbai, India, from 2008 to 2010.

Dr. Hafiz has published over 40 research articles in leading journals and conferences. His research interests include Computational Intelligence (CI), Machine Learning (ML), Multi-Objective Optimization (MOO), and their applications in systems and control, healthcare, and business analytics.

Akshya Kumar Swain (Senior Member, IEEE) received the B.Sc. degree in Electrical Engineering, the M.E. degree in Electronics and Communication Engineering, and the Ph.D. degree in Control Engineering from The University of Sheffield, Sheffield, UK, in 1985, 1988, and 1996, respectively.

He has authored over 200 papers in international journals and conferences. His research interests include non-linear system identification and control, Machine Learning (ML), and big data analytics.

Dr. Swain serves as an Associate Editor of the *IEEE Sensors Journal* and is a member of the editorial boards of the *International Journal of Automation and Control* and the *International Journal of Sensors, Wireless Communications and Control*.

Acronyms

AHU	Air Handling Unit
ANL	Argonne National Laboratory
ANN	Artificial Neural Network
ASCII	American Standard Code for Information Interchange
BDT	Building Digital Twin
BEM	Building Energy Modeling
BEMS	Building Energy Management System
BMS	Building Management System
CI	Computational Intelligence
CPES	Cyber-Physical Energy System
CPP	Critical Peak Pricing
DM	Decision Maker
DR	Demand Response
DSM	Demand-Side Management
DT	Digital Twin
EA	Evolutionary Algorithm
EPW	EnergyPlus Weather File
EV	Electric Vehicle
HCU	Heating Coil Unit
HPC	High-Performance Computing
HVAC	Heating, Ventilation, and Air Conditioning
i-BEMS	Intelligent Building Energy Management System
IDD	Input Data Dictionary (EnergyPlus)
IDF	Input Data File (EnergyPlus)
KPI	Key Performance Indicator
LBNL	Ernest Orlando Lawrence Berkeley National Laboratory
MBC	Model-Based Control
MCDM	Multi-Criteria Decision Making
MFC	Model-Free Control

ML	Machine Learning
MOO	Multi-Objective Optimization
MPC	Model Predictive Control
NeSI	New Zealand eScience Infrastructure
NREL	National Renewable Energy Laboratory
ORNL	Oak Ridge National Laboratory
PNNL	Pacific Northwest National Laboratory
PV	Photovoltaics
QoS	Quality of Service
RL	Reinforcement Learning
ToU	Time-of-Use

Part I
Background

Chapter 1
Introduction

Abstract Buildings are rapidly transforming from passive energy consumers into intelligent, Cyber-Physical Systems (CPS) capable of sensing, learning, and Decision-Making. This chapter introduces the motivation, scope, and structure of the book, which is inspired by and extends the author's doctoral (PhD) research on Computational Intelligence (CI)-enabled Building Energy Management Systems (BEMSs). The book targets senior undergraduate students, graduate students, doctoral researchers, and practicing engineers working at the intersection of electrical engineering, clean energy systems, and intelligent control. Emphasis is placed on bridging theory with practice by combining physical modeling, data-driven intelligence, and real–world implementation considerations. The chapter outlines the vision of the book, its pedagogical objectives, and its organization into three coherent parts which progressively guide the reader from fundamentals to advanced Digital Twin (DT)-enabled optimization and control frameworks.

Keywords Book scope · Book structure · Cyber physical systems · Objectives

1.1 Motivation of Research

The global transition toward low-carbon, resilient, and intelligent energy systems has placed buildings at the center of the clean energy challenge. Buildings account for a substantial fraction of global electricity consumption and Greenhouse Gas (GHG) emissions, yet they also represent one of the most promising domains for efficiency improvements, demand flexibility, and intelligent energy coordination. Advances in sensing, communication, and computation have enabled unprecedented access to building operational data, creating new opportunities for intelligent Decision-Making across multiple time scales.

This book stems directly from the author's doctoral (PhD) research conducted in the School of Electrical, Computer & Software Engineering (ECSE) at the University of Auckland (2018–2023). The work explored how Computational Intelligence (CI), optimization, and control techniques can be applied to Building Energy Management Systems (BEMSs), with a particular emphasis on integrating physics-based

M. H. Wani et al., *Digital Twins for Building Optimization: Enhancing Energy and Comfort*, Studies in Infrastructure and Control,
https://doi.org/10.1007/978-3-032-23274-8_1

models and data-driven learning methods. The research demonstrated that meaningful energy and comfort improvements are not achieved through algorithms alone, but through the careful integration of physical understanding, control theory, and practical engineering constraints [1]. This realization forms the intellectual backbone of the present monograph.

Despite significant progress in Artificial Intelligence (AI) and Machine Learning (ML), their application in building energy systems often remains fragmented, overly abstract, or disconnected from real operational realities. Many existing texts either emphasize theoretical algorithmic development without sufficient physical grounding, or focus on building simulation tools without exploiting modern Computational Intelligence (CI) techniques. This book is motivated by the need to close this gap by presenting a unified, engineering-driven perspective which treats buildings as Cyber-Physical Energy Systems (CPESs) rather than isolated simulation artifacts.

At its core, this monograph contends that *intelligent buildings are best understood as controlled dynamical systems operating under uncertainty, where Computational Intelligence (CI) serves as an enabler rather than a replacement for physical modeling and control principles*. By embedding learning, optimization, and Decision-Making within physically informed frameworks, buildings can evolve from reactive energy consumers into proactive, adaptive participants in future energy systems.

1.2 Objectives and Scope

The primary objective of this book is to provide a structured, technically rigorous, yet accessible treatment of Computational Intelligence (CI) applications in Building Energy Management (BEM) and optimization. The book is intended for a broad audience which includes senior undergraduate students, master's students, doctoral researchers, and practicing engineers in electrical engineering, energy systems, and related disciplines.

Specifically, the book aims to

- establish strong foundational understanding of building energy systems, control architectures, and performance metrics;
- introduce Computational Intelligence methods—including Evolutionary Algorithms (EAs), swarm intelligence, and learning-based approaches—from an application-driven perspective;
- demonstrate how physical modeling and data-driven intelligence can be systematically integrated within BEMS;
- present Digital Twin (DT) concepts as an evolution of conventional simulation models, emphasizing real-time interaction, adaptation, and decision support;
- bridge academic research and industrial practice through implementation-oriented discussions, case studies, and architectural frameworks.

Rather than positioning Computational Intelligence (CI) as a `black-box` solution, the scope of this book emphasizes interpretability, robustness, and engineering relevance. Mathematical formulations are introduced where necessary, but always in service of practical understanding and real-world applicability. Throughout the book, emphasis is placed on scalable architectures, control hierarchies, and optimization frameworks which can be scaled and eventually deployed in realistic building and campus-scale environments.

1.3 Structure of the Book

The book is organized into three coherent parts, each building progressively on the concepts introduced in the previous sections.

Part I establishes the foundational principles of building energy systems, Building Energy Management Systems (BEMSs), and the role of modeling and control in intelligent operation. This part introduces the physical and operational context within which Computational Intelligence (CI) methods are later applied, ensuring that readers develop a strong system-level intuition.

Part II focuses on Computational Intelligence (CI) techniques for building energy optimization and control. Evolutionary computation, swarm intelligence, and learning-based methods are introduced from an application-centric perspective, with emphasis on objective formulation, constraint handling, and performance evaluation within building energy contexts.

Part III advances toward Digital Twin (DT)-enabled intelligent BEMS, where simulation models, real-time data, optimization, and control are integrated into closed-loop decision frameworks. Topics such as surrogate modeling, consensus-based coordination, and scalable architectures are explored, highlighting the transition from offline analysis to online intelligent operation.

To visually summarize the conceptual flow of the book and its *three-part* structure, a pictorial overview is recommended, see Fig. 1.1, at the beginning of the monograph. Such a visual abstract helps the reader to focus and reinforces the narrative progression from fundamentals to advanced intelligent systems.

Together, these three parts reflect the central philosophy of the book: intelligent building energy systems emerge not from isolated algorithms, but from the disciplined integration of physical insight, Computational Intelligence (CI), and practical engineering design. The chapters which follow invite the reader to engage with this integration and to view buildings as active, intelligent participants in the future energy ecosystem.

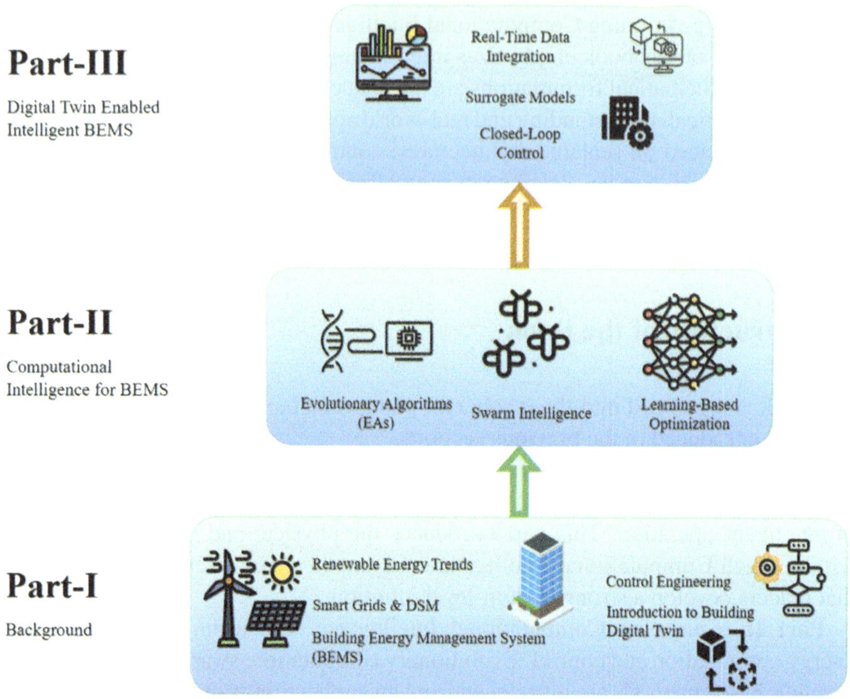

Fig. 1.1 Conceptual structure of the book and progression of themes across the three parts

Reference

1. Wani MH (2022) Intelligent control strategies for efficient building energy management systems (BEMS). Department of Electrical and Electronic Engineering, University of Auckland, PhD thesis

Chapter 2
Renewable Energy, Smart Grids, and Demand-Side Management (DSM)

Abstract This chapter explores the transforming energy landscape with a particular emphasis on `renewable energy`, `smart grids`, and `Demand-Side Management (DSM)`. It begins by outlining global trends in renewable-energy deployment—highlighting the rapid expansion of solar and wind power—while acknowledging persistent challenges such as fossil-fuel subsidies, intermittency of renewable resources, and infrastructural constraints. The discussion then examines the structural and operational barriers which hinder large-scale renewable integration and underscores the role of energy efficiency as an immediate pathway toward sustainability. Subsequently, the chapter introduces DSM as a strategic framework for optimizing energy use on the demand side, classifying its key approaches—including Demand Response (DR) and intelligent load management. Recent research on decentralized and transactive DSM architectures is analyzed to illustrate how distributed intelligence and market-based coordination can enhance flexibility and reliability. The chapter concludes by highlighting the role of Building Energy Management Systems (BEMS) as the cyber-physical core of DSM, responsible for forecasting, optimization, and real-time control. By embedding predictive intelligence and interoperability, BEMS establish the conceptual and technological bridge to Building Digital Twins, which form the focus of the next chapter.

Keywords Demand Side Management · Demand response · Renewable energy · Smart grids

2.1 Renewable Energy—The Current Trends

The global energy sector stands at a decisive inflection point. Over the past two decades, renewable energy technologies have advanced from niche research interests to integral components of national power strategies. Solar Photovoltaic (PV) and wind energy have witnessed exponential growth, supported by falling technology costs, policy incentives, and international climate commitments. According to recent assessments by REN21 [1] and IRENA [2], renewables now account for more than one-third of global installed power capacity—a remarkable achievement considering

M. H. Wani et al., *Digital Twins for Building Optimization: Enhancing Energy and Comfort*, Studies in Infrastructure and Control,
https://doi.org/10.1007/978-3-032-23274-8_2

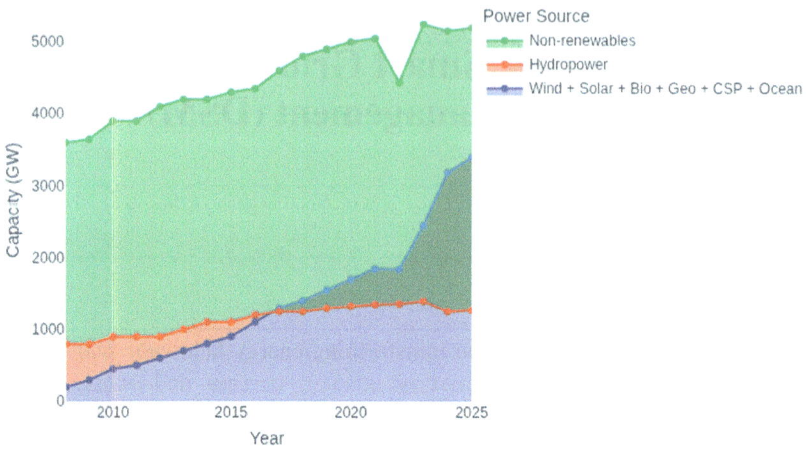

Fig. 2.1 Global power-generating capacity by source [1]. Non-renewables have dominated between 2008 and 2015, but renewables (solar, wind, and hydro) have grown steadily, especially after 2015, illustrating the accelerating global energy transition

their marginal contribution just a decade ago. In fact, IRENA reports that global renewable energy capacity experienced a record annual growth rate of 15.1% in 2024, expanding by 585 gigawatts and reaching a total of 4,448 GW [2]. These figures underscore the accelerating pace of the energy transition and the growing role of renewables in shaping future power systems.

Yet, this apparent progress conceals a slower and more complex reality. Despite rising installations, fossil fuels continue to dominate the world's primary energy supply, see Fig. 2.1, driven by legacy infrastructure, existing subsidies, and the intermittency of renewable resources. In 2022, global fossil-fuel subsidies were estimated at around $7 trillion USD, representing nearly 7% of global GDP [3]. These subsidies include both explicit consumption support and implicit costs linked to environmental and health externalities, revealing how deeply embedded conventional energy sources remain within national economies [4, 5]. Thus, despite the rapid increase in renewable energy deployment, significant investments still flow into carbon-heavy energy routes. This suggests that the shift toward a completely renewable and low-carbon energy economy will be a gradual process, likely taking multiple decades as power systems, markets, and technologies slowly evolve.

During this prolonged energy transition, enhancing efficiency and optimizing demand-side performance are just as critical as expanding renewable energy supply. Reflecting this evolving perspective, recent research underscores that effective decarbonization depends equally on intelligent Demand-Side Management (DSM) and the continued growth of renewable capacity. In this context, `buildings`—responsible for nearly 40% of global electricity consumption—emerge as a pivotal focus, offering the most immediate and scalable opportunity for impact through intelligent DSM [6].

Buildings as Catalysts for Intelligent Energy Transition

Accounting for nearly two-fifths of global electricity use, buildings represent one of the most immediate arenas for realizing measurable decarbonization. Through intelligent Demand-Side Management (DSM), they can act as dynamic energy nodes—adjusting consumption patterns, interacting with distributed renewable sources, and participating in grid-supportive Demand Response (DR) programs. Beyond passive efficiency improvements, emerging Building Energy Management Systems (BEMS) equipped with artificial intelligence, predictive control, and digital-twin capabilities can transform buildings into active, learning entities that contribute to system-wide stability and sustainability.

While managing demand efficiently is one pillar of a sustainable energy future, strengthening and diversifying the renewable supply side forms the other—ensuring that generation portfolios remain resilient to variability and geographic constraints. For instance, Lavidas et al. [7] emphasize that achieving 100 % renewable systems requires not only expanding solar and wind generation but also diversifying toward marine renewables such as wave and tidal energy. Using the extended PyPSA-Eur-MREL model with high-resolution climate data, the authors demonstrate that integrating marine sources can significantly reduce storage needs and system costs while improving demand matching and lowering curtailments. Their findings underline that resource diversification and accurate system modeling are essential for a balanced energy transition, aligning with the broader view that renewable expansion must advance alongside intelligent demand-side measures.

Mehmood et al. [8] present a comprehensive review of how Artificial Intelligence (AI) and Big Data (BD) are reshaping energy efficiency in buildings. The authors outline how machine learning, deep learning, and data-driven optimization are enabling energy-efficient building design, predictive operation, and comfort-oriented control, thereby transforming buildings into active participants in sustainable energy systems. Through examples such as data-driven predictive control, building energy dashboards, and large-scale calibration tools, the paper demonstrates that integrating AI and BD can substantially enhance both efficiency and occupant comfort while reducing operational costs and emissions. This perspective complements the broader narrative of this chapter—that digital intelligence within the built environment is indispensable to achieving effective decarbonization alongside renewable expansion.

Perr-Sauer et al. [9] explore how Explainable AI (XAI) enhances transparency and trust in renewable energy. They note that while traditional AI excels in forecasting, its "black-box" nature limits critical decision interpretability. By using XAI techniques like feature attribution and model visualization, stakeholders can better understand outputs related to solar irradiance and wind forecasting. This interpretability is crucial for responsible AI deployment, supporting the digitalization of demand-side and building-energy management.

Jagannath and Sharma [10] extend the discussion on intelligent Demand-Side Management by demonstrating how coordinated operation of electric-vehicle charging and renewable-energy integration can strengthen grid flexibility. Using mixed-integer optimization and simulation under high-penetration solar and wind scenarios, their work shows that EV load shifting and Vehicle-to-Grid (V2G) participation can mitigate renewable intermittency and reduce peak-load stress without significant investment in additional storage. The study reinforces that effective DSM, supported by data-driven scheduling and electrified transport, will be essential in ensuring stability during the renewable transition. Similarly, Wang and Trutnevyte [11] analyze the complementarity of electric-vehicle and heat-pump flexibility in a fully renewable Swiss electricity system for 2050. Their high-resolution EXPANSE model reveals that synchronized flexibility from these technologies can reduce national battery-storage needs by up to 95 % and improve renewable integration, particularly when coupled with Vehicle-to-Grid (V2G) strategies. However, the study also finds that such synergies depend strongly on policy and interconnection conditions, emphasizing that DSM must evolve in tandem with sector coupling and smart-grid infrastructure.

Collectively, these studies illustrate that the path toward a fully decarbonized energy system is not defined solely by the pace of renewable expansion but by the integration of intelligence, flexibility, and coordination across all layers of the energy network. Yet, despite the technological and analytical progress, large-scale renewable adoption continues to face structural, economic, and operational barriers. These enduring challenges—ranging from intermittency and grid constraints to policy and market design limitations—are discussed in the following section.

2.2 What is Holding Renewable Energy Back?

While the momentum toward renewable energy expansion appears irreversible, the transition remains riddled with structural, economic, and operational constraints which challenge its scalability. In addition to the concerns discussed in Sect. 2.1—including fossil-fuel subsidies, intermittency of renewable resources, and market inertia—the broader ecosystem of renewable energy faces persistent `bottlenecks` in its supply chains, regulatory frameworks, and technological interfaces. Recent literature emphasizes that addressing these barriers demands both systemic resilience and technical innovation across the entire *energy value chain*.

For instance, Labaran and Masood [12] highlight that sustainability and logistics barriers—particularly import dependency, inadequate recycling, and weak institutional frameworks—continue to undermine renewable project deployment in emerging markets. Their work underscores that strengthening the renewable energy supply chain through *green supply chain management* and digital transparency tools such as IoT and blockchain is critical to ensuring scalability and resilience.

At the technological frontier, Singh et al. [13] examine the persistent challenges of coupling renewable sources with non-isolated DC–DC converter topologies. The study identifies converter voltage stress, power-density limitations, and efficiency trade-offs as key bottlenecks in achieving seamless integration of solar and wind systems within hybrid microgrids. The authors assert that enhanced control strategies, high-frequency switching devices, and adaptive converter architectures are essential for bridging the reliability gap between variable renewable outputs and stable DC-bus operation. Such innovations form the electrical backbone of future `smart grids`, where both hardware and control intelligence determine overall system performance.

Grid-level challenges further compound these issues. Liang in [14] emphasizes that large-scale `renewable integration` introduces significant fluctuations in voltage and frequency, complicating grid stability and dispatch reliability. Their analysis of grid-connected systems demonstrates how weak interconnection networks and limited reactive power control lead to transient oscillations and power-quality degradation issues still relevant in contemporary high-penetration grids. Similarly, Elavarasan et al. [15] highlight that `grid integration` barriers persist even in nations with robust renewable capacity additions, attributing them to inadequate grid codes, transmission bottlenecks, and the absence of synchronized forecasting systems. Both studies converge on the conclusion that achieving higher renewable penetration necessitates dynamic grid planning, inverter-based control, and enhanced energy storage participation to maintain operational resilience.

From a system-planning perspective, Koltsaklis [16] presents an optimization-based framework for generation expansion in pursuit of 100% renewable systems. Their findings show that such a transition demands extensive spatial coordination of variable resources, large-scale storage deployment, and strong interregional interconnections to balance temporal mismatches between supply and demand. The study also cautions that without integrated policy mechanisms, the pursuit of full decarbonization may inadvertently increase overall system costs and carbon leakages through upstream supply-chain activities (e.g., sourcing raw materials, supplier negotiations, and production scheduling).

Complementing these insights, Bisht and Pandey [17] review the evolving role of `biofuels` as part of diversified renewable portfolios. While biofuels offer potential in addressing intermittency and transport-sector decarbonization, the authors note enduring challenges in feedstock availability, water demand, land-use competition, and the economic viability of third and fourth-generation biofuels. They advocate hybrid *biochemical–thermochemical routes* (e.g., combining anaerobic digestion with gasification or pyrolysis) integrated with *circular-economy* principles (such as waste-to-fuel valorization and nutrient recycling) and digital monitoring to improve efficiency and lifecycle sustainability.

Finally, Zhao et al. [18] contend that surmounting these multi-dimensional barriers requires a paradigm shift toward *cyber-physical integration*. Their editorial on `Digital-Twin methodologies` outlines how virtual replicas of renewable-dominant systems, synchronized through real-time data streams, can enhance observability, forecasting, and system-level coordination. Digital Twins equipped with AI-driven optimization can simulate contingencies, predict maintenance needs, and optimize renewable dispatch, thereby reinforcing the operational intelligence of future power systems.

Collectively, these studies reveal that the transition to renewable energy is not confined simply by capacity expansion but by the maturity of the technological, infrastructure and institutional components of the surrounding ecosystem. Overcoming these intertwined limitations necessitates an equally intelligent response from the `demand side`—through mechanisms which reshape consumption, enable flexibility, and integrate end-user systems within smart-grid architectures.

2.2.1 DSM as the Bridge: Decentralized Flexibility, Market Coordination, and Building-Centric Control

Before formally defining DSM (Sect. 2.3), *it is instructive to synthesize recent contributions that clarify how demand-side intelligence—particularly in decentralized contexts—can alleviate the integration bottlenecks outlined earlier.*

At the `building level`, early Demand-Side Management (DSM) studies in Nearly Zero-Energy Buildings (NZEBs) showed that treating the entire building as a controllable "*prosumer*"—both producing and consuming energy—can enable coordinated load shaping. This includes using the building's thermal inertia as virtual energy storage, reducing peak loads, and increasing self-consumption of solar power, all without needing centralized grid control [19]. In the proposed fully electric microgrid setup—combining heat pumps, solar panels, and building automation—aggregated load control helped ease stress on local feeders and encouraged user participation in Demand Response (DR) programs. These results demonstrate that local control can provide meaningful flexibility, even when upgrades to the wider grid are delayed (e.g., when utility companies postpone improvements to *transformers* or *substations* needed to support growing local energy demand).

When moving from single buildings to `networks of multiple energy users`, game-theoretic DSM approaches help model how prosumers and energy suppliers interact strategically. For example, Chouikhi et al. [20] proposed a scheduling game where multiple leaders and followers coordinate energy use. This method lowers overall energy costs and peak-to-average load ratios, while boosting the use of renewable energy. It supports decentralized decision-making and ensures that participants are motivated to cooperate, showing that even self-interested users can contribute to grid stability—without needing a central authority.

At the `district or community level`, Transactive Energy (TE) market models help coordinate energy use by linking local consumption with renewable generation through real-time pricing. Chandra et al. [21] introduced a multi-agent system where smart building controllers—called Energy Management Demand Agents (EMDAs)—schedule appliances based on changing electricity prices and submit bids to a local energy market. This market uses a clearing process (known as "*Walrasian*" clearing) to balance supply and demand, which helps reduce peak loads and lower energy costs—achieving average bill savings of 11.6%. The system works well both when connected to the main grid and in isolated community setups, showing how decentralized control can be both flexible and scalable.

At the `household and microgrid level`, combining load shifting with Peer-to-Peer (P2P) energy exchange has shown strong results. Barkouki et al. [22] demonstrated that when homes coordinate appliance use and share energy locally, they can improve efficiency, reduce grid imports, and increase self-consumption by up to 23%, with energy cost savings nearing 59%. This shows that DSM focused on individual prosumers, supported by local energy markets, can reshape demand patterns and ease pressure on distribution networks—without needing centralized control.

`Electric mobility` adds another layer to decentralized DSM. Sánchez-Loor and Ayala-Chauvin [23] studied Electric Vehicle (EV) charging in Quito, Ecuador using a large-scale data model. They analyzed over two million charging sessions and found that if EVs charge at the same time under fixed tariffs, peak demand could rise by 32%. However, using smart, randomized charging strategies could reduce that peak by about 18±2.5%, with a forecasting error below 5%. Their work shows how data-driven DSM can guide policy and infrastructure planning in fast-growing urban areas.

`Energy storage` also plays a key role in strengthening decentralized systems. Benavides et al. [24] developed a predictive storage management system using Redox-Flow Batteries (VRFBs) and machine learning to forecast demand. This setup reduces battery wear, improves short-term energy availability, and cuts down on wasted renewable energy. It proves that accurate local forecasting combined with smart storage can handle variability without needing backup from the main grid.

Bringing all these ideas together, decentralized DSM offers two major advantages over traditional, centralized approaches:

- **Local Control**—Buildings and homes can quickly adjust appliance schedules, setpoints, and storage use to shape their energy demand in real time [19, 22].
- **Market-Based Coordination**—Systems like transactive energy markets and game-theoretic models help align individual decisions with broader grid goals, without requiring heavy central oversight [20, 21].

Thus, when grid connections are strong, these methods can help reduce peaks and improve efficiency. Whereas, when connections are weak or expensive, they allow buildings and communities to operate flexibly and semi-independently. This makes DSM—especially building-focused DSM—a key operational link between today's infrastructure challenges and the smart, renewable-powered systems of the future.

2.2.2 Emerging Approaches and Research Direction in DSM

Some interesting approaches have been presented in recent studies which advance the technical and operational scope of Demand Side Management (DSM) across different scales. For instance, Roy and Das [25] introduce the *Class Topper Optimizer*, a metaheuristic designed for cost-efficient smart-grid operation under renewable uncertainty. Their results show that adaptive optimization can outperform traditional Particle Swarm Optimization (PSO) and Genetic Algorithm (GA) methods in maintaining cost stability and convergence under stochastic generation, highlighting the growing influence of intelligent solvers in DSM.

Bakare et al. [26] provide a comprehensive overview of DSM evolution toward smart grids, identifying key barriers such as interoperability, cybersecurity, and market readiness, while outlining strategies to scale DSM from pilot implementations to citywide deployment. Their synthesis reinforces DSM's dual identity as both a technical and policy mechanism within the broader smart-grid framework.

At the community level, Rollo et al. [27] examine various technological configurations of DSM in renewable-energy communities, demonstrating that hybrid PV-battery systems with coordinated load shifting achieve the highest self-consumption and minimize grid dependency. Similarly, Ouassaid et al. [28] propose a hierarchical optimization framework for Demand Response within smart microgrids, showing that coordinated scheduling of storage and flexible loads mitigates renewable intermittency and improves local system resilience.

From a building-level standpoint, Han and Yu [29] quantify the impact of battery-storage capacity on building energy flexibility within a PV-ESS-GSHP setup. Their findings reveal diminishing flexibility returns beyond certain storage thresholds, underscoring the importance of intelligent control over capacity expansion alone. Complementarily, Aboutalebi et al. [30] investigate cost-comfort trade-offs in implicit DSM for fully electrified Nordic households, illustrating how `adaptive control` can balance occupant comfort with operational economy.

Collectively, these studies demonstrate that *DSM research is converging toward multi-objective, comfort-aware, and optimization-driven paradigms—cementing its role as the operational core of decentralized, renewable-ready energy systems.*

Perspective: The AI-Energy Paradox and the Future of Renewable Power

The accelerating adoption of Artificial Intelligence (AI) is redefining the global energy landscape. While AI is widely recognized as an enabler of energy efficiency—driving smarter grids, predictive control, and intelligent building management—it simultaneously fuels an unprecedented surge in computational energy demand. Modern data centers, the physical backbone of AI, already account for nearly 2% of global electricity use and are projected to exceed 4% by 2030 [31]. Each new generation of large language models and high-density computing clusters amplifies this demand, challenging the very efficiency goals which AI seeks to advance.

This emerging "paradox" highlights a crucial inflection point: AI is both a *tool for* and a *driver of* energy consumption. On one hand, AI-enabled optimization in buildings, grids, and industry accelerates decarbonization through predictive analytics, fault detection, and dynamic control. On the other, the exponential growth of AI training and inference workloads—especially in data centers and edge devices—introduces massive and continuous baseload requirements. Unlike intermittent renewable generation, data-center operations demand constant, high-quality power with minimal downtime.

In this context, nations endowed with abundant renewable resources—particularly those capable of producing stable, low-cost baseload power through hydro, geothermal, or hybrid solar-wind configurations—are positioned to become strategic hubs for AI infrastructure. For example, Iceland, Canada, and the Nordic countries already attract hyperscale data-center investments by coupling clean energy portfolios with natural cooling advantages [32]. As the digital economy expands, such countries could play an outsized role in supporting the AI revolution sustainably, provided that renewable integration keeps pace with demand growth.

Ultimately, reconciling this paradox requires a symbiotic approach—deploying AI not only as an efficiency enabler but also as an active participant in its own sustainability loop. Integrating AI-based energy forecasting, thermal management, and real-time demand response within data centers can transform them from static energy consumers into flexible, grid-supportive assets. When powered by renewables and managed intelligently, these digital infrastructures can exemplify the very balance between innovation and responsibility which defines the next stage of the energy transition.

2.3 Buildings and Demand-Side Management (DSM)

Having reviewed the broader literature on DSM frameworks and their decentralized architectures, this section narrows the focus to buildings—the most tangible and controllable entities in the demand-side landscape.

The transition to a sustainable energy future cannot rely solely on expanding renewable generation. While technologies such as solar PV, wind, and energy storage are transforming the supply side, the demand side—particularly buildings—represents the most immediate and scalable frontier for system-wide optimisation.

Buildings account for nearly 40% of global electricity consumption, with a significant portion attributed to flexible loads such as HVAC, lighting, and equipment. As illustrated in Fig. 2.2, HVAC systems alone typically consume around 50% of electricity in commercial buildings, followed by lighting (33%), electric equipment (10%), and other uses (7%). These figures underscore the active role buildings can play in the energy ecosystem—not as passive consumers, but as dynamic participants.

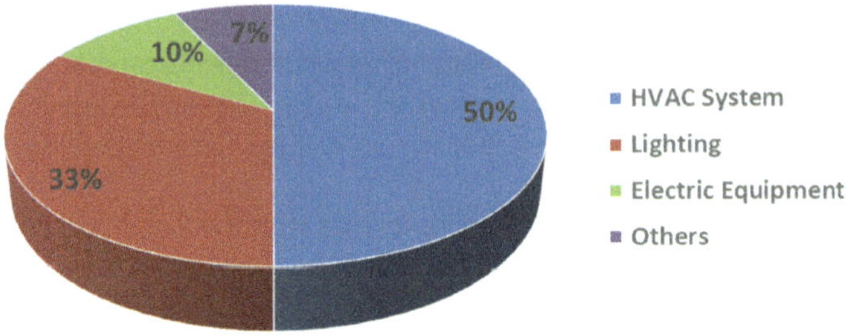

Fig. 2.2 Energy consumption proportion of building end uses in a typical *commercial* building [33]

Given that HVAC and lighting systems are inherently responsive to scheduling, set-point adjustments, and load shifting, buildings are ideally suited for integration into Demand-Side Management (DSM) frameworks. *By leveraging this flexibility, energy demand can be reshaped to better align with variable renewable supply*—enhancing grid stability, lowering energy costs, and improving overall system efficiency.

2.3.1 What is DSM?—A Concise Definition

Having reviewed, in Sect. 2.2, *how decentralized DSM and DR frameworks mitigate renewable integration challenges, it becomes necessary to formalize these concepts. The following sections define DSM and DR, clarifying their operational principles and their centrality to intelligent, building-centric energy systems.*

Demand-Side Management (DSM) is the coordinated application of strategies which *shape, shift, reduce, or schedule* end-use demand to improve efficiency, reliability, and economics. DSM actions generally fall into three complementary "*tiers*":

i. **Load Shaping**: adjusting consumption profiles through peak shaving, valley filling,[1] or load shifting.
ii. **Energy Efficiency**: permanent demand reduction via high-performance envelopes, equipment, and control.
iii. **Flexibility Orchestration**: dynamic coordination of distributed assets—PV, storage, EVs, and smart loads—to follow renewable generation.

[1] *Valley shifting* refers to the strategic increase of electricity consumption during periods of low demand (typically off-peak hours), helping to balance the load curve and utilize *base-load* or *renewable* generation more efficiently.

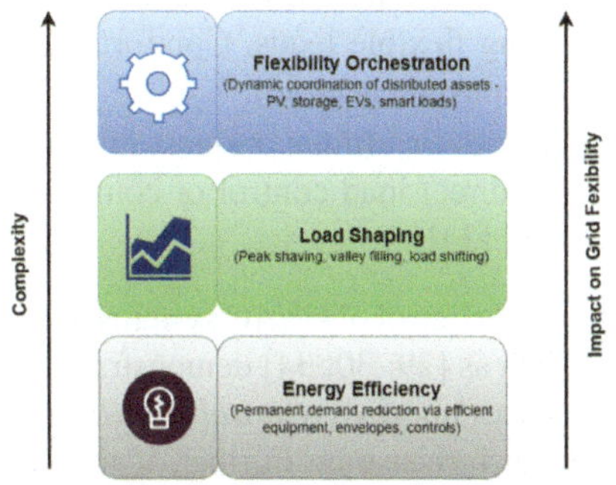

(a) Hierarchy of Demand-Side Management (DSM) measures showing progressive complexity and grid impact.

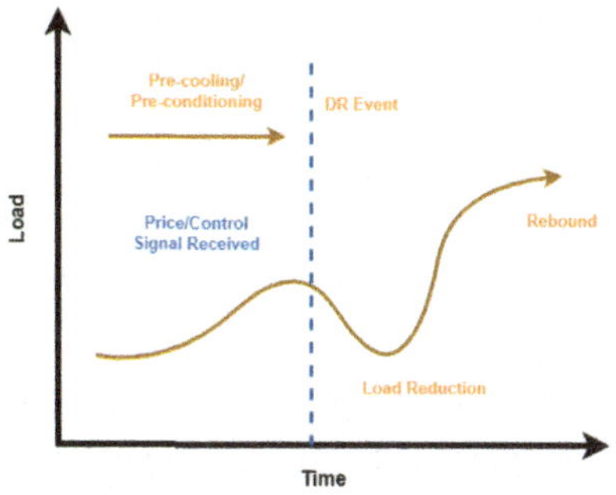

(b) Demand Response (DR) timeline illustrating pre-conditioning, load reduction, and rebound following a control signal.

Fig. 2.3 Visual illustration of DSM and DR concepts. **a** DSM hierarchy from efficiency to flexibility orchestration. **b** Indicative DR event timeline showing load variation in response to a control signal

DSM can be deployed through centralized (utility-led) programs or decentralized (prosumer/market-driven) control schemes, as demonstrated in NZEB microgrids [19], game-theoretic scheduling among smart buildings [20], and transactive markets [21].

As shown in Fig. 2.3, panel Fig. 2.3a depicts the hierarchical nature of DSM—progressing from long-term, efficiency-oriented measures to increasingly sophisticated strategies which dynamically coordinate distributed assets such as PV, batteries, and EVs. Each tier not only enhances system performance but also increases the level of control complexity and contribution to grid flexibility.

Panel Fig. 2.3b illustrates the operational side of this framework through a typical (indicative) Demand Response (DR) timeline. When a price or control signal is received, buildings or microgrids adjust their load profile by pre-conditioning (e.g., pre-cooling or pre-heating), reducing consumption during the event, and gradually recovering afterward. This real-time responsiveness exemplifies how DSM principles translate into actionable flexibility which supports renewable integration and grid stability.

2.3.2 *What is Demand Response (DR)? How it Relates with DSM?*

Demand Response (DR) is the *operational arm* of DSM which "*modulates*" demand in response to external signals (prices, incentives, grid constraints). Two widely considered categories are

- *Price/Tariff-Based DR* (Implicit): End-users react voluntarily to Time of Use (ToU) or Critical Peak Pricing (CPP) tariffs by shifting flexible loads. Comfort-aware controllers (algorithms) trade/balance cost and convenience [30].
- *Incentive/Event-Based DR* (Explicit): Aggregators or utilities request dispatchable load changes (reduction/shift), often with direct load control or committed flexibility; EV and storage coordination fits here [34].

In buildings, DR is implemented through flexible control of HVAC systems, lighting, EV charging, and energy storage. Studies such as [29, 30, 34] demonstrate that DR programs not only reduce *peak demand* but also improve *renewable utilization* when integrated with predictive control and local forecasting. Further, if the system(s) is integrated with *smart grid*, it could even sell energy back to the grid during peak times—another DR strategy. A good amount of information on DR for smart buildings is provided in Chap. 5 of [35], which discusses both implicit and explicit DR schemes, smart appliance coordination, and integration with smart grids.

Example: PV + Battery for Load Shifting (DSM) and Peak Support (DR)

Scenario: Midday PV surplus charges a battery (DSM: valley filling). During evening peak, the BEMS discharges the battery to meet loads (DSM: peak shaving) and, if an event signal arrives, reduces grid import or exports to the local market (DR).
Effect: Higher self-consumption, lower bills, reduced feeder stress, and availability of dispatchable flexibility for local markets or aggregators [21, 22]. This operational sequence is illustrated in Fig. 2.4, where building clusters coordinate energy flows and DR signals through a Central BEMS, enabling both local and community-scale energy balancing.

As DSM and DR mechanisms mature, their effectiveness increasingly depends on how intelligently loads can respond within real-time control environments. The growing integration of sensing, communication, and computation within buildings transforms traditional electrical loads into smart, context-aware entities. These "*intelligent energy systems*" form the adaptive layer of DSM—where decision-making, prediction, and learning take precedence over static control logic, paving the way for responsive and self-optimizing building operations.

2.3.3 Intelligent Energy Systems and Smart Loads

"Smart" loads expose controllable degrees of freedom (power, timing, set-points) and are orchestrated by a *BEMS equipped with forecasting, optimization and learning*. Examples include

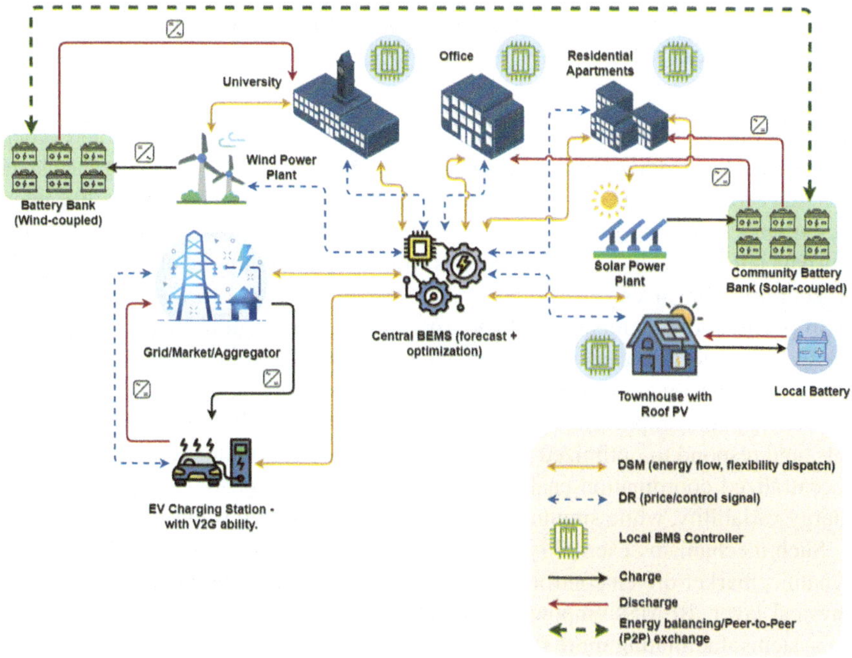

Fig. 2.4 Expanded illustration of decentralized community-level Demand-Side Management (DSM) and Demand Response (DR) integration. During off-peak hours, surplus generation from renewable sources (solar and wind) charges local and community batteries (DSM—valley filling). At evening peaks, stored energy is discharged to supply building and EV loads (DSM—peak shaving). When a grid or market signal is received, the Central BEMS coordinates flexibility dispatch and load curtailment across connected buildings (DR event)

- EV smart charging integrated with home EMS [34],
- comfort-aware appliance scheduling [30],
- coordinated PV-battery operation in energy communities [27].

As shown by [29], beyond a certain threshold, it is not the hardware—such as battery storage capacity—but rather the intelligence embedded in software (e.g., predictive control algorithms) which becomes the primary driver of energy flexibility.

Further, the intelligence of smart loads alone is insufficient without a unifying framework to sense, decide, and coordinate actions across diverse building systems. This is where Building Energy Management Systems (BEMS) become indispensable. Acting as the "brain" of DSM, the BEMS integrates data from sensors, forecasts, and occupant behaviors, executing optimal control strategies which align *local comfort objectives* with *global energy goals*. The following section outlines how BEMS functions as the operational and "cyber-physical nucleus" linking building-level control with broader smart-grid coordination.

2.3.4 Role of Building Energy Management Systems (BEMS) in DSM

Building Energy Management Systems (BEMS) serve as the operational and cyber-physical core of Demand-Side Management (DSM), providing the intelligence layer necessary to `predict`, `decide`, and `actuate` energy-related actions. By interfacing with local smart loads and distributed energy resources, BEMSs transform raw data into actionable control strategies which align building-level operations with broader grid objectives.

In practical deployments, as demonstrated by Chandra et al. [21], BEMS units engage with transactive energy markets through specialized entities known as Energy Management Demand Agents (EMDAs). These agents forecast their respective building's energy consumption and generation, submit flexibility bids to local energy markets, and respond to optimized dispatch instructions or dynamic price signals. This decentralized coordination enables clusters of buildings to self-balance renewable energy variability, while simultaneously contributing to grid stability.

Such mechanisms exemplify the transition from traditional static scheduling to dynamic, market-driven control paradigms. By embedding intelligence at the cyber-physical layer, BEMSs empower buildings to act as active participants in energy ecosystems, facilitating more resilient, efficient, and sustainable grid operations.

Functionally, BEMSs operate across *three* tightly coupled dimensions:

- **Forecasting**: predicting loads, PV generation, prices, and comfort parameters using data-driven or model-based approaches.
- **Optimization**: balancing multi-objective trade-offs (cost, comfort, carbon) and solving local market bids in real time.
- **Control**: applying optimal set-points to HVAC, lighting, storage, and EV systems through adaptive feedback loops.

Through layered coordination and cyber-physical integration, Building Energy Management Systems (BEMS) act as *aggregators of distributed intelligence*, aligning building-level autonomy with community-scale energy objectives. By interfacing with smart loads, distributed energy resources, and transactive energy markets, BEMS enable predictive control, decentralized decision-making, and real-time actuation. This positions them not only as the operational core of Demand-Side Management (DSM) and Demand Response (DR), but also as a strategic enabler of grid-interactive efficient buildings.

While DSM frameworks establish the strategic role of BEMS in energy-efficient buildings, practical implementation demands accurate modeling, simulation, and predictive optimization. The integration of the **`Building Digital Twin`** provides this critical capability—enabling dynamic representation of building systems (*virtual commissioning*), facilitating what-if analyses, and paving the way for supervisory control of `intelligent` buildings.

Together, these developments in DSM, DR, and BEMS signal a paradigm shift from reactive energy control to predictive digital representation. The next chapter extends this evolution by introducing the Building Digital Twin—a dynamic cyber-physical mirror which enables real-time simulation, forecasting, and optimization of building performance. In this context, BEMS serve as the natural bridge to the Digital Twin concept, embedding intelligence and interoperability that make such advanced modeling and control possible.

References

1. REN21 (2023) Renewables 2023 Global Status Report. Technical report, Renewable Energy Policy Network for the 21st Century (REN21). https://www.ren21.net/wp-content/uploads/2019/05/GSR2023_Demand_Modules.pdf. Accessed Oct 2025
2. International Renewable Energy Agency (IRENA) (2025) Renewable capacity statistics 2025. Statistical report, International Renewable Energy Agency, Abu Dhabi. https://www.irena.org/Publications/2025/Mar/Renewable-Capacity-Statistics-2025
3. Black S, Liu A, Parry I, Vernon N (2023) IMF fossil fuel subsidies data: 2023 update. https://www.elibrary.imf.org/view/journals/001/2023/169/001.2023.issue-169-en.xml
4. Black S, Liu A, Parry I, Vernon N (2023) Fossil fuel subsidies: data and trends. https://www.imf.org/en/Topics/climate-change/energy-subsidies
5. United Nations Environment Programme (processed by Our World in Data) (2024) Fossil-fuel subsidies (consumption and production) [dataset]. https://archive.ourworldindata.org/20250909-093708/grapher/fossil-fuel-subsidies.html. Accessed 1 Nov 2025. (data for 2010–2022)
6. Wani MH (2022) Intelligent control strategies for efficient building energy management systems (BEMS). Department of Electrical and Electronic Engineering, University of Auckland, PhD thesis
7. Lavidas G, Mezilis L, Alday M, Baki H, Tan J, Jain A, Engelfried T, Raghavan V (2025) Marine renewables in energy systems: impacts of climate data, generators, energy policies, opportunities, and untapped potential for 100% decarbonized systems. Energy p 138359
8. Mehmood MU, Chun D, Han H, Jeon G, Chen K et al (2019) A review of the applications of artificial intelligence and big data to buildings for energy-efficiency and a comfortable indoor living environment. Energy Build 202:109383. https://doi.org/10.1016/j.enbuild.2019.109383
9. Perr-Sauer J, Ugirumurera J, Gafur J, Bensen EA, Nguyen T, Paul S, Severino J, Nag A, Vijayshankar S, Gasper P et al (2025) Applications of explainable artificial intelligence in renewable energy research. Energy Rep 14:2217–2235
10. Jagannath GS, Sharma RB (2025) Enhancing demand-side management in smart grids with electric vehicle and renewable energy integration. J Energy Storage 140:118985

11. Wang Z, Trutnevyte E (2025) Demand-side flexibility of electric vehicles and heat pumps in the Swiss electricity system with high shares of renewable generation. Energy 275:138903
12. Labaran MJ, Masood T (2025) Technological frontiers: addressing renewable energy supply chain and sustainability challenges. Discover Sustain 6(1):100
13. Mishra S, Ahmad R, Srivastava A (2025) Bridging renewable energy sources with non-isolated DC-DC converters: challenges and innovations. Discover Electron 2(1):1–30
14. Liang X (2016) Emerging power quality challenges due to integration of renewable energy sources. IEEE Trans Ind Appl 53(2):855–866
15. Elavarasan RM, Shafiullah G, Padmanaban S, Kumar NM, Annam A, Vetrichelvan AM, Mihet-Popa L, Holm-Nielsen JB (2020) A comprehensive review on renewable energy development, challenges, and policies of leading indian states with an international perspective. IEEE Access 8:74432–74457
16. Koltsaklis NE, Knapek J (2025) Challenges and opportunities in generation expansion for 100% renewable energy systems: an optimization approach. Optim Eng 26(2):827–868
17. Bisht GS, Pandey S (2025) Biofuels for a sustainable future: generations, technologies, and key challenges in renewable energy. BioEnergy Res 18(1):43
18. Zhao H, Wang P, Palensky P, Filizadeh S, Liu Y, Chen Y, Zhu J, Li P, Lekić A, Terzija V (2025) Guest editorial: methodologies and applications of digital twin for renewable-dominant power systems. Int J Electric Power Energy Syst 171:111006
19. Martirano L, Habib E, Parise G, Greco G, Manganelli M, Massarella F, Parise L (2017) Demand side management in microgrids for load control in nearly zero energy buildings. IEEE Trans Ind Appl 53(3):1769–1779
20. Chouikhi S, Merghem-Boulahia L, Esseghir M, Snoussi H (2019) A game-theoretic multi-level energy demand management for smart buildings. IEEE Trans Smart Grid 10(6):6768–6781
21. Chandra R, Banerjee S, Radhakrishnan KK, Panda SK (2021) Transactive energy market framework for decentralized coordination of demand side management within a cluster of buildings. IEEE Trans Ind Appl 57(4):3385–3395
22. El Barkouki B, Mahir O, Laamim M, Rochd A, Ouassaid M, Oufettoul H (2024) Energy efficiency and optimal operation of a residential microgrid based on demand side management strategy. In: 2024 9th international youth conference on energy (IYCE). IEEE, pp 1–6
23. Sánchez-Loor I, Ayala-Chauvin M (2025) Modeling of electric vehicle energy demand: a big data approach to energy planning. Energies 18(20):5429
24. Benavides D, Arévalo-Cordero P, Ochoa-Correa D, Torres D, Ros A (2025) Predictive energy storage management with redox flow batteries in demand-driven microgrids. Sustainability 17(19). https://doi.org/10.3390/su17198915, https://www.mdpi.com/2071-1050/17/19/8915
25. Roy C, Das DK (2025) Class topper optimizer for cost-efficient smart grid operation under renewable energy uncertainties: C Roy and DK Das. Energy Effic 18(5):47
26. Bakare MS, Abdulkarim A, Zeeshan M, Shuaibu AN (2023) A comprehensive overview on demand side energy management towards smart grids: challenges, solutions, and future directions. Energy Inf 6(1):4
27. Rollo A, Serafini P, Aleotti F, Cilio D, Morandini E, Moneta D, Rossi M, Zulianello M, Angelucci V (2025) Load shifting and demand-side management in renewable energy communities: simulations of different technological configurations. Energies (19961073) 18(4)
28. Ouassaid M, El Idrissi RN, Azaroual M (2025) Demand response in smart microgrids: advanced and optimal energy management strategies
29. Han F, Yu S (2025) Quantifying the impact of energy storage capacity on building energy flexibility: a case study of the PV-ESS-GSHP system. Buildings 15(19):3536
30. Aboutalebi M, Bagherpour M, Noll J, Horn G (2025) Exploring cost-comfort trade-off in implicit demand response for fully electric solar-powered Nordic households. Energies
31. International Energy Agency (2024) Electricity 2024–analysis and outlook for data centre and ai energy use. https://www.iea.org/reports/electricity-2024. Accessed 5 Nov 2025
32. Statista Research Department (2025) Share of electricity consumption from renewable sources used by data centers worldwide in 2024, by region. https://www.statista.com/statistics/1405663/data-centers-renewable-electricity-share-worldwide/. Accessed 5 Nov 2025

33. US Energy Information Administration (2023) Use of energy in commercial buildings. https://www.eia.gov/energyexplained/use-of-energy/commercial-buildings.php. Accessed 7 Nov 2025
34. Etedadi F, Danté AW, Kelouwani S, Henao N, Agbossou K, Fournier M (2025) Distributed coordination of electric vehicles charging station and home energy management systems in residential neighborhood. Int J Electric Power Energy Syst 172:111142
35. Prusty BR, Arun SL, De Falco P (2022) Demand response in smart buildings. In: Tomar A, et al (eds) Control of smart buildings, studies in infrastructure and control. Springer, pp 121–131. https://doi.org/10.1007/978-981-19-0375-55, https://link.springer.com/chapter/10.1007/978-981-19-0375-55

Chapter 3
Introduction to Building Digital Twins

Abstract This chapter introduces the concept of the Building Digital Twin (BDT) as the natural evolution of modern Building Energy Management Systems (BEMS) and Control System Engineering. It establishes the foundational bridge from Demand-Side Management (DSM) and supervisory automation toward intelligent, data-driven, and continuously synchronized cyber-physical building systems. Beginning with the principles of feedback, prediction, and adaptation, the chapter explains how classical control theory—through both model-based and model-free paradigms—underpins the operational intelligence of Digital Twins. Model Predictive Control (MPC) and Reinforcement Learning (RL) are examined as complementary approaches, each offering unique advantages for handling the slow, nonlinear, and stochastic dynamics which characterize buildings. The chapter distinguishes simulation models from Digital Twins by emphasizing real-time synchronization, closed-loop intelligence, and continuous parameter calibration. Through multi-layered architectures, including perception, virtual modeling, and physical execution, it illustrates how Digital Twins integrate sensing, forecasting, optimization, and learning to support predictive and autonomous building operation. A contrast with fast, deterministic electromechanical systems—exemplified through a motor Digital Twin—highlights the additional complexity and long-memory behavior inherent in building environments. Toward the end, we provide an integrated overview of the challenges, advantages, and new considerations involved in putting BDTs into practical use. Key challenges include nonlinear dynamics, stochasticity, partial observability, model drift, data quality issues, multi-objective trade-offs, and computational burden. These challenges simultaneously justify the need for Computational Intelligence (CI) techniques—surrogate modeling, evolutionary optimization, Reinforcement Learning, and hybrid predictive-learning frameworks—which form the focus of Part II of the book. Ultimately, the chapter positions Digital Twins as predictive, adaptive, and control-oriented constructs which unify physics-based modeling with data-driven intelligence to achieve the next generation of autonomous, grid-interactive, and resilient building operation.

Keywords Building Energy Management System · Digital Twins · Model free control · Model based control · Model predictive control · Reinforcement learning

M. H. Wani et al., *Digital Twins for Building Optimization: Enhancing Energy and Comfort*, Studies in Infrastructure and Control,
https://doi.org/10.1007/978-3-032-23274-8_3

3.1 Prelude and Context

The preceding chapter concluded with an important realization: modern buildings are no longer passive energy sinks, but intelligent, responsive entities capable of shaping their demand in synchrony with renewable generation. Central to this transformation is the Building Energy Management System (BEMS)—the cyber-physical "brain" which integrates sensing, forecasting, optimization, and control. Yet, as these systems evolve, the question arises—*how can their behavior be represented, predicted, and optimized continuously, in real time?* The answer lies in the emergence of the **Building Digital Twin (BDT)**—a dynamic virtual counterpart of the physical building, continuously updated through data streams and analytical models.

While the term Digital Twin often evokes images of futuristic automation, its essence is rooted in the discipline of `Control System Engineering`. A twin is valuable only if it can sense, decide, and act—precisely the capabilities which define a feedback control system. Thus, before delving into the architecture and applications of BDTs, it is essential to understand the control logic which underpins them.

This chapter therefore commences by delineating the conceptual integration between Control System Engineering and Building Energy Management Systems (BEMS), demonstrating how feedback, prediction, and adaptation form the functional scaffolding of any digital twin. The discussion then differentiates between model-free and model-based control approaches, explores Model Predictive Control (MPC) and Reinforcement Learning (RL) as modern exemplars, and finally contextualizes these within the building domain. The chapter culminates by distinguishing simulation models from digital twins, clarifying why real-time connectivity and closed-loop intelligence mark the next frontier in building performance management.

It is worth mentioning here that the gradual shift from static simulation to adaptive, data-driven control has already been evidenced in early BEMS research. For instance, Yang et al. [1] demonstrated that data-mining-based prediction within Building Energy Management Systems (BEMSs) could learn consumption patterns, anticipate Demand-Response (DR) signals, and adjust control policies autonomously—an "*embryonic*" form of what is now recognized as the Digital Twin philosophy. Building on this continuum, recent surveys define a Digital Twin as a multi-layered construct encompassing physical, virtual, connection, data, and service domains, sustained by continuous synchronization and twin intelligence [2]. Within such frameworks, *feedback and predictive learning cease to be auxiliary functions; they become the operational core of intelligent buildings*. Real-world implementations now extend these principles beyond single assets to campus-scale infrastructures. The Smart Campus Digital Twin System reported in [3] illustrates this translation from concept to practice—integrating IoT sensing, analytics, and AI-based control into a dynamic ecosystem where the virtual and physical continuously inform one another.

Together, these developments underline that Digital Twins represent not a technological leap but a convergence of control theory, simulation, and computational intelligence—a natural progression from the data-driven BEMS architectures introduced a decade earlier. This understanding prepares the ground for Sect. 3.2, which examines how the fundamental tenets of Control System Engineering form the backbone of these intelligent, self-adapting systems.

3.2 Control System Engineering and BEMS

Control System Engineering provides the theoretical and mathematical foundation for every cyber-physical process which senses its state, compares it to a desired reference, and acts to minimize deviation. In buildings, these principles are embodied in the **Building Energy Management System (BEMS)**—which monitors environmental variables (temperature, humidity, illumination, occupancy) and manipulates actuators (HVAC dampers, valves, fans, blinds, lighting circuits) to maintain comfort and efficiency.

From a systems perspective, a building can be viewed as a complex Multi-Input Multi-Output (MIMO) plant, characterized by nonlinear, time-varying, and *stochastic dynamics*. Internal `heat gains, weather fluctuations, and occupant behavior` create continuous disturbances which require adaptive feedback rather than static control. The BEMS, therefore, performs the role of a *supervisory controller*—aggregating data from distributed subsystems and orchestrating them to satisfy energy and comfort objectives simultaneously.

Traditional control strategies—such as `PID` loops [4]—remain fundamental at the equipment level (for example, maintaining coil temperatures or fan speeds). However, at the supervisory layer, *classical feedback* alone becomes insufficient. As building systems become increasingly integrated with renewables, storage, and electric mobility, `predictive` and `optimal` control mechanisms are required to anticipate future disturbances and coordinate multiple objectives. This evolution transforms the BEMS from a rule-based automation platform into an intelligent decision-making engine.

Figure 3.1 might illustrate this conceptual transformation (at a high-level): sensors provide data to a control logic layer; the BEMS performs forecasting and optimization; actuators execute commands; and the resulting states continuously feed back, creating a digital loop that eventually integrates into the Digital Twin framework.

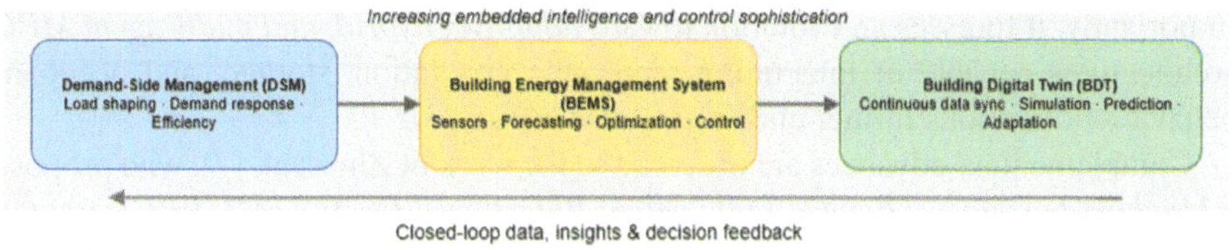

Fig. 3.1 Conceptual evolution of *intelligent building control* from DSM through BEMS to Digital Twin

Integration of Control Theory and BEMS

In essence, Control System Engineering and BEMS are two sides of the same coin:

- **Control theory** contributes the structure—feedback, stability, optimality.
- **BEMS** contributes the context—data acquisition, interoperability, and operational purpose.

Their integration is what enables buildings to transition from automated to autonomous, setting the stage for Digital Twins which learn, adapt, and evolve with their physical counterparts.

Recent literature has reinforced the notion that ***Control System Engineering*** remains the analytical core of the Building Energy Management System (BEMS), yet its role has expanded in scope and complexity. Traditional feedback and supervisory mechanisms are increasingly complemented by optimization and data-driven intelligence, reflecting the evolving dynamics of grid-interactive buildings and digital ecosystems.

Tomar et al. [5] provide a broad foundation linking *classical control theory* with the operation of grid-integrated buildings and local energy communities. Their work highlights that next-generation "smart" buildings must simultaneously maintain internal comfort and respond to external grid signals, establishing the BEMS as a cyber-physical bridge. Within this hierarchy, multi-layered control coordinates HVAC, lighting, and renewable subsystems while supporting Demand-Side Management (DSM) and flexibility provision—marking the transition from automation to "autonomy" and aligning directly with the control-centric framework described earlier in this section. Expanding on this concept, Michailidis et al. [6] offer an extensive review of *Model Predictive Control* (MPC) in the context of smart buildings, highlighting its role as the prevailing *model-based framework* in modern BEMSs. MPC frameworks optimize control actions over a *receding horizon* by predicting system states and enforcing comfort and operational constraints simultaneously. Their review documents the methodological shift from rigid *white-box* implementations toward *hybrid gray and black-box* structures enriched by machine-learning forecasts. Importantly, it foresees an evolution toward *adaptive, hybrid, and multi-agent MPC* architectures capable of integrating renewable generation, storage, and occupant behavior—directions further elaborated later in Sect. 3.2.1.

Complementary advances are observed in the work of Zhu et al. [7], who propose a *Machine-learning-enhanced Lightweight Rule-Based Control (ML-RBC)* strategy for Demand-Response (DR) applications. Their findings demonstrate that ML-RBC can emulate near-optimal MPC performance while reducing computational cost by several orders of magnitude, addressing a key barrier to real-time deployment in embedded BEMS hardware. Such hybrid schemes—blending explicit rule logic with data-driven adaptation—represent a pragmatic *path toward intelligent control implementation*.

Zhang et al. [8] further synthesize the field of *Machine Learning Control* (MLC) in building operations, classifying developments into (i) ML-based system or *component modeling* for control and (ii) learning of the control process itself. Their survey emphasizes that with the growing availability of operational data, learning algorithms can derive optimal policies without explicit physical models, enhancing adaptability under nonlinear and uncertain conditions. Reinforcement learning and hybrid neuro-fuzzy methods are identified as promising routes toward autonomous supervisory control, anticipating the model-free paradigms introduced in the next section.

Earlier work by Fang et al. [9] had already recognized that intelligent control techniques—neural networks, fuzzy inference, and adaptive algorithms—could augment conventional PID feedback loops by embedding self-tuning and learning capabilities. Collectively, these studies trace the progression from classical control toward *learning-augmented and predictive frameworks*, wherein control theory, optimization, and Artificial Intelligence (AI) converge within the BEMS hierarchy.

In summary, contemporary literature portrays BEMS as a "maturing" control ecosystem: hierarchical in structure, data-driven in operation, and increasingly participatory in grid coordination. The convergence of MPC, MLC, and hybrid control approaches does not signify a replacement of control theory, but its natural evolution—extending its principles of feedback, stability, and optimality into high-dimensional `stochastic energy environments`. The following sections therefore differentiate **Model-Free** and **Model-Based** control approaches, illustrating how each contributes to the emergence of intelligent Digital Twin frameworks in buildings.

3.2.1 Model Free Control

Model-free control (MFC) represents a departure from the long-standing paradigm of developing explicit physical or empirical models of building dynamics. Instead, it enables control actions to be learned directly through interaction with the environment, relying on historical or real-time operational data. *The system* "`learns by doing`", *improving its policy based on measured outcomes rather than predefined differential equations*. This characteristic makes MFC particularly attractive for complex and stochastic systems such as buildings, whose thermal and occupancy dynamics are difficult to capture precisely.

At its core, MFC leverages the principles of *Reinforcement Learning* (RL) and *approximate dynamic programming*, where a learning agent interacts with an environment, observes its current state "s_t", executes an action "a_t", and receives a reward "r_t" which reflects the quality of the decision. The aim is to discover an optimal control policy "π*"–a mapping from states to actions—which maximizes the expected long-term cumulative reward:

$$\pi^*(s) = \arg\max_{\pi} \mathbb{E}_{\pi}\left[\sum_{t=0}^{\infty} \gamma^t r_t \,\middle|\, s_0 = s\right] \tag{3.1}$$

where "γ" is the discount factor ($0 < \gamma < 1$) and "r_t" the instantaneous reward at time "t". The value of this optimal policy can be expressed recursively through the *Bellman optimality equation*:

$$Q^*(s_t, a_t) = \mathbb{E}\left[r_t + \gamma \max_{a_{t+1}} Q^*(s_{t+1}, a_{t+1})\right] \tag{3.2}$$

where $Q^*(s_t, a_t)$ is the optimal action—*value function*. Consequently, the `optimal policy` can be obtained as

$$\pi^*(s_t) = \arg\max_{a_t} Q^*(s_t, a_t) \tag{3.3}$$

meaning that, for each observed state "s_t", the controller selects the action "a_t" yielding the highest expected return.

Historical Emergence. Claessens et al. [10] demonstrated one of the earliest large-scale applications of MFC in energy systems by using Reinforcement Learning (RL) to control clusters of thermostatically controlled loads in a district heating network. Their approach—based on fitted Q-iteration and market-based coordination—achieved over 65% of the theoretical optimization potential within a learning period of two months, illustrating the viability of learning control without explicit plant models. Importantly, they showed that embedding minimal domain knowledge (e.g., safe policy shapes or temperature limits) can accelerate convergence and stabilize learning—lessons which remain vital for building-scale applications today.

Recent Advances in Building Energy Systems. Building on this foundation, Xiong et al. [11] applied a Deep Q-Network (DQN) algorithm to optimize the operation of a building's cooling water system. Their study highlighted *two major obstacles* of purely model-free learning—*slow convergence* and the possibility of *unsafe actions*. To mitigate these, the authors refined the *action space* by dividing equipment control variables (pumps and towers) into safe and unsafe regions and then restricting exploration to the safe subset. This adjustment shortened the convergence period by one cooling season while maintaining a stable Coefficient Of Performance (COP) improvement of approximately 6%, proving that structured exploration can make MFC practical in HVAC domains.

Li et al. [12] extended this direction by embedding hyper-parameter optimization within Deep Reinforcement Learning (DRL) controllers, demonstrating that automatic tuning of learning rate, exploration factor, and neural-network depth can substantially enhance stability and energy efficiency during online training. Their work reinforces that the success of MFC hinges not only on algorithm selection but also on the meta-learning of its tuning parameters.

Transfer and Generalization. Because model-free controllers rely on trial-and-error interaction, one key limitation has been the discomfort and inefficiency during their early learning phase. Esrafilian-Najafabadi and Haghighat [13] proposed an elegant solution by combining unsupervised learning and transfer learning. Their framework clustered residential occupancy patterns using dynamic time warping and k-means, then transferred pretrained DQN agents between households with similar occupancy signatures. The approach improved the controller's jump-start performance by 25% and reduced mean temperature deviation by 68%, showing that knowledge reuse can make model-free Reinforcement Learning far more deployable in practice.

Coraci et al. [14] further advanced this idea by introducing heterogeneous transfer learning for DRL controllers, enabling policies trained on one building typology to be adapted to another with minimal retraining. They demonstrated improved scalability and real-time adaptability across heterogeneous building datasets, emphasizing that transferable and federated learning schemes may form the backbone of next-generation model-free Digital Twins.

Model Formulation and Conceptual Summary. In Model-Based Control (further explained in Sect. 3.2.2) formulations, the system is typically represented in the state-space form:

$$\begin{cases} \dot{x}(t) = f(x(t), u(t), w(t)) \\ y(t) = g(x(t), u(t)) \end{cases} \tag{3.4}$$

where "$x(t)$" denotes the state variables, "$u(t)$" the control inputs, "$w(t)$" the external disturbances (e.g., weather or occupancy), and "$y(t)$" the measured outputs. The functions "$f(\cdot)$" and "$g(\cdot)$" capture the system dynamics and output mapping, respectively. Model-Based Controllers such as MPC rely on these mathematical descriptions to forecast system evolution and compute optimal inputs. In contrast, Model-Free Control completely bypasses this identification step by learning the input–output relationship directly from operational data.

Accordingly, Model-Free Control (MFC) can be expressed as a data-centric learning paradigm in which the control law "$u(t)$"[1] is synthesized directly from observed *input–output trajectories*:

$$u(t) = \mathcal{L}_\theta\big(y(t), y(t-1), \ldots, r(t)\big) \tag{3.5}$$

[1] "a_t" and "$u(t)$" denote the same control action—the former (called as "action") follows the RL notation, whereas the latter (called as "control input") follows the classical control theory notation.

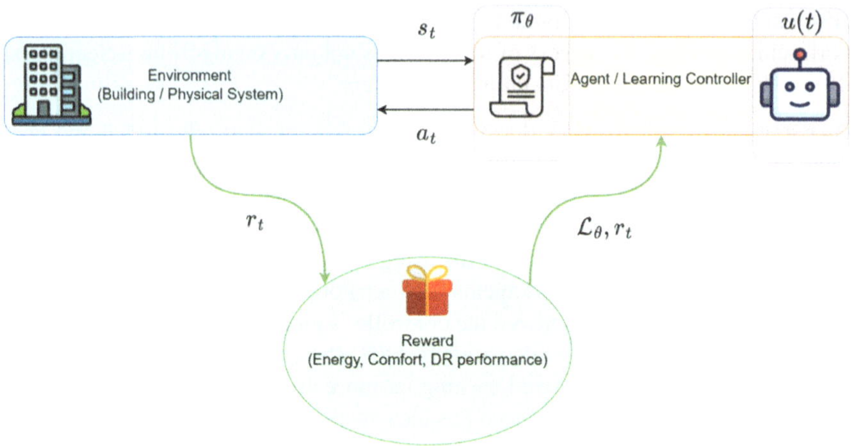

Fig. 3.2 Conceptual schematic of Model-Free Control: The agent maintains a learned control law "$\mathcal{L}_\theta$" (policy, "π_θ") which maps the observed building (environment) state, "s_t", to the control action, "a_t" (*control input*, "$u(t)$"). The environment returns a reward, "r_t", based on criteria such as energy use, comfort deviation, and Demand-Response performance. This reward is used to update the parameters, "θ", of the control law, "$\mathcal{L}_\theta$", enabling the control policy to improve over time without requiring an explicit model of the building (environment)

where "$\mathcal{L}_\theta$" is a learned policy parameterized by "θ",[2] "$y(t)$" the measured outputs, and "$r(t)$"[3] the desired reference. This bypasses explicit identification of system dynamics "$f(x, u)$" or "$g(x)$", as required in model-based schemes. A simplified schematic of the MFC process is presented in Fig. 3.2, showing how state observations, reward evaluation, and policy updates occur in closed loop through continual interaction with the environment.

Building on these formulations, a comparative perspective helps contextualize their practical implications.

Comparative Perspective. While MFC demonstrates growing promise for adaptive and data-driven operation, its strengths and weaknesses are best understood relative to Model-Based Control (MBC). Nagy et al. [15] provide a comprehensive comparison between the two paradigms, highlighting that MBC excels in transparency, constraint enforcement, and short-term predictability, whereas MFC offers superior scalability and robustness against model uncertainties. They further advocate `hybrid control architectures` which *couple model-based prediction with model-free adaptation*—an approach which bridges analytical rigor and learning flexibility. This perspective reinforces that MFC and MBC are complementary rather than

[2] "θ" here denotes the internal parameters of the learned policy, e.g., neural-network weights, value-function parameters, or other tunable components adjusted, for example, through Reinforcement-Learning updates.

[3] In this chapter, "r_t" denotes the RL reward at time "t" and is distinct from the reference signal "$r(t)$" used in classical control formulations.

competing frameworks, together forming the foundation for intelligent Digital Twin control in buildings.

In essence, the philosophy of Model-Free Control (MFC) is aligned with the spirit of Digital Twins (DT) which "learn by synchronization". Both (MFC and DT) rely on real-time feedback from the physical system, continuous learning from data, and iterative refinement of decision policies. As such, MFC provides the operational intelligence which allows a Digital Twin to move beyond static simulation toward adaptive and self-optimizing behavior—a theme which will further unfold in the following section.

3.2.2 *Model Based Control*

While Model-Free Control derives its behavior solely from interaction with the physical environment, *Model-Based Control* (MBC) relies explicitly on a mathematical representation of the building's dynamics. In this paradigm, the controller uses a `predictive model` of the system—typically expressed in state-space form, see Sect. 3.2.1, Eq. (3.4)—to forecast future states and compute the optimal control action. This approach preserves the classical principles of control engineering, offering transparency, constraint handling, and formal guaranties of stability and feasibility. As mentioned earlier, "$x(t)$" denotes the system states (which can be variables such as, zone temperatures or internal loads) "$u(t)$" the control inputs (e.g., airflow, chilled-water valve position and so on), "$w(t)$" external disturbances such as weather and occupancy, and "$y(t)$" the measurable outputs available to the controller. The functions "$f(\cdot)$" and "$g(\cdot)$" describe the physics-based or data-driven model upon which the controller relies.

Using these predictive dynamics, the controller computes the action "$u(t)$" by minimizing a performance index which penalizes deviation from a reference trajectory "$r(t)$" and *control effort* (which may correspond to energy use in building applications). A general formulation is

$$u(t) = \arg\min_{u} \int_{t}^{t+T} \left(\|y(\tau) - r(\tau)\|_Q^2 + \|u(\tau)\|_R^2 \right) d\tau \tag{3.6}$$

where "Q" and "R" (in this context) are positive semi-definite weighting matrices, and "T" is the `prediction horizon`. This formulation captures the essence of MBC: control signals are computed by *predicting* how the building will evolve under different actions.

The **term**:

- $\|y(\tau) - r(\tau)\|_Q^2$: Penalizes deviation of the output, "$y(\tau)$" from the reference, "$r(\tau)$", weighted by "Q".
- $\|u(\tau)\|_R^2$: Penalizes control effort, weighted by "R".

In contrast to Model-Free Control—where the policy "$\mathcal{L}_\theta$" is iteratively updated through the reward signal "r_t"—Model-Based Control incorporates the model "$f(\cdot)$" directly in the computation of "$u(t)$". The selection of "$u(t)$" therefore arises from *optimization over predicted trajectories*, rather than trial-and-error learning. Figure 3.3 conceptually illustrates this process.
A typical MBC formulation evaluates a sequence of candidate control inputs:

$$\mathcal{U}(t) = \{u(t), u(t+1), \ldots, u(t+N)\} \tag{3.7}$$

The predictive model, "$f(\cdot)$", generates the corresponding state trajectory:

$$\hat{s}(t+k+1) = f\left(\hat{s}(t+k), u(t+k)\right), \qquad k = 0, \ldots, N-1 \tag{3.8}$$

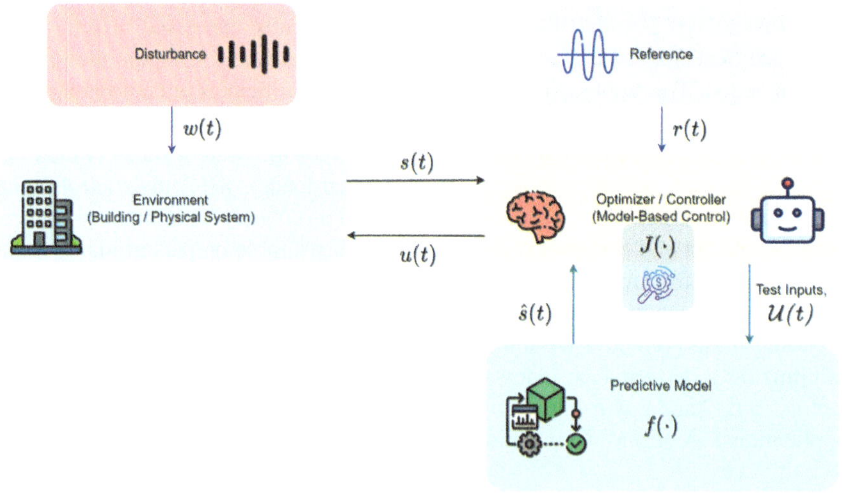

Fig. 3.3 Conceptual schematic of Model-Based Control (MBC): the controller receives the current building state, "$s(t)$", and a reference trajectory "$r(t)$", and uses a predictive model, "$f(\cdot)$", to simulate how the system would evolve under a sequence of candidate control inputs, "$\mathcal{U}(t)$". The predictive model returns the corresponding predicted states, "$\hat{s}(t)$", which the controller evaluates using a cost function, "$J(\cdot)$", which penalizes tracking error and control effort. The control action, "$u(t)$", which minimizes "$J(\cdot)$" is applied to the environment, while external disturbances, "$w(t)$", influence the true building response

These predictions are assessed through a cost function[4] of the general form:

$$J(t) = \sum_{k=0}^{N} \left(\|y(t+k) - r(t+k)\|_Q^2 + \|u(t+k)\|_R^2 \right) \tag{3.9}$$

where the term:

- "$y(t+k) - r(t+k)$", weighted by "Q", represents the *tracking error.*
- "$u(t+k)$", weighted by "R", represents the *control effort.*

The cost function, "$J(\cdot)$",[5] penalizes tracking error and control effort. *Note that, this expression is representative and may vary across applications*: different MBC or MPC formulations may incorporate nonlinear penalties, comfort bounds, equipment constraints, economic objectives, or multi-objective performance indicators.

The optimal input "$u(t)$" is the first element of the sequence "$\mathcal{U}(t)$" which minimizes "$J(\cdot)$" over the prediction horizon.

Model-Based Control encompasses several approaches, ranging from classical PID control tuned around simplified linear models, to advanced methods such as Linear Quadratic Regulators (LQR), robust control, and the widely used *Model Predictive Control* (MPC). While these methods differ in complexity, they share three foundational characteristics:

- **They require an explicit model** describing the building's response to control inputs.
- **They optimize control actions** based on predicted behavior, often enforcing constraints on temperature, energy, or equipment limits.
- **They provide stability and feasibility guarantees** under well-defined assumptions, offering transparency and interpretability—valued in building operations.

An important distinction thus emerges between Model-Based and Model-Free Control. In MBC, the model serves as the engine for computing optimal decisions, and the controller's performance largely depends on the fidelity of the model—its ability to capture thermal inertia, nonlinear responses, and the time-varying effects of occupancy and weather. In MFC, by contrast, the controller bypasses the modeling step entirely and learns a control policy directly from data. As demonstrated earlier, both paradigms are complementary, and their hybridization forms an active research frontier in intelligent building control.

From a practical standpoint, recent studies show that Model-Based and Learning-Based controllers are not competing entities, but complementary tools within a

---

[4] Although "$J(\cdot)$" is shown in a minimization form, MBC formulations do not necessarily require minimizing a cost; some applications instead *maximize* a performance index (e.g., thermal comfort, renewable-energy use, economic benefit, or service-level metrics). The choice of minimization or maximization depends on the application-specific definition of "$J(\cdot)$".

[5] Here, "Q" and "R" are positive semi-definite weighting matrices which tune the relative penalties on tracking deviation and control effort. Their specific form depends on the application and may reflect comfort priorities, energy considerations, or equipment constraints.

broader design space. Fu et al. [16] compare MPC and Deep Reinforcement Learning (DRL) for a single-zone fan coil unit and quantify how closely each can approximate a "true" dynamic-programming optimum under fixed computation budgets. Their results indicate that, in an idealized setting with perfect models and offline training, the best DRL agent can reach up to about 96.5% of the estimated optimality, while the best MPC configuration approaches roughly 90%. However, when stochasticity and deployment variability are taken into account, the expected performance of DRL and MPC becomes almost indistinguishable. *This nuance reinforces a key message of this chapter: Model-Based Control offers a transparent, optimization-ready backbone, while Model-Free methods contribute adaptivity and exploration capabilities—and a Digital Twin can act as the integrative layer where both co-exist.*

At the building-portfolio scale, Aslam et al. [17] present an Adaptive Model-Based Control (AMBC) scheme for thermal energy storage in multi-building networks, where the controller continuously updates a state-space model and feedback gains at event times to maximize a Requisition Time (RT) metric. Their framework illustrates how MBC can remain effective even when plant parameters drift or are only approximately known: an adaptive estimator refines the model online, and an event-triggered law preserves stability and comfort while responding to grid requests. This is strongly aligned with the Digital Twin philosophy, where the "virtual" model is continuously calibrated to the real building and used to orchestrate demand shifting, storage, and comfort.

In parallel, Morcillo-Jiménez et al. [18] demonstrate how Deep Learning (DL) based forecasting can feed into Model-Based strategies. Using sensor data from an office building, they show that memory-based architectures such as Long Short-Term Memories (LSTMs) yield superior energy-consumption predictions compared to simpler multilayer perceptrons and other baselines. Since, their work is explicitly framed within the context of Model-Based Control, Demand Response (DR), and energy optimization, it highlights a practical design pattern: data-driven predictors (e.g., LSTM, Seq2Seq) can be embedded as surrogate models inside MPC or other MBC formulations, thereby enhancing predictive accuracy without abandoning the interpretability and constraint handling of model-based schemes.

Redder et al. [19] shift the focus from algorithms to infrastructure, arguing that the real barrier to advanced model-based building control is often not the controller itself but the surrounding ICT stack. In their large-scale Living Lab Energy Campus, they deploy a flexible IoT-enabled ICT architecture which links metering, HVAC control, user interfaces, and data services across 13 mixed-use buildings. Within this environment, they demonstrate real-world MPC operation while emphasizing user engagement, scalability, and robust data exchange as prerequisites for "intelligent" building operation.

This perspective is crucial for Digital Twins: without an ICT backbone to support bidirectional data flow, even the most elegant MBC formulation cannot be deployed at scale.

Finally, cross-domain work by Vinh et al. [20] on self-learning Model-Based Control for sensor-less induction motor drives provides a forward-looking template for buildings. Their approach couples a backpropagation Neural Network (NN) based estimator with robust nonlinear control to maintain high precision under sensor faults and parameter uncertainties. Although the application domain is motor drives rather than HVAC, the conceptual structure—a physics-informed model augmented by learning components which tune themselves online—mirrors the type of hybrid MBC architectures which Digital Twins in buildings will likely adopt.

Taken together, these studies indicate a clear trajectory from static, hand-tuned model-based controllers toward adaptive, data-enriched, and infrastructure-aware Model-Based Control (MBC). Among these, *Model Predictive Control (MPC)* has emerged as the most prominent and practically deployed realization of this paradigm in buildings. MPC formulates control as a `rolling (receding) horizon` optimization problem which encodes a cost function, "$J(\cdot)$", prediction horizon, and operational constraints, enabling constraint-aware and anticipatory decision-making. Its predictive capabilities allow seamless integration of advanced forecasts, adaptive models, and Digital Twin representations, thereby translating high-fidelity simulations into real-time supervisory actions. This makes MPC the backbone of many advanced building control strategies and a primary vehicle for operationalizing Digital Twin concepts.

3.3 What is Model Predictive Control (MPC)?

Model Predictive Control (MPC) is the most widely adopted and practically deployed realization of Model-Based Control (MBC) in modern cyber-physical systems. *In buildings—where dynamics are slow, disturbances are stochastic, and comfort-energy trade-offs are inherently multi-objective—MPC offers a principled framework for anticipatory, constraint-aware decision-making.* Its foundations align naturally with the Digital Twin (DT) paradigm: both rely on prediction, optimization, and continual feedback from real-time data streams.

At its core, MPC is a discrete time, Multi-InputâŁ"Multi-Output (MIMO) control architecture which uses an internal model of the process to predict future states over a finite *receding horizon*. Based on these predictions, the controller solves an optimization problem which determines the optimal sequence of `future control inputs`. These inputs are evaluated using a cost function, "$J(t)$", which encodes objectives such as comfort, energy use, renewable utilization, and equipment constraints. Although "$J(t)$" is often minimized, MPC can also be formulated as a maximization problem depending on the desired performance index [21].

Only the first control move of the optimized sequence is applied[6]; at the next sampling instant, updated measurements are obtained, the horizon shifts forward, and the optimization is solved again. This rolling (receding) formulation equips MPC with adaptiveness—one of the core requirements for Digital Twin synchronization.

In the context of intelligent buildings, MPC has been used for HVAC scheduling, Demand-Response (DR) coordination, hybrid thermal storage optimization, and predictive comfort control [22, 23]. Its predictive nature enables seamless integration of forecasts (weather, occupancy, pricing), surrogate models, and learning components, making MPC a natural supervisory layer for Building Digital Twins.

Figure 3.4 provides an expanded block diagram of MPC, illustrating its prediction, optimization, and feedback structure.

3.3.1 Expanded View of Basic Block Diagram of MPC

The MPC structure can be decomposed into key components: (i) the prediction model; (ii) the optimizer which computes optimal moves; (iii) constraint-handling mechanisms; and (iv) the feedback loop which synchronizes the model with real-time measurements. Such decomposition not only clarifies the computational workflow but also mirrors how a Digital Twin operates—continuously predicting, comparing, and updating its internal representations.

At each sampling instant, "t", the controller collects the current state (or measurement), "$s(t)$", and any exogenous inputs such as weather disturbances, "$w(t)$", or occupancy signals. These inputs feed into the predictive model, "$f(\cdot)$", which generates predictions of the future outputs: $\hat{y}(t+1), \ldots, \hat{y}(t+P)$ for a candidate input sequence:

$$\mathcal{U}(t) = \{u(t), u(t+1), \ldots, u(t+M-1)\} \tag{3.10}$$

where "P" is the prediction horizon and "M" is the control horizon with $M \leq P$.

These predicted trajectories are evaluated using a cost function, "$J(t)$", which—in its representative form—penalizes tracking error and control effort, see Eq. (3.9):

$$J(t) = \sum_{k=0}^{P} \left(\|\hat{y}(t+k) - r(t+k)\|_Q^2 + \|u(t+k)\|_R^2 \right) \tag{3.11}$$

Note: The above form is representative; MPC formulations may instead maximize economic objectives, renewable utilization, or thermal storage benefit [22, 23]. Weighting matrices "Q" and "R" reflect comfort priorities, energy cost, equipment limits, or risk preferences.

---

[6] Only the first control input is implemented because future predictions may deviate from reality due to disturbances or model inaccuracies. Re-optimizing at each step ensures adaptiveness and robustness—this is the essence of receding horizon control.

```
+----------------------------------------------------------+
|                    Perception Layer                      |
+----------------------------------------------------------+
            |                               |
+----------------------+        +-------------------------+
|  Measurements s(t)   |        |      Forecasts          |
|      (sensors)       |        | (weather, occupancy,    |
|                      |        |   pricing, etc.)        |
+----------+-----------+        +------------+------------+
           |                                 |
           +---------------+-----------------+
                           |
                           v
+----------------------------------------------------------+
|           Virtual Layer (MPC Controller)                 |
|                                                          |
|  +-------------------+   +---------------------------+   |
|  | Predictive Model  |--> |   Optimization / Solver  |<--+
|  |      f(.)         |   +---------------------------+   |
|  +-------------------+              ^                    |
|                                     |                    |
|                           Cost Function J(t)             |
|                                     |                    |
|                                     v                    |
|                        +---------------------------+     |
|                        |       Constraints         |---->+
|                        |  (limits, bounds, etc.)   |     |
|                        +---------------------------+     |
+----------------------------------------------------------+
                           |
                    Control Action u(t)
                           |
                           v
+----------------------------------------------------------+
|      Physical Layer (Building / System)                  |
|  Environment / Building / Equipment (HVAC, TES, ...)     |
+----------------------+-----------------------------------+
                       |
                Disturbance w(t)
                       |
                       v
        Affects real building dynamics
                       |
                       v
      Measurements / Feedback s(t) ------------> Back to MPC
```

Fig. 3.4 Conceptual schematic of Model Predictive Control (MPC), showing the interaction between perception (sensor feedback and forecasts), the virtual decision-making layer (predictive model, optimization, cost evaluation, and constraints), and the physical building system. This layered structure reflects the Digital Twin logic underpinning MPC, where predictions and real-time measurements continuously inform optimal control actions

A unique feature of MPC—highlighted in process industries and increasingly in building applications—is the coordination between *target calculation* (steady-state economic optimization) and *move calculation* (dynamic optimization). This two-layer architecture allows MPC controllers to pursue long-term operational goals (e.g., cost minimization or load shifting) while simultaneously respecting equipment and comfort constraints.

The expanded block perspective is essential for Digital Twin environments: the `prediction model` corresponds to the `virtual twin`, real-time measurements synchronize physical and virtual states, and the optimizer uses both information streams to produce actionable supervisory decisions.

3.3.2 Basic Concept (Receding Horizon)

The fundamental principle underlying MPC is the *receding horizon*. MPC operates in discrete time, optimizing future control actions at each sampling instant "k" based on the current measurement "$y(k)$" and the *internal model*.

At `every time step`, MPC performs three key computations:

1. **Prediction**: Using "$f(\cdot)$" and the latest measurements, the controller predicts the system's future evolution over the prediction horizon "P".
2. **Optimization**: It computes the optimal sequence of future control moves over the control horizon "M"[7]:

$$\mathcal{U}(k) = \{u(k), u(k+1), \ldots, u(k+M-1)\} \tag{3.12}$$

   by *minimizing* or *maximizing* "$J(\cdot)$" subject to constraints.
3. **Application:** Only the first action "$u(k)$" is implemented. At the next time step "$k+1$", new measurements become available and the optimization is repeated.

The prediction horizon "P" determines how far ahead the controller anticipates system behavior. `Longer horizons benefit systems with slow dynamics, such as HVAC and energy storage.` The control horizon "M" determines the number of explicitly optimized moves. Its choice balances:

- **Computational Feasibility**: larger "M" increases the dimensionality of the optimization problem.
- **System Behavior**: systems with slow or delayed dynamics often benefit from longer horizons.
- **Actuator Health and Over-Actuation Risks**: smaller "M" mitigates aggressive transients.

[7] In this context, "k" represents the discrete sampling index. As previously noted in Eq. (3.10), "$\mathcal{U}(t)$" is a general reference to a continuous-time perspective.

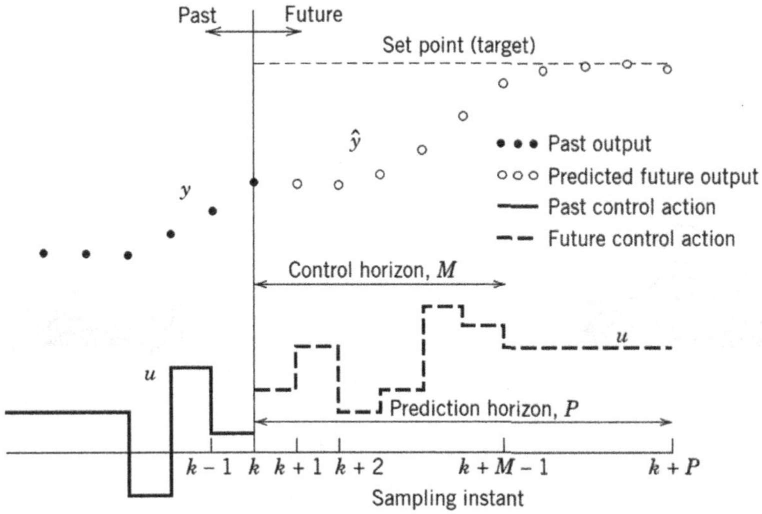

Fig. 3.5 Illustration of the *Receding Horizon* Principle in Model Predictive Control (MPC): At each sampling instant, future trajectories are predicted over a finite horizon, and an optimal sequence of control moves is computed. Only the first move is implemented, and the horizon shifts forward at the next time step. Typically, $M \leq P$, where "M" (control horizon) and "P" (prediction horizon) are chosen based on several factors (such as Computational Constraints, System Dynamics, and Over-Actuation Concerns)—a compromise between computational complexity and control performance. *Note that, the diagram is an exemplar I/O response for a Single-Input-Single-Output (SISO) system*

Figure 3.5 illustrates the receding horizon strategy, showing predicted trajectories, implemented actions, and the shift of the horizon at every sampling step.

The receding horizon formulation equips MPC with adaptability: the controller always incorporates the latest real-time measurements, disturbance realizations, forecast updates, and Digital Twin synchronization signals. This continual re-optimization is what sets MPC apart from static controllers and makes it particularly suited for intelligent, data-rich building environments.

Typical Example of MPC

A canonical example frequently used to teach MPC is *Adaptive Cruise Control* (ACC) in autonomous driving, see Fig. 3.6. The ACC system uses radar sensors to measure the relative distance and velocity to the leading vehicle. At each sampling instant, the MPC predicts future trajectories of both vehicles and optimizes throttle or brake commands to maintain a safe distance while respecting comfort and acceleration limits. If the leading vehicle is far enough, MPC prioritizes `speed tracking`; if it is too close, the controller prioritizes `spacing`. This demonstrates how MPC switches control objectives based on real-time context—an ability central to building Digital Twin orchestration.

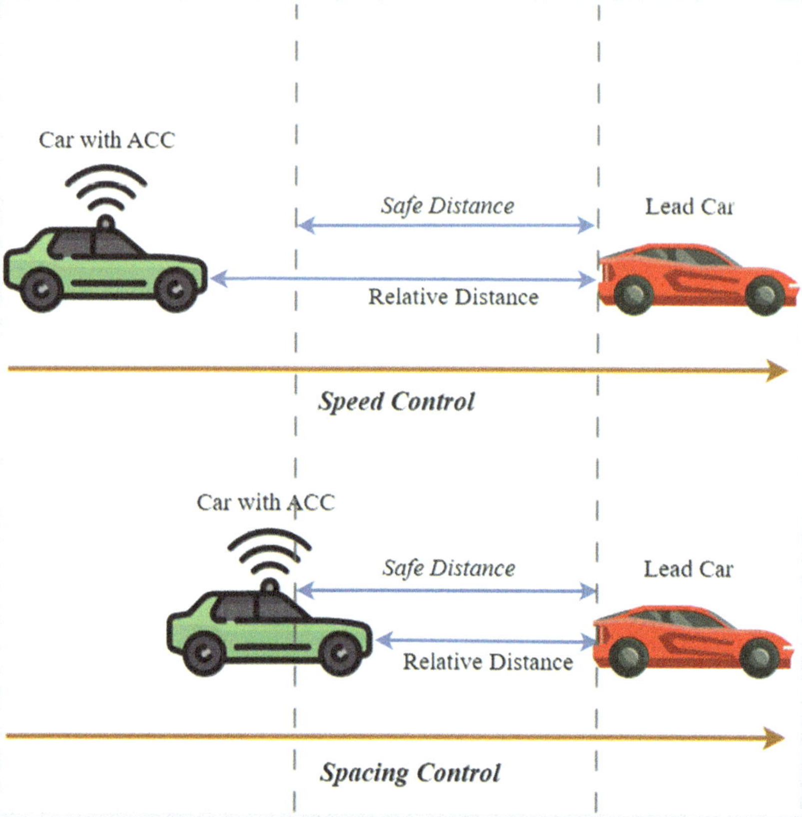

Fig. 3.6 Adaptive Cruise Control (ACC) as an example of MPC. The controller maintains safe spacing or tracks the desired speed by optimizing future inputs while respecting constraints

In buildings, MPC has been deployed for a wide variety of supervisory tasks:

- **Thermal Storage Optimization**: Habib et al. [22] demonstrate a data-driven MPC framework for hybrid thermal energy systems which reduces energy consumption and peak demand through optimal charge/discharge coordination.
- **Real Building Operation**: Behzadi et al. [23] apply MPC to a high-performance building and achieve measurable reductions in cooling energy while maintaining comfort, illustrating the benefits of anticipatory control under real-world disturbances.
- **NN-Enhanced MPC**: Wald et al. [24] integrate a Neural Network based parameter-varying model with MPC, enabling adaptation to time-varying dynamics and highlighting the convergence of MBC and learning—a cornerstone of Digital Twin intelligence.

These examples illustrate a broader theme: MPC is not merely a controller but a computational paradigm which unifies prediction, optimization, and constraint integration. This makes it a natural supervisory backbone for Building Digital Twins, where high-fidelity simulation models, forecasts, adaptive components, and real-time data streams converge to guide optimal decision-making.

The next section introduces Reinforcement Learning (RL), which complements MPC through adaptive, data-driven decision-making, allowing Digital Twins to operate under uncertainty and partial observability.

3.4 Reinforcement Learning (RL)

Reinforcement Learning (RL) represents the most general formulation of Model-Free Control, where an agent learns control behavior through repeated interaction with the physical (or simulated) environment. Unlike Model-Based Control—which requires explicit knowledge of system dynamics—RL discovers optimal control policies by evaluating the long-term consequences of its actions through a scalar reward signal. *This paradigm makes RL particularly suitable for buildings, whose dynamics are nonlinear, stochastic, and influenced by delayed thermal effects which cannot be captured easily using static (e.g.,* `white or gray-box`*) or empirical (e.g.,* `black-box`*) models.*

Formally, RL frames control as a sequential decision-making problem. At each discrete time step, "t", the agent observes a state, selects an action, and receives a reward describing the quality of that decision. The objective is to learn a policy which maximizes the expected discounted return, a formulation introduced earlier in Eq. (3.1). Likewise, the recursive structure describing how optimal action values are related across time follows from the *Bellman optimality principle* presented in Eq. (3.2). In RL, however, these relationships are not solved analytically; instead, they are *learned* through sampled transitions (s_t, a_t, r_t, s_{t+1}), enabling the agent to improve its policy without requiring any explicit building model.

This temporal coupling is what allows RL to capture "long-horizon effects" such as thermal inertia, peak pricing, or Demand-Response (DR) participation—phenomena which are essential when controlling modern buildings but are difficult to anticipate using short-sighted or purely "myopic" control strategies. *A key feature of RL is that learning occurs entirely through interaction: the agent improves its policy by continuously evaluating how its actions influence future outcomes.* This interaction-driven adaptation forms the conceptual basis for the analogy introduced next.

Reinforcement Learning (RL)—A Simple Analogy

A useful analogy for understanding RL is dog training, see Fig. 3.7. The environment represents the context (e.g., the room); the agent is the dog; and the actions include standing, jumping, or shaking hands. Rewards reinforce desirable behaviors, and over time the dog learns a *policy*[8] mapping situations to the best actions.

In buildings, the analogy holds directly. The environment is the building; the agent is the controller; the actions might be modulating airflow or dispatching storage; and the reward encodes energy cost, comfort deviation, or Demand-Response (DR) performance. Through repeated interaction, the agent gradually identifies control strategies that improve overall building performance across time.

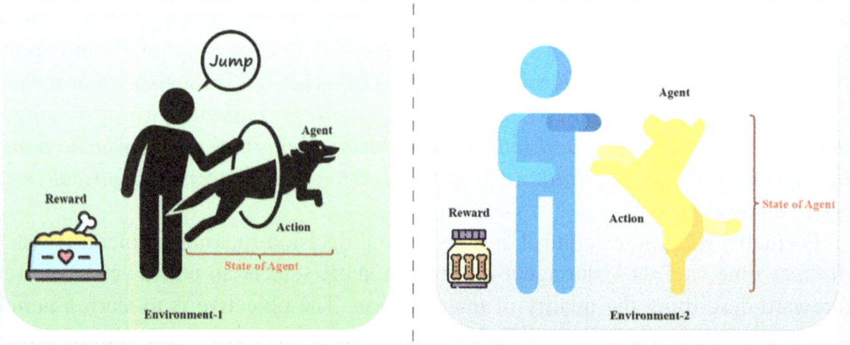

Fig. 3.7 Illustration of Reinforcement Learning concepts using dog training. Both environments show the core RL loop: the agent (dog) takes an action, the environment provides a reward or response, and the agent updates its behavior based on this feedback

3.4.1 RL for Control System Applications: A Control-Theoretic View

Although RL originates from Artificial Intelligence (AI), its structure aligns closely with classical control theory. The policy, "π_θ", serves as the *controller*, the environment contains the *plant*, disturbances, sensors, and actuators; observations, o_t, correspond to measurable building states ($o_t \approx s_t$); and the reward, r_t, encodes the performance objective. From this perspective, RL can be interpreted as an adaptive

[8] This is like the dog's strategy or way of deciding which *action* to take based on what it knows. Over time, with enough rewards and punishments, the dog will hopefully develop a policy which makes it to do a certain action every time one commands it to.

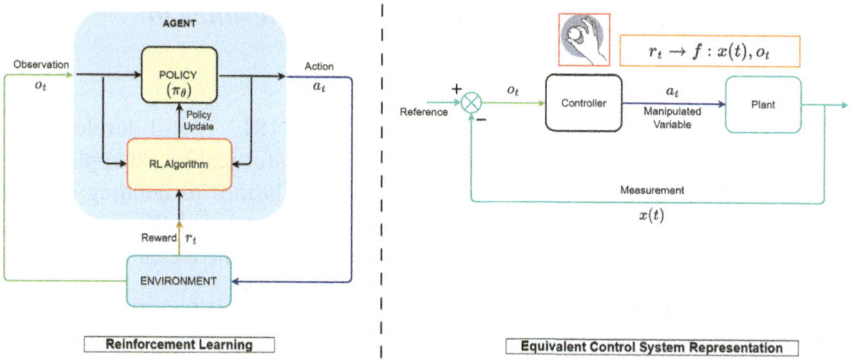

Fig. 3.8 **Left**: The canonical RL structure in which an agent selects an action, "a_t", receives an observation, "o_t", and evaluates the reward, "r_t", for policy improvement. **Right**: Equivalent control-theoretic interpretation where the RL policy functions as an adaptive controller, the environment corresponds to the plant and disturbances, and the reward synthesizes performance criteria such as comfort, energy cost, or constraint satisfaction

Table 3.1 Conceptual mapping between reinforcement learning (RL) components and classical control system elements

RL	Control systems
Policy	Controller
Environment	Everything which is not the controller
Observation	Any measurable output from the environment which is visible to the agent
Action	Manipulated variables or control actions
Learning algorithm	Adaptation mechanism of an adaptive controller

controller whose internal parameters are continuously updated. Figure 3.8 illustrates this equivalence. Further, Table 3.1 highlights how RL can be interpreted through a control-theoretic lens.

This interpretation has enabled RL to address increasingly complex building control problems. Ding et al. [25] formulated the coordinated scheduling of a multi-energy *Electric- Hydrogen-Heat Trigeneration* system as a *Markov decision process* and applied an improved Deep Deterministic Policy Gradient (I-DDPG) controller, achieving measurable reductions in operating cost and improved robustness under renewable uncertainty. Li et al. [26] further demonstrated that appropriate hyper-parameter optimization substantially increases the stability of deep RL controllers in HVAC applications.

These building-level insights extend foundational work by Nian et al. [27], who highlighted RL's suitability for industrial process control under noise, disturbances, and *nonlinear dynamics—a set of characteristics strongly mirrored in buildings*.

3.4.2 Some Famous Examples and Their Relevance to Buildings

It is important to note that Reinforcement Learning.[9] (RL) is still developing in the built-environment sector; however, its most successful real-world applications can be found in industries which have structural similarities to building control. These include environments characterized by uncertainty, delayed effects, nonlinear dynamics, and high- dimensional decision spaces. Representative examples include

- **Gaming (e.g., AlphaGo, Atari)**: RL excels in environments requiring long-term planning and strategic exploration—analogous to buildings where thermal inertia and pricing signals create delayed and temporally coupled effects.
- **Robotics**: Robots must operate under sensor noise, partial observability, and continuous control signals. Buildings exhibit similar characteristics, with uncertain occupancy, weather disturbances, and continuous HVAC control actions.
- **Healthcare**: Treatment sequencing requires balancing immediate and long-term outcomes, similar to balancing comfort and energy cost in building operations.
- **Finance and Portfolio Optimization**: Portfolio management involves risk-reward trade-offs and optimal sequential decision-making—paralleling Demand-Response (DR) participation and energy-cost optimization in smart buildings.
- **Autonomous Vehicles**: These systems must act under uncertainty while satisfying safety and comfort constraints—mirroring the safety, comfort, and operational limits imposed in building control systems.
- **Recommendation Systems**: Learning user preferences over time aligns closely with buildings which adapt to occupant comfort patterns and usage behaviors.

Within the built environment itself, RL is beginning to show substantial promise. Homod et al. [28] developed a cooperative multi-agent RL controller for Phase-Change-Material (PCM) storage sequencing, achieving over 28% energy savings by coordinating thermal storage with chiller COP characteristics. Moshari et al. [29] demonstrated real-world RL deployment in district energy control, showing that RL can operate robustly under measurement noise, partial observability, and dynamic disturbances.

These examples illustrate a central point: Reinforcement Learning (RL) thrives in complex, uncertain, and multi-timescale decision problems—precisely the characteristics of modern buildings. This makes RL a natural control paradigm for Digital Twin (DT) environments, where closed-loop learning, predictive adaptation, and continuous data synchronization enable autonomous refinement of operational behavior. By continuously evolving policies with incoming data, RL bridges the gap between Model-Free Control (MFC) and intelligent Digital Twins (DTs), allowing them not only to mirror the physical building but also to optimize its strategy dynamically.

[9] While promising, RL deployment in real buildings remains constrained by safety and sample-efficiency concerns.

This adaptive capability forms the foundation for hybrid control frameworks which integrate RL and MPC, discussed in the following sections.

3.5 Applications of MPC and RL—Relevance to Building Services Engineering

Having examined the structures and operating principles of Model Predictive Control (MPC) and Reinforcement Learning (RL), it is now important to situate these two control paradigms within the practical context of Building Services Engineering. Modern buildings are complex cyber-physical systems whose operation spans multiple timescales, objectives, and uncertainty sources. HVAC dynamics evolve slowly; occupancy and weather introduce stochastic disturbances; equipment limits impose hard constraints; and grid-interactive programs such as Demand Response (DR) require coordinated, anticipatory action. No single control strategy can satisfy all of these demands in isolation, which is why MPC and RL have emerged as complementary tools within the contemporary BEMS ecosystem.

From an engineering perspective, MPC excels in scenarios where reliable models exist and constraints must be respected explicitly. Its strengths include transparency, predictable behavior, and the ability to incorporate forecasts—weather, occupancy, solar irradiance, or electricity pricing—`within a unified optimization problem`. These capabilities make MPC particularly well-suited for supervisory HVAC control, thermal storage coordination, and DR scheduling, as demonstrated in the works of Habib et al. [22] and Behzadi et al. [23]. When integrated into a Digital Twin (DT) architecture, the MPC model acts as the *virtual* representation of the building, continuously calibrated through real-time measurements and used to generate informed, constraint-aware actions.

RL, in contrast, offers advantages in settings where models are uncertain, costly to maintain, or difficult to derive from *first principles*. Buildings exhibit nonlinear and partially observable thermal behavior, occupancy patterns which vary seasonally, and interaction effects among zones which defy low-order physics-based approximation. In these cases, RL can learn effective control strategies directly from data. Recent studies in buildings (e.g., Homod et al. [28], Moshari et al. [29]) show that RL agents can adapt to unmodeled dynamics, respond to disturbances, and refine their policies over time—attributes which align naturally with DT environments where continuous learning is a central feature.

Importantly, MPC and RL need not be viewed as competing alternatives. Instead, they occupy different regions of a broader design space:

- **MPC Provides Structure**: explicit constraint handling, interpretability, and the ability to embed domain knowledge through physics-based models or surrogate representations.

- **RL Provides Adaptability**: the capability to handle unknown dynamics, evolve with new data, and explore novel operating regimes—particularly valuable under uncertainty, variability, and *long-horizon effects*.
- **Hybrid MPC-RL Approaches**: increasingly explored in building research, combine prediction with learning, using MPC for safe or constraint-critical regimes and RL for adaptation, exploration, or online refinement.

This complementary viewpoint is especially relevant to Building Services Engineering (BSE), where operational conditions vary significantly across seasons, occupancy types, equipment age, and utility-market signals. A Digital Twin can orchestrate these complementary controllers: the MPC component ensuring reliable and constraint-compliant operation; the RL component enabling long-term performance improvement through interaction; and the twin itself mediating between the two using synchronized data.

In practice, choosing between MPC and RL therefore depends not on algorithmic preference, but on the nature of the building problem, the availability of models, the tolerance for exploration, and the criticality of constraints.

Which One to Choose?—Selecting between MPC and RL is ultimately a question of operational priority rather than theoretical superiority. Table 3.2 summarizes the core differences and helps identify which paradigm is better suited to specific building-control scenarios.

In numerous practical applications within the building sector, a *hybrid approach* is advantageous by integrating the optimal features of multiple methodologies. Model Predictive Control (MPC) is capable of ensuring baseline performance which is aware of constraints, whereas Reinforcement Learning (RL) functions within a "*learning layer*", refining policies, exploring strategies for energy conservation, or adapting to dynamic changes which are not encapsulated by the model. The Digital Twin shall serve as a secure environment conducive to experimental exploration, allowing RL to acquire knowledge in a virtual sandbox prior to implementation in the actual physical structure.

Thus, the decision between MPC and RL is not binary; it is architectural. Buildings which demand stability, compliance, and transparency benefit from MPC-dominant structures. Buildings operating in uncertain, data-rich, or rapidly evolving contexts benefit from RL-driven adaptability. Digital Twins may unify both by providing a synchronized, real-time platform where predictive models and learning agents work cooperatively.

Table 3.2 Comparison of MPC and RL for building control applications

Criterion	Model Predictive Control (MPC)	Reinforcement Learning (RL)
Model requirements	Requires accurate white, grey, or surrogate models; performance depends on model fidelity.	No explicit model required; learns behavior directly from interaction or data
Handling constraints	Naturally enforces constraints (comfort, equipment limits, safety) through optimization.	Does not guarantee constraint satisfaction unless enhanced with safe-RL mechanisms
Adaptability to change	Less adaptive to unforeseen dynamics; requires model updates or re-identification.	Highly adaptive; policy evolves with new data and changing building conditions
System/building dynamics	Best for systems with predictable and moderately nonlinear dynamics.	Well-suited for strongly nonlinear, uncertain, or partially observable systems
Operational stability (predictability)	Provides stable, transparent, and interpretable control decisions.	Policies may be opaque and require monitoring, especially during exploration phases
Long-horizon effects (Thermal inertia, demand response)	Handles anticipation via forecasts, but accuracy depends on model quality.	Captures long-term effects naturally through repeated interaction and reward structure
Computational complexity	Requires solving optimization problems at each step; complexity grows with model size.	High training cost; inference is fast once policy is learned
Practical use cases	Supervisory HVAC control, thermal storage coordination, schedule-based Demand Response.	Adaptive HVAC control, Demand Response with variable occupancy/weather, online optimization
Integration in Digital Twins	Serves as the predictive engine using the DT's calibrated virtual model.	Serves as the learning engine, improving performance through data-driven adaptation

The interaction between MPC, RL, the Digital Twin, and the physical building is summarized in Fig. 3.9, which illustrates how predictive and learning-based components could operate cooperatively within a unified DT architecture.

3.6 Understanding Digital Twins

Digital Twins (DTs) have rapidly emerged as a cornerstone of modern cyber-physical systems, yet the concept itself is not new. Long before buildings adopted the terminology, engineering domains such as aerospace, electric motors, and precision manufacturing had already embraced the idea of maintaining a continuously updated virtual representation of a physical asset. In these domains, the Digital Twin serves as a

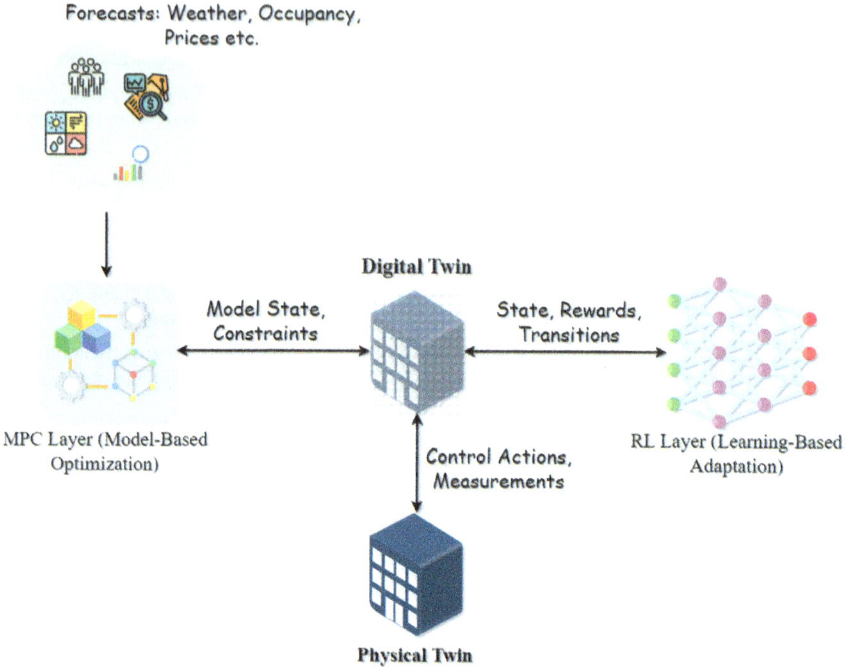

Fig. 3.9 The DT receives measurements from the physical building and provides synchronized state estimates to both the MPC layer (model-based optimization) and the RL layer (learning-based adaptation). MPC computes constraint-aware optimal actions using forecasts and virtual-model predictions, while RL refines policies through interaction data (state, reward, transitions). The DT fuses both outputs into coordinated control actions delivered to the physical building

"`computational replica`" capable of predicting behavior, detecting anomalies, and informing optimal control strategies in real time.

At its simplest, a Digital Twin can be understood as a dynamic virtual counterpart of a physical system, continuously synchronized through real-time data. The essential requirement is *bidirectional coupling*: the physical system provides measurements to the virtual model, and the virtual model generates predictions, diagnostics, or control decisions which influence the physical system. This continuous data-model-action loop distinguishes a Digital Twin from a traditional simulation model.

Digital Twin Versus Simulation Model: The Core Distinction

A simulation model can reproduce how a system behaves under specified inputs, but it is typically *static*: once built, its parameters do not evolve automatically when the real system changes. A Digital Twin, by contrast, must satisfy three additional properties:

1. **Continuous Synchronization**: Model states and parameters are regularly updated using sensor data.
2. **Predictive Capability**: The twin can forecast future states, not merely replicate current ones.
3. **Closed-Loop Intelligence**: The twin can recommend or generate control actions in real time.

Thus, while every Digital Twin contains a simulation model, not every simulation model qualifies as a twin. It is the integration of prediction, feedback, and adaptation which elevates a model into a Digital Twin.

Digital Twins Beyond Buildings

The concept of a Digital Twin is well established in domains with fast and largely deterministic dynamics. A widely studied example is the Digital Twin of an `electric motor` [30–33], used extensively for monitoring, fault detection, real-time interface development, and performance optimization. Previous works show the development of motor twins by utilizing empirical performance models. These works combine analytical and data-driven approaches. Further, these also employ high-fidelity implementations which integrate Unity3D/ROS environments for real-time visualization and controller-in-the-loop testing. The studies show that motor-based twins operate over compact electromechanical dynamics with dense, high-rate sensing—conditions which make their virtual counterparts tightly synchronized and highly predictive. In such systems, the governing physics is compact, measurable, and described by relatively simple electromechanical equations. For instance, a basic motor twin may rely on the following differential relationships:

Example: Core Equations Behind a Motor Digital Twin

The dynamics governing the rotor of a simple Direct Current (DC) motor can be mathematically expressed as

$$J\frac{d\omega(t)}{dt} = T_m(t) - T_L(t) - B\omega(t)$$

where $\omega(t)$ denotes the angular velocity, $T_m(t)$ represents the electromagnetic torque, $T_L(t)$ is the load torque, J is the moment of inertia, and B is the viscous damping coefficient.

The electromagnetic torque $T_m(t)$ is described by

$$T_m(t) = K_t i(t)$$

where K_t is the torque constant and $i(t)$ is the armature current. The current dynamics are given by

$$L\frac{di(t)}{dt} = V(t) - Ri(t) - K_e\omega(t)$$

where L and R represent inductance and resistance, respectively, and K_e is the back electromotive force (EMF) constant.

A Digital Twin framework utilizes real-time measurements of $i(t)$ and $\omega(t)$ to iteratively update these model parameters, diagnose deviations, and forecast future system behavior.

The differential equations above form the mathematical core of a simple motor Digital Twin. Listing 3.1 illustrates how these equations can be implemented in Python™[10] to simulate the motor's behavior in real time. This implementation is then augmented with Digital Twin capabilities, including sensor-noise emulation, parameter-drift tracking, and online error evaluation. Listing 3.2 encapsulates the supporting utility functions which enable these enhancements, providing a modular and reusable structure for advanced Digital Twin applications. In this example, the Digital Twin evolves its internal states (speed and current) according to the governing dynamics and the applied voltage, thereby closely mirroring the response of the physical motor.

Figure 3.10 illustrates the simulated response of this virtual motor, showing how the Digital Twin evolves its internal states over time. Even this simple implementation demonstrates the essential characteristic of a Digital Twin: a computational model capable of predicting how the physical system behaves under a given set of inputs. *In its current form, the simulation is an "open-loop" twin*[11] *which mirrors the underlying physics; however, the same structure can be transformed into a fully predictive Digital Twin by synchronizing the model with real measurements, updating parameters online, and validating predictions against incoming data.*

[10] The Python™ implementation in Listing 3.1 relies on three widely used scientific libraries: (i) `NumPy` for numerical operations, (ii) `SciPy` for solving ordinary differential equations via `solve_ivp`, and (iii) `Matplotlib` for visualization. All three libraries are part of the standard scientific Python™ ecosystem and are freely available.

[11] An *open-loop Digital Twin*: predicts motor behavior from physics and fixed inputs without feedback, yet remains a valid twin as a virtual representation.

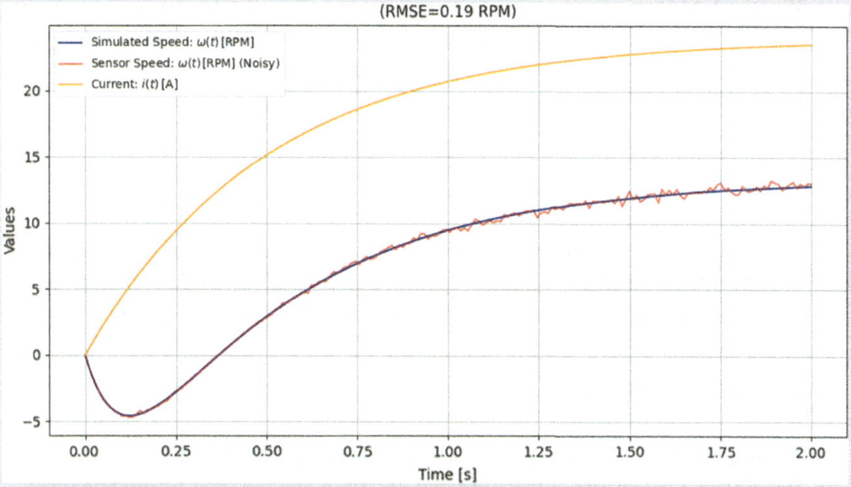

Fig. 3.10 Simulated behavior of the motor Digital Twin. Using the dynamic model in Listing 3.1, the twin predicts the motor's speed $\omega(t)$ and armature current $i(t)$ in response to an applied voltage. This response underpins a predictive Digital Twin, which can be calibrated and enhanced with real sensor data

Parameters Which Make the Motor Digital Twin Predictive

A motor Digital Twin becomes *predictive* when its dynamic model parameters are continuously updated using real sensor data. In the DC motor model, the key parameters are

- **Mechanical**: J (inertia), B (damping)
- **Electrical**: R (resistance), L (inductance)
- **Electromechanical**: K_t (torque constant), K_e (back-EMF constant)
- **External inputs**: $V(t)$ (applied voltage), $T_L(t)$ (load torque)

These parameters govern acceleration, damping, current evolution, torque generation, and load interaction. Over time, factors such as friction changes, winding resistance drift, or aging can cause deviations between the simulated and actual motor behavior. A predictive Digital Twin compensates by recalibrating these parameters online using measurements of speed, current, temperature, or vibration.

Important Clarification: When we say "update B," we mean that the Digital Twin updates the value of B in its simulation model—not that the physical damping of the motor changes. The distinction is

- **Physical Motor**: The actual damping (friction) in the hardware is determined by mechanical wear, lubrication, and load conditions. This cannot be changed by the Digital Twin; it is a physical property.
- **Digital Twin**: The virtual model uses B in its differential equation

$$\frac{d\omega}{dt} = \frac{K_t i - T_L - B\omega}{J}$$

If the real motor exhibits more friction (e.g., due to aging), measured speed will deviate from the simulated speed. The Digital Twin detects this and adjusts its internal value of B so that the simulation matches reality.

Why Is This Beneficial?

- **Accuracy**: Keeps the simulation aligned with the real motor.
- **Predictive Capability**: Once updated, the twin can forecast future behavior under different loads or voltages.
- **Maintenance Insight**: A persistent increase in the estimated B indicates rising friction → potential bearing wear → need for scheduled maintenance.
- **Control Optimization**: Updated parameters enable better tuning of controllers (e.g., PID gains).

Through continuous estimation, the Digital Twin remains synchronized with the physical motor, enabling accurate forecasting under varying control inputs. This capability supports real-time monitoring, anomaly detection, and control optimization.

```python
from motor_twin import add_sensor_noise, apply_parameter_drift, compute_rmse, plot_dashboard
    , log_data
import numpy as np
from scipy.integrate import solve_ivp
import matplotlib.pyplot as plt

# Original motor parameters
J, B = 0.01, 0.1           # inertia, damping
R, L = 1.0, 0.5            # resistance, inductance
Kt, Ke = 0.01, 0.01        # torque and back-EMF constants
V = 24                     # applied voltage
T_load = 0.1               # load torque

# Differential equations
def motor_dynamics(t, y):
    omega, i = y
    domega_dt = (Kt * i - T_load - B * omega) / J
    di_dt = (V - R * i - Ke * omega) / L
    return [domega_dt, di_dt]

# Initial conditions
y0 = [0.0, 0.0]            # omega, i
t_span = (0, 2)
```

```
t_eval = np.linspace(*t_span, 200)

# Solve
sol = solve_ivp(motor_dynamics, t_span, y0, t_eval=t_eval)
rpm = sol.y[0] * 60 / (2 * np.pi)

# Digital Twin Enhancements: Sensor Noise, Parameter Drift, and Online Error Evaluation
rpm_noisy = add_sensor_noise(rpm)            # Emulate noisy sensor readings
R_drift = apply_parameter_drift(R, sol.t)    # Simulate gradual resistance drift
error = compute_rmse(rpm, rpm_noisy)         # Calculate RMSE between simulated and noisy data

# Plot
plot_dashboard(sol.t, rpm, rpm_noisy, sol.y[1], error)
log_data('motor_twin_log.csv', sol.t, rpm, sol.y[1])
```

Listing 3.1 Python implementation of the core motor equations (with noise injection) used in a simple Digital Twin simulation.

```
import numpy as np
import matplotlib.pyplot as plt
import pandas as pd

# ==============================
# SENSOR NOISE SIMULATION
# ==============================
def add_sensor_noise(data, noise_level=0.02):
    """
    Adds Gaussian noise to simulated sensor data.
    :param data: Original data array
    :param noise_level: Noise intensity (default 2%)
    :return: Noisy data array
    """
    return data * (1 + noise_level * np.random.randn(len(data)))

# ==============================
# PARAMETER DRIFT MODEL
# ==============================
def apply_parameter_drift(R, t, drift_rate=0.05):
    """
    Simulates gradual drift in resistance over time.
    :param R: Initial resistance value
    :param t: Time array
    :param drift_rate: Percentage drift over full time span
    :return: Array of drifted resistance values
    """
    return R * (1 + drift_rate * t / max(t))

# ==============================
# ERROR METRIC CALCULATION
# ==============================
def compute_rmse(actual, noisy):
    """
    Computes Root Mean Square Error between actual and noisy data.
    :param actual: Original data array
    :param noisy: Noisy data array
    :return: RMSE value
    """
    return np.sqrt(np.mean((actual - noisy)**2))

# ==============================
# PLOTS
# ==============================
def plot_dashboard(t, rpm, rpm_noisy, current, error):
    """
    Plots simulated RPM, noisy RPM, and current on a dashboard.
    Displays RMSE in the title.
    :param t: Time array
```

```
    :param rpm: Simulated RPM array
    :param rpm_noisy: Noisy RPM array
    :param current: Motor current array
    :param error: RMSE value
    """
    plt.figure(figsize=(10, 5))
    plt.plot(t, rpm, label=r'$\text{Simulated Speed: } \omega(t)\,\text{[RPM]}$', color='blue')
    plt.plot(t, rpm_noisy, label=r'$\text{Sensor Speed: } \omega(t)\,\text{[RPM] (Noisy)}$',
        color='red', alpha=0.6)
    plt.plot(t, current, label=r'$\text{Current: } i(t)\,\text{[A]}$', color='orange')
    plt.title(f'(RMSE={error:.2f} RPM)')
    plt.xlabel('Time [s]')
    plt.ylabel('Values')
    plt.legend()
    plt.grid(True)
    plt.show()

# ==============================
# DATA LOGGING TO CSV
# ==============================
def log_data(filename, t, rpm, current):
    """
    Logs time, RPM, and current data to a CSV file.
    :param filename: Output CSV file name
    :param t: Time array
    :param rpm: RPM array
    :param current: Current array
    """
    df = pd.DataFrame({'time': t, 'rpm': rpm, 'current': current})
    df.to_csv(filename, index=False)
```

Listing 3.2 Utility functions for the motor Digital Twin: sensor noise model, parameter drift, RMSE computation, plots, and data logging.

Extended Example: Real-Time Synchronization in a Motor Digital Twin

The open-loop model in Listing 3.1 provides the backbone of a Digital Twin by reproducing the dynamical response of the physical motor. A true predictive twin, however, must also process *real-world measurements*, compensate for *sensor noise*, and reason about *model drift*. These effects are commonly expressed as

$$\tilde{\omega}(t) = \omega(t)\big(1 + \eta_\omega(t)\big), \qquad \tilde{i}(t) = i(t)\big(1 + \eta_i(t)\big)$$

where $\eta_\omega(t), \eta_i(t) \sim \mathcal{N}(0, \sigma^2)$ denote Gaussian sensor noise. The multiplicative form reflects realistic speed and current measurement errors.

In some cases, gradual parameter drift—such as an increase in resistance due to winding temperature—may be represented as

$$R(t) = R_0\left(1 + \alpha \frac{t}{t_{\max}}\right)$$

Here, R_0 is the nominal resistance and α a drift factor. $t_{\max}$ is a time constant representing the maximum (or characteristic) time used to normalize t, setting the time scale over which $R(t)$ changes. Although the synchronized twin used here focuses on noise and state estimation rather than explicitly modeling this drift, the expression reflects how Digital Twins typically represent long-term component changes.

A synchronized Digital Twin continuously evaluates the discrepancy between the simulated speed $\omega(t)$ and the *estimated* speed $\widehat{\omega}(t)$ obtained via filtering. A widely used metric is the root-mean-square error:

$$\text{RMSE} = \sqrt{\frac{1}{N}\sum_{k=1}^{N}\Big(\omega(k) - \widehat{\omega}(k)\Big)^2}$$

These additional layers—sensing, estimation, and deviation quantification—transform the simulation from a passive open-loop model into an actively synchronized Digital Twin capable of tracking measurements, identifying mismatch, and supporting predictive control.

Listing 3.3 shows how the above equations can be implemented in Python™, and Listing 3.4 encapsulates the corresponding utility functions along with the *Kalman filtering*. Figure 3.11 illustrates the simulated response.

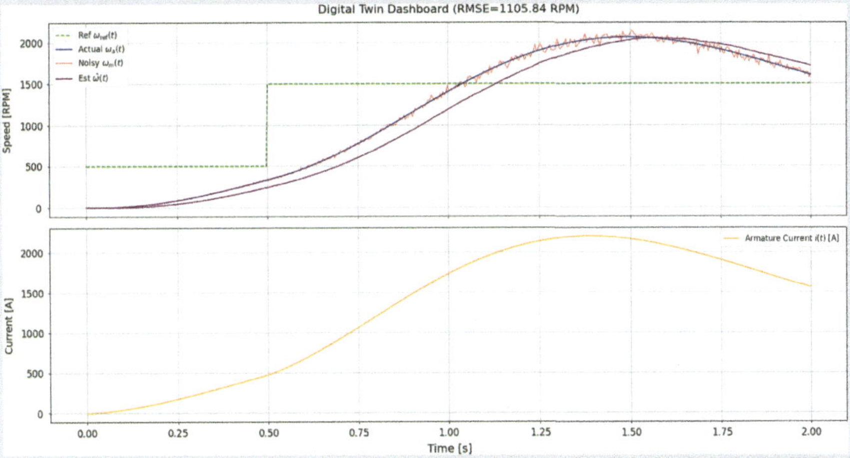

Fig. 3.11 Synchronized Digital Twin dashboard for the DC motor example. The upper subplot compares the reference trajectory, $\omega_{\text{ref}}(t)$, "actual" (simulated) motor speed, $\omega_a(t)$, emulated noisy measurements, $\omega_m(t)$, and Kalman-filtered estimate, $\hat{\omega}(t)$. The lower subplot shows armature current, $i(t)$. The Digital Twin computes an RMSE value which quantifies model-measurement deviation, enabling real-time monitoring and drift detection

```python
# motor_twin_main.py
import numpy as np
from motor_twin_utils import add_sensor_noise, kalman_filter, compute_rmse, plot_dashboard,
    log_data

# ==============================
# MOTOR PARAMETERS
# ==============================
J, B = 0.01, 0.1           # Inertia and damping
R, L = 1.0, 0.5            # Resistance and inductance
Kt, Ke = 0.01, 0.01        # Torque and back-EMF constants
T_load = 0.1               # Load torque

# PI controller gains
Kp, Ki = 2.0, 50.0

# Reference speed trajectory (rad/s)
def reference_speed(t):
    return (1500 * 2 * np.pi / 60) if t > 0.5 else (500 * 2 * np.pi / 60)


# ==============================
# SIMULATION SETTINGS
# ==============================
t_eval = np.linspace(0, 2, 500)
dt = t_eval[1] - t_eval[0]

# Initial states
omega, i, integral_error = 0.0, 0.0, 0.0

# Logging arrays
omega_list, ref_list, noisy_list, est_list, current_list = [], [], [], [], []

# Kalman Filter initialization
x_est = np.array([[0.0]])  # Estimated omega
P = np.array([[1.0]])      # Covariance
A = np.array([[1]])        # State transition
H = np.array([[1]])        # Measurement matrix
Q = np.array([[0.001]])    # Process noise covariance
R_kf = np.array([[0.5]])   # Measurement noise covariance

# ==============================
# MAIN SIMULATION LOOP
# ==============================
for t in t_eval:
    # Reference speed
    ref = reference_speed(t)
    error = ref - omega

    # PI control law
    V = Kp * error + Ki * integral_error

    # Motor dynamics (Euler integration)
    domega_dt = (Kt * i - T_load - B * omega) / J
    di_dt = (V - R * i - Ke * omega) / L
    omega += domega_dt * dt
    i += di_dt * dt
    integral_error += error * dt

    # Sensor noise
    noisy_measurement = add_sensor_noise(np.array([omega]))[0]

    # Kalman filter update
    x_est, P = kalman_filter(noisy_measurement, A, H, Q, R_kf, x_est, P)

    # Log data
    omega_list.append(omega)
```

```
    ref_list.append(ref)
    noisy_list.append(noisy_measurement)
    est_list.append(x_est[0])
    current_list.append(i)

# ==============================
# POST-PROCESSING
# ==============================
rpm_actual = np.array(omega_list) * 60 / (2 * np.pi)
rpm_ref = np.array(ref_list) * 60 / (2 * np.pi)
rpm_noisy = np.array(noisy_list) * 60 / (2 * np.pi)
rpm_est = np.array(est_list) * 60 / (2 * np.pi)
current_array = np.array(current_list)

# Compute RMSE
rmse = compute_rmse(rpm_actual, rpm_est)

# Visualization
plot_dashboard(t_eval, rpm_ref, rpm_actual, rpm_noisy, rpm_est, rmse, current_array)

# Save data
log_data('motor_twin_kalman.csv', t_eval, rpm_actual, rpm_est, rpm_noisy, current_array)
```

Listing 3.3 Interactive Digital Twin for a DC motor with PI control, sensor noise emulation, Kalman-based state estimation, and online RMSE evaluation.

```

# motor_twin_utils.py
import numpy as np
import matplotlib.pyplot as plt
import pandas as pd

# ==============================
# SENSOR NOISE SIMULATION
# ==============================
def add_sensor_noise(data, noise_level=0.02):
    """Adds Gaussian noise to simulated sensor data."""
    return data * (1 + noise_level * np.random.randn(len(data)))

# ==============================
# KALMAN FILTER FOR STATE ESTIMATION
# ==============================
def kalman_filter(z, A, H, Q, R, x_est, P):
    """
    Applies one step of Kalman Filter.
    z: measurement
    A: state transition matrix
    H: measurement matrix
    Q: process noise covariance
    R: measurement noise covariance
    x_est: previous state estimate
    P: previous covariance
    Returns: updated state estimate and covariance
    """
    # Prediction
    x_pred = A @ x_est
    P_pred = A @ P @ A.T + Q

    # Update
    K = P_pred @ H.T @ np.linalg.inv(H @ P_pred @ H.T + R)
    x_est = x_pred + K @ (z - H @ x_pred)
    P = (np.eye(len(P)) - K @ H) @ P_pred
    return x_est, P

# ==============================
# ERROR METRIC CALCULATION
# ==============================
```

```python
def compute_rmse(actual, estimated):
    """Computes RMSE between actual and estimated data."""
    return np.sqrt(np.mean((actual - estimated)**2))

# ==============================
# DASHBOARD VISUALIZATION (Speed + Current)
# ==============================
def plot_dashboard(t, rpm_ref, rpm_actual, rpm_noisy, rpm_est, error, current):
    """
    Plots reference, actual, noisy, and estimated RPM in one subplot,
    and armature current in another subplot.
    """
    fig, axs = plt.subplots(2, 1, figsize=(10, 8), sharex=True)

    # Speed tracking
    axs[0].plot(t, rpm_ref, '--', label=r'Ref $\omega_{\text{ref}}(t)$', color='green')
    axs[0].plot(t, rpm_actual, label=r'Actual $\omega_{a}(t)$', color='blue')
    axs[0].plot(t, rpm_noisy, label=r'Noisy $\omega_{m}(t)$', color='red', alpha=0.5)
    axs[0].plot(t, rpm_est, label=r'Est $\hat{\omega}(t)$', color='purple')
    axs[0].set_ylabel('Speed [RPM]', fontsize=14)
    axs[0].set_title(f'Digital Twin Dashboard (RMSE={error:.2f} RPM)', fontsize=16)
    axs[0].legend(fontsize=12)
    axs[0].grid(True)

    # Current evolution
    axs[1].plot(t, current, label=r'Armature Current $i(t)$ [A]', color='orange')
    axs[1].set_xlabel('Time [s]', fontsize=14)
    axs[1].set_ylabel('Current [A]', fontsize=14)
    axs[1].legend(fontsize=12)
    axs[1].grid(True)

    # Increase tick label size
    for ax in axs:
        ax.tick_params(axis='both', labelsize=12)

    plt.tight_layout()
    plt.show()

# ==============================
# DATA LOGGING TO CSV
# ==============================
def log_data(filename, t, rpm_actual, rpm_est, rpm_noisy, current):
    """Logs time, actual RPM, estimated RPM, noisy RPM, and current to CSV."""
    df = pd.DataFrame({
        'time': t,
        'actual_rpm': rpm_actual,
        'estimated_rpm': rpm_est,
        'noisy_rpm': rpm_noisy,
        'current_A': current
    })
    df.to_csv(filename, index=False)
```

Listing 3.4 Utility functions for Digital Twin enhancements: sensor noise emulation, Kalman filtering, RMSE computation, dashboard visualization, and data logging.

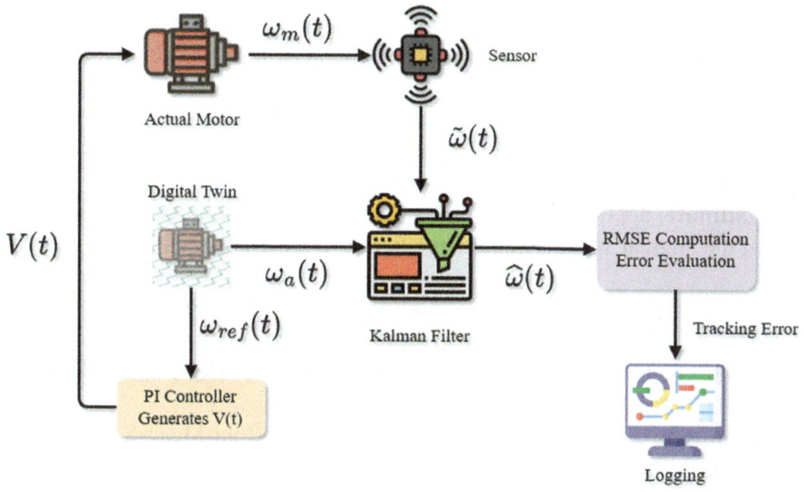

Fig. 3.12 Conceptual structure of a synchronized Digital Twin: sensor data, estimation, physics-based prediction, error evaluation, and supervisory control

Key Parameters for a Predictive DC Motor Digital Twin

The synchronized Digital Twin implementation[12] in Listings 3.3 and 3.4 extends the open-loop example by adding measurements, noise, online estimation, and deviation tracking. Figure 3.12 provides a visual illustration of this concept. Together, these elements show how a Digital Twin evolves from a purely physics-based simulation into an operational model which continuously aligns with its physical counterpart.

At each time step, the twin performs four functions:

1. **Simulate**: integrates the motor dynamics using the physical model:

$$\dot{\omega}(t) = \frac{K_t\, i(t) - T_L(t) - B\,\omega(t)}{J}, \qquad \frac{i(t)}{dt} = \frac{V(t) - R\,i(t) - K_e\,\omega(t)}{L}$$

as in the previous example.

---

[12] The synchronized Digital Twin implementation in Listings 3.3 and 3.4 uses four widely available scientific Python™ libraries: (i) `NumPy` for numerical operations, (ii) `SciPy` for differential equation solvers, (iii) `Matplotlib` for dashboard visualization, and (iv) `pandas` for structured data logging. All four are part of the standard scientific Python ecosystem and freely available.

2. **Sense**: generates a noisy speed measurement by injecting Gaussian sensor noise:
$$\tilde{\omega}(t) = \omega(t)\big(1 + \eta(t)\big)$$
where $\eta(t) \sim \mathcal{N}(0, \sigma^2)$ captures sensor imperfections.
3. **Estimate**: applies a `Kalman filter` to obtain a filtered estimate $\widehat{\omega}(t)$ of the angular speed using the prediction-correction scheme:
$$\widehat{\omega}^-(t) = A\,\widehat{\omega}(t - \Delta t), \qquad \widehat{\omega}(t) = \widehat{\omega}^-(t) + K\big(\tilde{\omega}(t) - H\,\widehat{\omega}^-(t)\big)$$
where A and H define the state and measurement models, and the Kalman gain, K, balances model and sensor information.
4. **Compare**: quantifies predictive accuracy via the Root-Mean-Square Error (RMSE):
$$\text{RMSE} = \sqrt{\frac{1}{N}\sum_{k=1}^{N}(\omega(k) - \widehat{\omega}(k))^2}$$
measuring deviation between simulated and estimated behavior.

These steps yield a *synchronized virtual model* rather than a static, open-loop replica, able to track the physical motor under measurement noise and parameter uncertainty. The `Proportional-Integral (PI) controller` in Listing 3.3 further shows how the Digital Twin functions in closed loop, mirroring both the motor dynamics and the supervisory control actions.

The dashboard-style view shown in Fig. 3.11 reflects typical Digital Twin operation in cyber-physical systems: real sensors supply imperfect data, the virtual model predicts behavior, and an estimation layer reconciles both to preserve fidelity. The same framework can be extended to update parameters such as R, B, or K_t online for fault detection and adaptive calibration. Buildings, as discussed later in this chapter, follow the same principles but with slower, noisier dynamics, sparser sensing, and more interacting subsystems.

A Digital Twin is, therefore, more than a simulation; it is a dynamic model which continuously synchronizes with its physical counterpart through real-time data. For an electric motor, this means that sensor measurements periodically update the model states or parameters, ensuring accuracy and enabling predictive capabilities such as evaluating future trajectories under alternative control inputs. When coupled with a closed-loop mechanism, the twin can influence operational decisions, transforming it into a tool for forecasting, anomaly detection, and real-time optimization.

The same principles apply to buildings, but with slower thermal dynamics, higher uncertainty, and broader measurement needs. *Unlike motors, which exhibit fast and largely deterministic behavior, Building Digital Twins often combine physics-based models with machine learning and data assimilation to remain predictive* [2, 3]. In essence, whether for motors or buildings, a true Digital Twin integrates continuous data, predictive modeling, and decision-making to bridge the physical and virtual worlds.

For motors, the twin typically operates at millisecond time scales with dense, high-resolution sensing. The dynamics are fast, deterministic, and well understood, making the Digital Twin relatively compact and easier to maintain.

`Buildings`, by contrast, exhibit

- slow thermal dynamics,
- heterogeneous materials,
- uncertain occupancy,
- nonlinear HVAC behavior,
- partial observability.

Consequently, a Building Digital Twin must rely more heavily on statistical models, surrogate models, machine learning, or hybrid AI components to maintain predictive fidelity. Table 3.3 illustrates this contrast by comparing Digital Twins of fast electromechanical machines with slow, stochastic building systems.

Table 3.3 Comparing Digital Twin complexity: fast deterministic machines versus slow stochastic building systems

Aspect	Motors/machines	Buildings/facilities
Dynamics	Milliseconds to seconds; rapid electromechanical states	Minutes to hours; slow thermal and occupancy dynamics
Dimensionality	Few, well-measured states	Many zones, materials, loads, and latent factors
Determinism	Largely deterministic	Highly stochastic (weather, occupancy)
Sensing	Dense, high-rate measurements	Sparse, heterogeneous sensing; partial observation
Physics	Compact ODE/PDE models	Hybrid thermo-electrical behavioral coupling
Control loop	Real-time $(kHz - Hz)$	Supervisory $(minutes - hours)$
Synchronization latency	Near real-time; minimal lag	Asynchronous; frequent misalignment
Calibration	Infrequent, batch updates	Continuous re-anchoring to new data
Drift risk	Low; relatively stable operation	High; usage patterns, sensor drift, and aging effects
Data exchange	High-rate, deterministic	Lower-rate; e.g., 15-min comfort and 5-min control intervals
Modeling effort	Moderate	Extensive multi-zone and schedule data required
Validation KPIs	Efficiency, torque ripple, EMI	Energy, comfort, IAQ, flexibility, and cost
Typical tools	SPICE, Simulink, Thermal FE solvers	EnergyPlus™ with Python™/BMS/EMS co-simulation
Twin approach	Single-rate, fixed model	Multi-rate, adaptive, surrogate-assisted

Supervisory Control and Data Acquisition (SCADA) Versus Digital Twin (DT)

At this point, it is also imperative to distinguish Supervisory Control and Data Acquisition (SCADA) systems from Digital Twins. Table 3.4 highlights the key differences between the two.

Table 3.4 Comparison between SCADA and Digital Twin across key aspects

Aspect	SCADA	Digital Twin
Purpose	Real-time monitoring and basic supervisory control of industrial processes	Virtual replica for simulation, prediction, and optimization
Nature	Primarily descriptive and reactive	Predictive and often prescriptive
Data usage	Collects, logs, and displays sensor data	Uses sensor data to update model parameters and run simulations
Modeling	No explicit physics-based model; focuses on visualization and alarms	Physics-based or data-driven model (e.g., differential equations or ML surrogates)
Forecasting	Not inherently supported	Predicts future states under varying conditions
Control	Allows operator commands (start/stop, setpoints)	Suggests or generates optimal control strategies based on predictions
Maintenance insight	Relies on thresholds and alarms	Detects trends (e.g., friction increase) for predictive maintenance
Integration	Often the data source for a Digital Twin	Consumes SCADA/IoT data for continuous calibration

The Digital Twin as a Control-Oriented Construct

Despite differences across domains, the unifying idea is that a Digital Twin is fundamentally a *control system construct*. It is not merely a visualization or a geometric model, but a computational entity which:

- predicts the future behavior of the physical asset,
- evaluates alternative control actions,
- assimilates real-time data to update its internal representation,
- generates or supports optimal supervisory actions.

Thus, the Digital Twin extends the logic of control engineering into a continuously learning, data-driven environment. In buildings, this integration becomes essential: slow dynamics, delayed effects, and stochastic disturbances require blending physics-based models with adaptive intelligence—precisely the combination discussed in Sect. 3.5.

Why Digital Twins Are Gaining Traction in Buildings

Buildings differ from motors, aircraft, or manufacturing assets in one fundamental respect: they exhibit *long memory*. Thermal inertia, occupancy variability, and equipment cycling introduce "`long-horizon effects`" which cannot be captured through instantaneous measurements. Digital Twins enable buildings to

- track drifting parameters (e.g., insulation degradation, equipment efficiency),
- anticipate demand peaks and thermal behavior,
- coordinate storage, solar, HVAC, and grid signals,
- provide a safe environment for testing AI-driven controllers,
- support supervisory control through MPC and RL (as established in Sect. 3.5).

Viewed this way, *Digital Twins are not futuristic constructs but a natural evolution of building energy management*—linking sensing, simulation, prediction, and control in a unified operational loop. The next section formalizes the distinction between simulation models and Digital Twins and situates these concepts within the broader context of modern building automation and cyber-physical intelligence.

3.7 Simulation Models Versus Digital Twins

Simulation models occupy a foundational role in building performance analysis, yet their purpose and behavior differ fundamentally from that of Digital Twins (DTs). A simulation model represents a mathematical or computational abstraction of a physical system under *assumed* conditions, whereas a Digital Twin represents a *continuously evolving* virtual counterpart that remains anchored to reality through incoming data. The distinction is subtle but crucial: simulation models help us understand and predict behavior under predefined scenarios, while Digital Twins help us operate, control, and *optimize* behavior under actual, time-varying conditions.

Simulation environments allow engineers to explore "what-if" scenarios, evaluate design alternatives, and study system behavior before deployment. However, they remain static unless manually recalibrated. By contrast, a Digital Twin operates as a *living model*—updating its internal parameters, learning from new measurements, and adapting its forecasts to reflect the current state of the physical asset. This recurring data-model-action loop gives the Digital Twin an operational character which a traditional simulation model inherently lacks.

In buildings, this distinction has profound implications. A detailed simulation of a room, zone, or entire facility may be accurate at the time of creation but gradually diverges from reality as occupancy patterns shift, equipment ages, architectural modifications occur, or sensors drift. A Digital Twin prevents such divergence by continuously synchronizing itself with new measurements, thereby maintaining predictive fidelity. Simulation models therefore remain indispensable for design and planning, while Digital Twins extend these capabilities into *operations*, ensuring the model remains relevant throughout the building's lifecycle.

Simulation Models

Simulation models approximate the physical behavior of buildings using physical laws, empirical relationships, or data-driven abstractions. They form the computational backbone of early-stage design and retrofit analysis. Examples include

- thermal simulations for HVAC sizing and energy demand estimation,
- day-lighting simulations for visual comfort and glare assessment,
- airflow and contaminant dispersion modeling,
- electromechanical equipment models for chiller, pump, or generator optimization.

These models typically rely on fixed parameter sets—such as envelope thermal resistances, equipment performance curves, occupancy schedules, and predefined control strategies—all of which are assumed constant over time. Once constructed, the simulation proceeds with no intrinsic link to real-world measurements unless manual calibration is performed.

In practice, simulation models excel at evaluating design choices under controlled assumptions, performing sensitivity analysis, exploring scenarios, and satisfying compliance or code-driven requirements. However, they do *not* operate synchronously with the real building.

Digital Twins

Digital Twins extend the concept of simulation by embedding it within an ecosystem of continuous sensing, prediction, and control. Their value lies not in reproducing the "ideal" building but in understanding and anticipating the *actual* building as it evolves. Unlike simulation models which rely on static assumptions, Digital Twins continually update their internal representations using:

- sensor data (temperature, flow, occupancy, weather),
- equipment performance metrics,
- actuator feedback (valves, dampers, setpoint behavior),
- historical and forecasted disturbances (weather, schedules, pricing).

Where simulation models answer *what might happen*, Digital Twins answer *what is happening now* and *what is likely to happen next*. This distinction is what makes Digital Twins indispensable for supervisory control strategies such as MPC and RL.

3.8 Room or Building Simulation Models as Digital Twins

Not every simulation model qualifies as a Digital Twin—but with appropriate extensions, many room or building-level models can *become* one. This transformation is particularly relevant in buildings, where simulation engines such as EnergyPlus™, Modelica™, TRNSYS™, or custom MATLAB™ and Python™-based thermal models already capture much of the governing physics.

A room-level or whole-building simulation becomes a Digital Twin when three conditions are met:

1. Continuous Data Assimilation
The simulation model must ingest real-world measurements and update its internal states or parameters accordingly. Examples include

- updating *thermal capacitances* or *conductances* based on measured drift,
- adjusting HVAC efficiency curves using observed performance,
- correcting occupancy or lighting schedules based on sensor data,
- re-anchoring envelope parameters using weather-driven discrepancies.

Without this data assimilation, the model remains a simulation, not a twin.

2. Predictive and Scenario-Driven Intelligence
A Digital Twin does not merely reflect the present state; it forecasts future trajectories under variable conditions such as

- anticipated weather variations,
- expected occupancy trends,
- dynamic electricity tariffs,
- peak load or Demand-Response (DR) events.

These predictive capabilities make the Digital Twin a forward-looking decision engine.

3. Integration with Control Logic
The final requirement is that the Digital Twin participates in—or meaningfully supports—supervisory control. This may involve

- Model Predictive Control (MPC) using the DT to compute and optimize future trajectories,
- Reinforcement Learning (RL) agents interacting with the DT to learn policies safely,
- fault or anomaly detection via DT–sensor deviations,
- automated setpoint recommendations or constraint-aware advisory control.

Once integrated with control, the simulation model transitions from a passive tool to an *active participant* in building operation.

Example: EnergyPlus™ as a Digital Twin Backbone
EnergyPlus™, traditionally used for performance simulation, can serve as a Digital Twin engine when:

- model parameters are recalibrated using real operational data,
- it receives real-time inputs via co-simulation (e.g., Python™ API, FMI, BOPTEST),

- prediction cycles are executed at supervisory intervals (e.g., 5–15 min),
- the model exchanges information with a BMS/EMS for control decisions.

This elevates the simulation model from a design-analysis tool to a continuously synchronized cyber-physical asset, aligning directly with the hybrid MPC–RL architecture illustrated earlier in Fig. 3.9.

As buildings transition toward grid-interactive, flexible, and autonomous operation, the distinction between "`simulation`" and "`Digital Twin`" becomes increasingly significant. Simulation models support design. Digital Twins support operation. And hybrid cyber-physical systems (MPC + RL + DT) support *intelligent, adaptive* operation across the *building lifecycle*.

3.9 Challenges, Benefits, and Emerging Considerations

Building Digital Twins (BDTs) hold immense promise as cyber-physical platforms for prediction, diagnostics, optimization, and control. Yet, the same characteristics which make buildings complex to operate also make their Digital Twins challenging to maintain. *These include nonlinear dynamics, stochastic disturbances, high-dimensional state spaces, calibration drift, data quality issues, and the need to balance multiple conflicting objectives.* A growing body of literature documents these challenges in detail [34–36].

Importantly, these limitations also establish the natural motivation for Computational Intelligence (CI) techniques—surrogate modeling, Evolutionary Optimization, Reinforcement Learning, and data-driven state estimation—which are explored in detail in Part II of this book. This section positions those developments by examining the key challenges and emerging considerations in modern BDT practice.

3.9.1 The Challenge of Nonlinearity in Building Energy Dynamics

Among all challenges, nonlinearity remains the most fundamental. Buildings exhibit highly nonlinear thermo-electrical interactions: radiative and convective heat transfer, HVAC equipment cycling, humidity dynamics, saturation effects in actuators, Variable Air Volume (VAV) flow relationships, and envelope-weather coupling [37, 38]. Unlike electromechanical systems—such as DC motors—where nonlinearities are limited and can be reliably linearized over operating ranges, building physics lack such convenience. Even small changes in occupancy, solar gains, or internal loads can create large deviations in indoor states.

This has direct implications for Digital Twins. While state observers such as the Kalman Filter are effective for low-dimensional, weakly nonlinear systems (e.g., device-level twins or small thermal zones), their performance deteriorates in whole-building contexts due to non-Gaussian disturbances, multi-modal dynamics, and partial observability. For this reason, Digital Twins of motors can rely on compact stateâŁ"space models with high accuracy, whereas buildings require hybrid approaches which combine physics-based models with Machine Learning (ML), surrogate modeling, or Reinforcement Learning (RL).

These nonlinear effects also motivate the need for CI-based optimization (e.g., NSGA-II, surrogate-assisted search), which does not assume convexity or smoothness—properties rarely satisfied in building dynamics.

3.9.2 Stochasticity and Unpredictability

Occupancy, weather, window blinds, internal equipment, and behavioral patterns introduce stochasticity at multiple scales. The uncertainty spectrum ranges from short-term disturbances (e.g., door openings, solar fluctuations) to long-term uncertainties (e.g., seasonal variations, retrofits, equipment degradation). Traditional deterministic models cannot fully anticipate this variability, resulting in reduced predictive accuracy and suboptimal control.

Modern BDTs shall incorporate stochastic modeling and uncertainty-aware approaches—probabilistic forecasts, ensemble learning, and RL-based policies—to maintain robustness under such conditions. This challenge lays the foundation for data-driven control methods discussed in Part II.

3.9.3 High-Dimensionality and Partial Observability

Large buildings and campuses contain hundreds of thermal zones, sensors, actuators, and operational degrees of freedom. Only a subset of internal states is monitored, and sensing infrastructures may be sparse or heterogeneous. This renders many building states effectively unobservable.

Device-level systems (like motors) are fully instrumented and highly observable; their Digital Twins can estimate hidden states with classical observers. In contrast, building-scale Digital Twins require:

- state estimation via Machine Learning (ML) or probabilistic filters,
- surrogate models to reduce dimensionality,
- decomposition techniques to handle subsystem-level observability.

3.9.4 Long-Memory and Delayed Dynamics

Buildings possess long thermal memory: HVAC actions may take minutes or hours to manifest in zone temperatures. This leads to delayed feedback, slow convergence, and strong cross-coupling between zones.

Such long-horizon effects demand predictive approaches which reason about future consequences. Model Predictive Control (MPC) and Reinforcement Learning (RL) are natural fits here, as they explicitly internalize predictions over extended horizons [22].

3.9.5 Model Drift and Degradation Over Time

Equipment aging, fouling, sensor bias, seasonal weather patterns, and changes in building usage cause progressive deviation between the physical building and its Digital Twin. This *twin-building divergence*, commonly referred to as `model drift`, can degrade control performance and predictive accuracy.

Motor systems also experience drift (e.g., friction increase, winding resistance change), but the underlying physics remains simpler and easier to recalibrate. Buildings require continuous parameter adaptation, online learning, and periodic re-anchoring of model parameters—strategies which motivate machine-learning-assisted and hybrid Digital Twins.

3.9.6 Data Quality, Noise, and Heterogeneity

Real-world building data are often noisy, missing, delayed, or inconsistent across platforms (BMS, IoT sensors, meters). Data fusion and cleaning remain labor-intensive.

Unlike DC motors—where measurements are dense, clean, and high frequency—buildings frequently suffer from sparse and infrequent sampling intervals (e.g., 5–15 min). This makes Kalman-type estimation partially effective only at the subsystem levels.

3.9.7 Multi-objective and Conflicting Operational Goals

Buildings must simultaneously optimize

- energy consumption,
- thermal comfort,
- indoor air quality,

- equipment health,
- cost and tariff signals,
- and grid-interactivity (Demand Response).

These objectives are often conflicting, non-convex, and time-varying. CI-based Multi-Objective Optimization (e.g., NSGA-II, SA-NSGA-II) thus becomes essential for navigating the operational trade-space [39]. Part II of the book is dedicated to this aspect.

3.9.8 Computational Burden and the Need for Surrogates

High-fidelity building simulations (e.g., EnergyPlus™) can be computationally expensive, especially when embedded in real-time supervisory control loops. Iterative optimization, sequential prediction, and Monte Carlo evaluation may become infeasible without suitable approximations.

> Machine Learning (ML)-based surrogate models dramatically reduce this burden and form a key pillar of CI-enhanced Digital Twins. Surrogates enable near-real-time control, optimization, and scenario evaluation.

3.9.9 Benefits and Opportunities Enabled by Digital Twins

While the preceding subsections have emphasized the challenges inherent to BDT development, it is equally important to recognize the significant benefits they unlock when implemented successfully. A well-calibrated Digital Twin shall provide a continuously updated representation of the building, enabling predictive insight, faster decision-making, and improved operational intelligence. Key advantages include: (i) enhanced supervisory control through model-based or learning-based optimization; (ii) improved fault detection, diagnostics, and predictive maintenance; (iii) integration of occupant-centric comfort models; (iv) scenario testing without disrupting actual operations; and (v) seamless participation in grid-interactive efficient building programs and Demand-Response (DR) strategies.

These benefits implicate why Digital Twins are emerging as a foundational technology for modern BEMS. They create an operational platform where Computational Intelligence (CI) methods—discussed in detail in Part II—can be deployed to achieve performance levels that would be impractical using traditional control approaches alone.

These benefits, however, come with the responsibility of ensuring that increasingly autonomous Digital Twins remain safe, trustworthy, and robust during operation. This naturally leads to emerging considerations such as runtime assurance and cyber-physical integrity.

3.9.9.1 Runtime Assurance and Cyber-Physical Integrity

Ensuring runtime assurance in Building Digital Twin (BDT) environments is fundamental to maintaining safety, trust, and cyber-physical integrity. As Digital Twins increasingly generate supervisory or optimal control commands—often through machine learning, optimization, or hybrid AI algorithms—the risk of issuing unsafe or anomalous actuation signals becomes non-negligible. These risks may stem from model inaccuracies, data corruption, unexpected system drift, or, in the worst case, deliberate `cyber-physical attacks`.

A useful conceptual parallel can be drawn from the work of Pearce et al. [40], who developed *Smart I/O Modules* to mitigate cyber-physical attacks in industrial control systems. Their framework, though designed for PLC and relay-based grid environments, demonstrates a broader principle: embedding a hardware-level *runtime enforcer* between the controller and the actuation layer to validate commands against a set of safety envelopes. Their approach uses Valued Discrete Timed Automata (VDTA) to ensure that malicious or erroneous signals never reach the physical plant unless they satisfy predefined temporal and operational constraints.

While buildings differ substantially from industrial plants, the underlying lesson translates meaningfully. In BEMS and BDT architectures, similar edge-level enforcers—implemented via embedded controllers, smart gateways, or safety wrappers—can act as a final safeguard. They can validate AI-generated or optimization-generated setpoints (e.g., temperature, airflow, valve positions, DR signals) against acceptable operational bounds, comfort constraints, and equipment protection limits. This *enforcer-at-the-edge* paradigm therefore represents a compelling strategy for achieving self-assured, cyber-resilient Digital Twins, capable of maintaining safe operation even if supervisory controllers, cloud analytics, or learned policies malfunction, drift, or are compromised.

3.9.9.2 Resilience and Fault-Tolerant Control

Modern buildings are increasingly operated under continuous sensing and automated supervisory control, yet the physical assets themselves remain susceptible to hardware faults, communication interruptions, and unexpected degradations. Sensors may drift or fail; actuators may stick, jam, or exhibit deadband nonlinearities; and supervisory control networks may experience delays or packet loss. A resilient Digital Twin must therefore recognize and compensate for such anomalies, ensuring that control performance degrades gracefully rather than abruptly.

Fault-tolerant Digital Twins incorporate mechanisms for early anomaly detection, fault diagnosis, and model reconfiguration. Machine-learning-based diagnostics, such as classification models or autoencoder-based reconstruction errors, play an increasingly important role in identifying deviations from expected behavior. When paired with robust or adaptive controllers—such as fault-tolerant MPC or fault-aware RL—BDTs can maintain safe operation even when parts of the sensing or actuation infrastructure degrade. Indeed, this capability is essential for real-world deployment.

3.9.9.3 Human-in-the-Loop and Explainability

Despite increasing autonomy in BEMS, human operators—facility managers, commissioning engineers, energy analysts—remain central to decision-making. Digital Twins are intended to generate optimal control recommendations or predictive insights that must be reviewed, validated, or occasionally overridden by humans. This creates a need for transparency and explainability in the underlying algorithms.

`Black-box` CI methods, especially Deep Learning (DL) and Reinforcement Learning (RL), may produce control actions which are difficult to interpret in operational terms. To ensure trust and adoption, BDT frameworks must present predictions, recommended actions, and inferred internal states in a "human-understandable" manner. Human-in-the-loop interfaces allow operators to explore "what-if" scenarios, visualize model uncertainties, and enforce overrides when necessary. Explainability also plays a critical role in verifying that CI-based controllers operate within equipment constraints, comfort boundaries, and regulatory requirements.

These considerations reinforce the importance of interpretable surrogate models, transparent optimization, and explainable RL.

3.9.10 Summary

In summary, buildings present a unique combination of nonlinear, stochastic, high-dimensional, and slow-moving dynamics which challenge traditional modeling and control approaches. These factors highlight both the limitations of physics-only Digital Twins and the necessity of integrating Computational Intelligence (CI) for prediction, estimation, optimization, and autonomous control. Part II of this book explores these CI methods in depth—demonstrating how hybrid modeling, Evolutionary Algorithms (EAs), surrogate learning, and reinforcement learning can enable the next generation of `intelligent Building Energy Management Systems.`

References

1. Yang C, Létourneau S, Guo H (2014) Developing data-driven models to predict BEMS energy consumption for demand response systems. In: International conference on industrial, engineering and other applications of applied intelligent systems. Springer, pp 188–197
2. Roda-Sanchez L, Cirillo F, Solmaz G, Jacobs T, Garrido-Hidalgo C, Olivares T, Kovacs E (2023) Building a smart campus digital twin: system, analytics, and lessons learned from a real-world project. IEEE Internet Things J 11(3):4614–4627
3. Zhang R, Wang F, Cai J, Wang Y, Guo H, Zheng J (2022) Digital twin and its applications: a survey. Int J Adv Manuf Technol 123(11):4123–4136
4. Wani MH (2022) Intelligent control strategies for efficient building energy management systems (BEMS). Department of Electrical and Electronic Engineering, University of Auckland, PhD thesis
5. Tomar A, Nguyen PH, Mishra S (2022) Control of smart buildings: an integration to grid and local energy communities. Springer. https://link.springer.com/chapter/10.1007/978-981-19-0375-5_8
6. Michailidis P, Michailidis I, Minelli F, Coban HH, Kosmatopoulos E (2025) Model predictive control for smart buildings: applications and innovations in energy management. Buildings 15(18). https://doi.org/10.3390/buildings15183298. https://www.mdpi.com/2075-5309/15/18/3298
7. Zhu J, Tian Z, Niu J, Lu Y, Cheng B, Zhou H (2025) Machine learning-enhanced lightweight rule-based control strategy for building energy demand response. In: Building simulation. Springer, pp 1–20. https://link.springer.com/article/10.1007/s12273-025-1275-1
8. Zhang L, Chen Z, Zhang X, Pertzborn A, Jin X (2023) Challenges and opportunities of machine learning control in building operations. In: Building simulation, vol 16. Springer, pp 831–852. https://link.springer.com/article/10.1007/s12273-023-0984-6
9. Fang Q, Zhu Q, Qiao F (2019) Advancements in smart city and intelligent building. Springer https://link.springer.com/chapter/10.1007/978-981-13-6733-5_5
10. Claessens BJ, Vanhoudt D, Desmedt J, Ruelens F (2018) Model-free control of thermostatically controlled loads connected to a district heating network. Energy Build 159:1–10
11. Xiong Q, Li Z, Cai W, Wang Z (2023) Model free optimization of building cooling water systems with refined action space. Building simulation, vol 16. Springer, pp 615–627
12. Li S, Su S, Lin X (2025) Optimizing the hyper-parameters of deep reinforcement learning for building control. Building simulation, vol 18. Springer, pp 765–789
13. Esrafilian-Najafabadi M, Haghighat F (2023) Transfer learning for occupancy-based HVAC control: a data-driven approach using unsupervised learning of occupancy profiles and deep reinforcement learning. Energy Build 300:113637
14. Coraci D, Brandi S, Hong T, Capozzoli A (2024) An innovative heterogeneous transfer learning framework to enhance the scalability of deep reinforcement learning controllers in buildings with integrated energy systems. Building simulation, vol 17. Springer, pp 739–770
15. Nagy Z, Henze G, Dey S, Arroyo J, Helsen L, Zhang X, Chen B, Amasyali K, Kurte K, Zamzam A et al (2023) Ten questions concerning reinforcement learning for building energy management. Build Environ 241:110435
16. Fu Y, Xu S, Zhu Q, O'Neill Z, Adetola V (2023) How good are learning-based control vs model-based control for load shifting? Investigations on a single zone building energy system. Energy 273:127073
17. Aslam MS, Kaplun D, Romanov S, Yahya A, Badruddin IA, Ayouni S (2025) Adaptive model-based control for optimized energy of complex building network requisition time in thermal energy storage. Alex Eng J 130:208–226
18. Morcillo-Jimenez R, Mesa J, Gómez-Romero J, Vila MA, Martin-Bautista MJ (2024) Deep learning for prediction of energy consumption: an applied use case in an office building. Appl Intell 54(7):5813–5825

19. Redder F, Althaus P, Ubachukwu E, Mork M, Johnen S, Küpper C, Lieberenz P, Oden M, Westphal L, Storek T et al (2025) Information and communication technologies (ICT) for the intelligent operation of building energy systems: Design, implementation and evaluation in a living lab. Energy Inf 8(1):77
20. Vinh NQ, Le TL (2025) Self-learning model-based control for sensorless induction motor drives. Electr Eng 1–14
21. Schwenzer M, Ay M, Bergs T, Abel D (2021) Review on model predictive control: an engineering perspective. Int J Adv Manuf Technol 117(5):1327–1349
22. Habib M, Palomba V, Frazzica A, Wang Q (2025) Optimizing hybrid thermal energy storage in building management systems using data-driven model predictive control. Energy Rep 14:2092–2109
23. Behzadi A, Goudarzi N, Ploskic A, Thorin E, Sadrizadeh S (2026) Advancing an already high-performance smart building with model predictive control: multi-layer optimization under forecast uncertainty in a real building case. Appl Energy 402:126999
24. Wald D, Doronina O, Johnson K, King R, Sinner M, Griffin K, Chintala R, Vaidhynathan D, Sanyal J, Day M (2025) A neural-network-enhanced parameter-varying framework for multi-objective model predictive control applied to buildings. In: Energy and AI, p 100566
25. Ding J, Chen B, Lei Y, Zhang W (2025) Coordinated scheduling of electric-hydrogen-heat tri-generation system for low-carbon building based on improved reinforcement learning. Energy Eng 122(11)
26. Li J, Xie H, Zhang J, Li L, Song G, Fu H, Chen P, Liu C, Zhang L, Shi Z, et al (2025) A systematic review of reinforcement learning in building-integrated photovoltaic (BIPV) optimization. Advances in applied energy, p 100239
27. Nian R, Liu J, Huang B (2020) A review on reinforcement learning: introduction and applications in industrial process control. Comput Chem Eng 139:106886
28. Homod RZ, Dhaidan NS, Mohammed HI, Al-Mousawi F, Alawi OA, Mahdi JM, Yadav KK, Togun H, Albadr MA, Sadeq AM et al (2025) Flattening power curves in smart buildings: deep reinforcement learning for enhancing chiller performance and binary phase change materials. J Energy Storage 140:119163
29. Moshari A, Javanroodi K, Nik VM (2026) Real-world deployment of model-free reinforcement learning for energy control in district heating systems: enhancing flexibility across neighboring buildings. Appl Energy 402:12699. https://doi.org/10.1016/j.apenergy.2025.126997, https://www.sciencedirect.com/science/article/pii/S0306261925017271
30. Venkatesan S, Manickavasagam K, Tengenkai N, Vijayalakshmi N (2019) Health monitoring and prognosis of electric vehicle motor using intelligent-digital twin. IET Electr Power Appl 13(9):1328–1335
31. Rassõlkin A, Rjabtšikov V, Vaimann T, Kallaste A, Kuts V, Partyshev A (2020) Digital twin of an electrical motor based on empirical performance model. In: 2020 XI International conference on electrical power drive systems (ICEPDS). IEEE, pp 1–4
32. Rassõlkin A, Rjabtšikov V, Kuts V, Vaimann T, Kallaste A, Asad B, Partyshev A (2022) Interface development for digital twin of an electric motor based on empirical performance model. IEEE Access 10:15635–15643
33. Hu W, Wang T, Chu F (2023) Novel Ramanujan digital twin for motor periodic fault monitoring and detection. IEEE Trans Industr Inf 19(12):11564–11572
34. Attaran M, Celik BG (2023) Digital twin: benefits, use cases, challenges, and opportunities. Decis Anal J 6:100165
35. Botín-Sanabria DM, Mihaita AS, Peimbert-García RE, Ramírez-Moreno MA, Ramírez-Mendoza RA, Lozoya-Santos JJ (2022) Digital twin technology challenges and applications: a comprehensive review. Remote Sens 14(6):1335
36. Rasheed A, San O, Kvamsdal T (2020) Digital twin: values, challenges and enablers from a modeling perspective. IEEE Access 8:21980–22012
37. Wani M, Swain A, Ukil A (2019) Control strategies for energy optimization of HVAC systems in small office buildings using EnergyPlusTM. In: 2019 IEEE innovative smart grid technologies-Asia (ISGT Asia), pp 2698–2703. https://doi.org/10.1109/ISGT-Asia.2019.8880806

38. Wani M, Hafiz F, Swain A, Ukil A (2021) Estimating thermal parameters of a commercial building: a meta-heuristic approach. Energy Build 231:11053. https://doi.org/10.1016/j.enbuild.2020.110537, https://www.sciencedirect.com/science/article/pii/S0378778820316832
39. Wani M, Swain A, Ukil A, Ploder M, Koole R (2022) Optimizing the overall performance of forced extraction systems: a multi-objective framework. Build Environ 221:109214.https://doi.org/10.1016/j.buildenv.2022.109214, https://www.sciencedirect.com/science/article/pii/S0360132322004504
40. Pearce H, Pinisetty S, Roop PS, Kuo MM, Ukil A (2019) Smart I/O modules for mitigating cyber-physical attacks on industrial control systems. IEEE Trans Industr Inf 16(7):4659–4669

Part II
Application of Computational Intelligence (CI) to BEMS

Chapter 4
Foundations of Computational Intelligence (CI) for Building Digital Twins

Abstract This chapter establishes Computational Intelligence (CI) as the conceptual bridge between physics-based building simulation and the emerging paradigm of Building Digital Twins (BDTs). Building energy systems are shown to exhibit nonlinear, stochastic, and multi-objective behavior, rendering classical optimization and control approaches insufficient in isolation. Through an illustration of complex search landscapes, global versus local optimization behavior, and Pareto-based trade-offs, optimization is reframed as a process of structured exploration under uncertainty rather than the pursuit of a single optimal solution. The chapter clarifies how CI complements physics-based models by enabling adaptive search, learning, and multi-objective reasoning, thereby addressing fundamental challenges associated with Digital Twin development. Rather than implementing BDTs directly, this chapter prepares the analytical and conceptual foundations upon which subsequent chapters build, motivating the CI methodologies, simulation infrastructures, and surrogate-assisted optimization frameworks developed in the rest of this book.

Keywords Artificial Neural Networks · Black-box models · Convolutional Neural Networks · Computational intelligence · Grey-box models · Hybrid systems · Machine learning

4.1 Prelude: From Control to Computational Intelligence (CI)

The preceding chapters established the foundations of Building Energy Management Systems (BEMS), feedback control, Model Predictive Control (MPC), Reinforcement Learning (RL), and the evolution of these ideas into modern Building Digital Twins (BDTs). As shown in Chap. 3, Digital Twins integrate physics-based modeling, real-time sensing, prediction, and decision-making, but their effectiveness is ultimately constrained by the inherent complexities of buildings—nonlinear dynamics, stochastic disturbances, partial observability, long-memory effects, multi-objective conflicts, and computational burden. These characteristics expose the limitations of

M. H. Wani et al., *Digital Twins for Building Optimization: Enhancing Energy and Comfort*, Studies in Infrastructure and Control,
https://doi.org/10.1007/978-3-032-23274-8_4

traditional control methods and motivate the transition toward more adaptive, data-driven, and optimization-rich methodologies.

This chapter forms a conceptual bridge between Part I and Part II of the book. Its purpose is to provide a clear, accessible, and unified introduction to the field of *Computational Intelligence* (CI) and to outline why CI has become indispensable for developing predictive, resilient, and autonomous Digital Twins. Rather than focusing on any single technique, the chapter presents the broader CI landscape—encompassing learning systems, evolutionary computation, swarm intelligence, fuzzy reasoning, and stochastic search—so that the reader may appreciate the underlying principles which bind these methods together. These foundations are crucial because the author's PhD research [1] (presented in Chaps. 5 and 6) relies directly on CI-driven Multi-Objective Optimization (MOO) and surrogate-assisted modeling within *first-principles* simulation settings. By the end of this chapter, the reader will be well equipped to grasp how CI augments classical control, why it is vital for Digital Twins, and how it lays the groundwork for the advanced frameworks introduced later in the book.

4.2 What Is Computational Intelligence?—Scope and Philosophy

Computational Intelligence (CI) should not be viewed as a single technique or algorithmic family, but rather as a unifying paradigm for building adaptive, perception-driven, and optimization-oriented systems capable of operating in uncertain, nonlinear, and partially observable environments. In the context of buildings and Building Digital Twins (BDTs), where dynamics are slow, disturbances are stochastic, sensors are imperfect, and the underlying thermal-electrical-occupant interactions are difficult to model explicitly, CI provides the methodological foundation needed to go beyond the limitations of purely analytical or classical control approaches.

CI has been interpreted differently across the literature, but the underlying ideas share substantial alignment. Early concepts emphasized the distinction between CI and symbolic Artificial Intelligence (AI). Bezdek's foundational works position CI as a form of "low-level cognition" concerned with numerical pattern recognition, adaptation, and fault-tolerant computation—while deliberately avoiding the symbolic knowledge structures which characterize expert systems and classical AI [2]. Pedrycz [3] expands the notion of CI into a cohesive computational framework built around fuzzy sets, neurocomputing, and evolutionary computation, highlighting their shared capacity for learning, approximation, and decision-making under uncertainty.

Duch [4] reframes CI as a practical "`toolbox for intelligent behavior`," emphasizing nonlinear mapping, classification, memory organization, and heuristic optimization. Meanwhile, Hanne and Dornberger [5] provide a contemporary (modern) and operational definition: *CI is a family of heuristic, adaptive, and optimization-centric methods—including Evolutionary Algorithms (EAs), swarm*

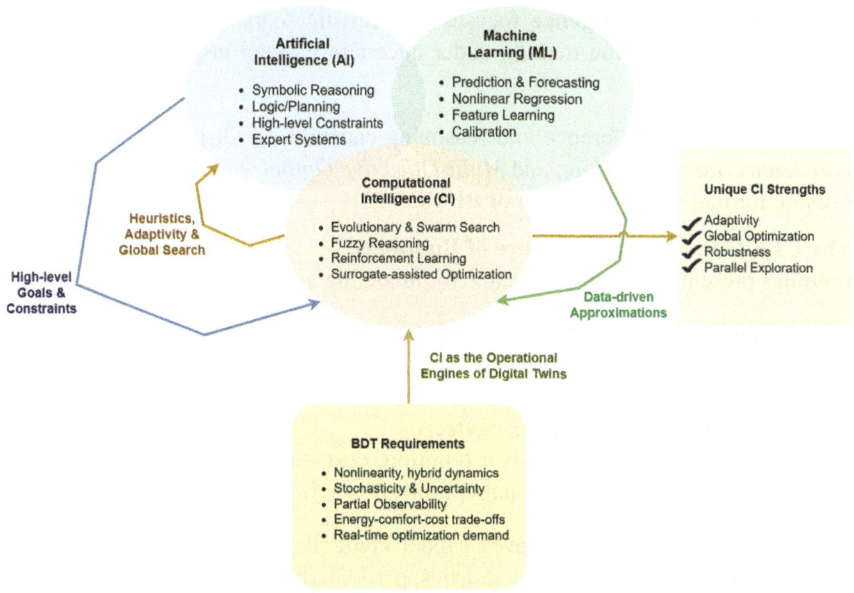

Fig. 4.1 Conceptual relationship between AI, ML, and CI. CI forms the optimization-centric, adaptive, and uncertainty-aware computational core which complements symbolic AI and leverages statistical learning from ML. *For Building Digital Twins, CI provides the global search, adaptivity, and robustness required to operate under nonlinear and stochastic real-world conditions*

intelligence, fuzzy reasoning, and neural learning—which solve complex real-world problems for which exact solutions are either `analytically intractable` *or* `computationally prohibitive.`

Despite their stylistic differences, all four perspectives converge on several defining principles of CI:

- CI methods are **numerical**, **data-driven**, and inherently **adaptive**.
- CI does not require a fully specified analytical model, but can leverage one when advantageous.
- CI thrives in environments characterized by **nonlinearity**, **noise**, and **high uncertainty**.
- CI complements (rather than replaces) both classical AI and modern Machine Learning (ML).

CI Versus AI Versus Machine Learning (ML)

Although CI, AI, and ML often overlap (see Fig. 4.1), distinguishing them clarifies their roles in Digital Twin development:

- **AI** (classical/symbolic) focuses on reasoning, logic, planning, rule-based decision-making, and explicit knowledge representation.
- **Machine Learning** focuses on statistical generalization, predictive modeling, and data-driven structure extraction.

- **Computational Intelligence** focuses on heuristic search, global optimization, adaptivity, and decision-making under uncertainty, often inspired by *biological or natural systems*.

ML and AI provide inference and reasoning capabilities; CI provides the *global search*, *uncertainty-handling*, and *Multi-Objective Optimization (MOO)* machinery essential for real-world operation of BDTs.

Why CI Fits the Complex Nature of Buildings?
Buildings present a difficult landscape for modeling and control [6]:

- strong nonlinearities and hybrid continuous—discrete behavior
- stochastic disturbances (weather, occupancy, equipment faults)
- partial observability and sensor uncertainty
- slow dynamics and long-memory effects
- conflicting performance objectives (comfort, cost, energy, IAQ)
- non-convex and high-dimensional optimization surfaces

Classical control—including convex Model Predictive Control (MPC)—struggles when confronted with these characteristics, particularly when the true optimum lies in a highly non-convex region or when the building model is imperfect. CI provides what these systems lack: **global, model-free (or model-light) search** capabilities which do not depend on convexity, linearity, or perfect models [7–9].

Bezdek's criteria for CI—adaptivity, fault tolerance, human-comparable error bounds, and pattern-recognition ability—align almost exactly with what operational Digital Twins require to function reliably in real time [2]. Evolutionary Algorithms (EAs), swarm intelligence, fuzzy reasoning, and Reinforcement Learning (RL) become natural complements to physics-based simulation, which could enable BDTs to [10]:

- explore non-convex search spaces
- adapt to *model drift* and changing conditions
- calibrate and refine models through data assimilation
- generate surrogate models (leveraging ML methods) for real-time prediction
- achieve multi-objective trade-offs between energy and comfort
- operate robustly even under sensor faults and uncertainty

Thus, CI could augment the "physics-based layer" of the Digital Twin with an *adaptive intelligence layer*, enabling Digital Twins to behave not merely as simulators but as predictive, resilient, and self-improving decision-making systems [11].

4.3 The Pillars of Computational Intelligence

The previous section established Computational Intelligence (CI) as a unifying paradigm for adaptive, optimization-oriented, and uncertainty-aware computation in complex environments such as buildings. While that discussion was mostly conceptual, operational Digital Twins require concrete algorithmic tools. This section therefore introduces the main "pillars" of CI which will appear throughout the remainder of the book: learning systems, evolutionary computation, swarm intelligence, and fuzzy systems.

At an abstract level, a CI-based decision maker can be viewed as a mapping:

$$u = \mathcal{F}(x, \theta) \tag{4.1}$$

where "x" denotes the input state (for example, indoor temperature, occupancy, and weather variables), "u" is the resulting action or decision (for example, a control signal or setpoint), and "θ" represents the adaptable parameters of the system. In contrast to purely model-based or analytical approaches, the parameters "θ" in CI are updated through heuristic search, learning, or approximate reasoning, often in the presence of noisy or incomplete information.

These pillars are not mutually exclusive silos (divisions). Instead, they frequently coexist and interact within a single framework—for example, Evolutionary Algorithms (EAs) tuning the hyper-parameters of a Reinforcement Learning (RL) controller, or fuzzy rules combined with neural networks in a neuro-fuzzy inference system.

Nevertheless, it is pedagogically useful to describe each "family" separately before discussing their integration in later chapters. A high-level taxonomy of these components is shown in Fig. 4.2, which complements the conceptual AI–ML–CI diagram in Fig. 4.1 introduced earlier.

Another subtle but important point concerns the boundary between CI and modern Machine Learning (ML). In current usage, the distinction is often blurred: many authors, for instance, treat Reinforcement Learning (RL) as part of CI, while others consider it primarily a subfield of ML and, broader, AI. In this book, RL is discussed as a *learning system* which is frequently used in conjunction with CI techniques (for example, using ANNs for function approximation in RL), but it is not treated as a core pillar of CI in the original, strict sense—which focused on systems that learn from data or adapt without prior assumptions about the underlying probability distributions.

4.3.1 Learning Systems (Model-Free, Model-Based, and Hybrid)

Learning systems constitute one of the foundational pillars of Computational Intelligence (CI) and provide the approximation, prediction, and adaptation capabilities required for modern Building Digital Twins (BDTs). While Chap. 3 focused on the control-theoretic perspective of *model-based* (e.g., MPC) and *model-free* (e.g., RL)

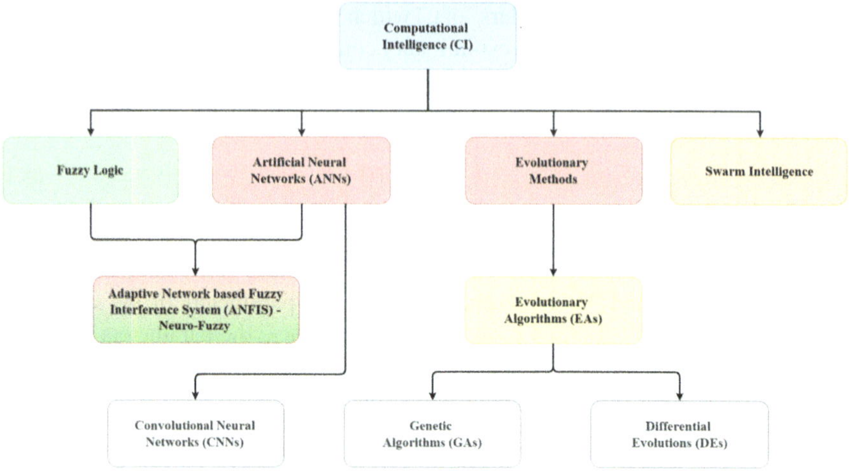

Fig. 4.2 Pillars of Computational Intelligence: The core families—fuzzy logic, neural systems, and evolutionary-swarm methods—form the adaptive engine which empowers CI to learn, reason, and optimize under uncertainty

decision-making, the present section introduces their complementary representation within CI: neural learning systems which approximate nonlinear mappings, update internal parameters online, and support calibration, forecasting, and decision-making under uncertainty.

Consistent with the generic CI mapping introduced in Eq. (4.1) learning systems primarily shape the parameter vector "θ" through data-driven adaptation. Depending on the structural relationship between the system model and the learning mechanism, these approaches are broadly grouped as

- **Model-Free Learning Systems**: approximate $u = \mathcal{F}(x, \theta)$ directly from data without requiring an explicit physics model.
- **Model-Based Learning Systems**: incorporate a known or partially known model into the learning process.
- **Hybrid Systems**: combine data-driven learning with physics- or rule-based structures.

In BDTs, these learning paradigms are employed for model calibration, state estimation, fault detection, surrogate modeling, and predictive control. The following subsections introduce core neural structures, illustrate how they are used in BDT workflows, and provide small, self-contained examples which complement—rather than repeat—the discussions of RL and MPC from Chap. 3.

Artificial Neural Networks (ANNs): Universal Function Approximators

Artificial Neural Networks (ANNs) are archetypal *model-free* systems. They learn a nonlinear mapping $u = \mathcal{F}(x, \theta)$ purely from data by adjusting internal weights through gradient-based optimization, consistent with the adaptable parameters introduced in Chap. 3 and Eq. (4.1). For a feed-forward ANN with L layers, the forward pass is

$$h^{(1)} = \sigma\left(W^{(1)}x + b^{(1)}\right), \quad h^{(\ell)} = \sigma\left(W^{(\ell)}h^{(\ell-1)} + b^{(\ell)}\right), \ \ell = 2, \ldots, L-1 \tag{4.2}$$

$$\hat{u} = W^{(L)}h^{(L-1)} + b^{(L)} \tag{4.3}$$

Symbol Definitions

- $x \in \mathbb{R}^{n_x}$: Input feature vector (e.g., sensor readings, occupancy data)
- $u \in \mathbb{R}^{n_u}$: True output variable (e.g., energy consumption, comfort index)
- $\hat{u}$: ANN-predicted output
- $h^{(\ell)}$: Activation vector of hidden layer ℓ
- $W^{(\ell)}$: Weight matrix for layer ℓ
- $b^{(\ell)}$: Bias vector for layer ℓ
- $\sigma(\cdot)$: Nonlinear activation function (e.g., ReLU, tanh)
- L: Total number of layers in the network
- $\theta = \{W^{(\ell)}, b^{(\ell)}\}$: Set of trainable parameters

The parameters "θ" are learned by minimizing a suitable loss function (e.g., mean squared error) using gradient-based optimization methods such as stochastic gradient descent or Adam (Adaptive Moment Estimation). By the Universal Approximation Theorem, ANNs can approximate any continuous function on a compact domain, making them powerful surrogates for complex building dynamics.

ANNs are widely researched in BDT pipelines to [12–15]:

- learn surrogate models of EnergyPlus™-based dynamics,
- approximate comfort, load, or occupancy patterns,
- support parameter re-identification or drift detection through online retraining,
- replace expensive physical simulations in MPC loops (hybrid MPC–ANN schemes).

Figure 4.3 presents a conceptual ANN structure consistent with notation used in previous diagrams.

Convolutional Neural Networks (CNNs)

CNNs originated as an advanced form of ANNs, introducing convolutional layers to exploit spatial correlations in data. Table 4.1 highlights the differences between the two. While ANNs treat all inputs as independent features, CNNs apply learnable filters across local neighborhoods, making them highly effective for structured data such as images or spatial grids. This architectural evolution enables CNNs to capture translation-invariant patterns and hierarchical features.

Although CNNs originated in image processing [16], their convolutional filters are well-suited to spatially distributed building datasets [17–19]. For instance:

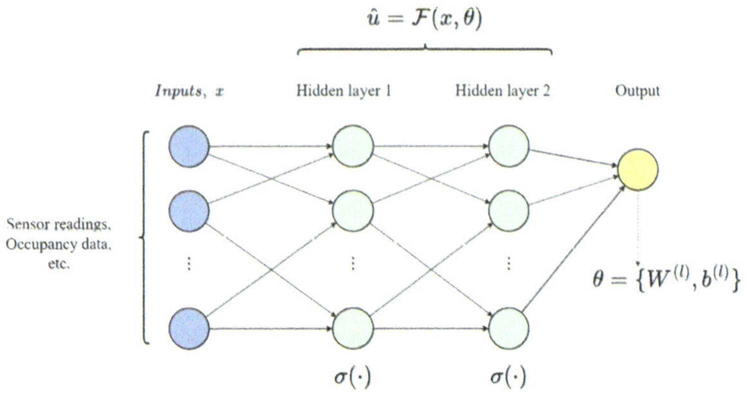

Fig. 4.3 Conceptual structure of a feed-forward ANN approximating the nonlinear mapping, $\hat{u} = \mathcal{F}(x, \theta)$, where nonlinear activation functions, $\sigma(\cdot)$, are applied at hidden layers. The input vector, x, may include sensor measurements, occupancy information, and environmental data in Building Digital Twins context

Table 4.1 Key differences between ANN and CNN

Aspect	Artificial Neural Network (ANN)	Convolutional Neural Network (CNN)
Origin	Base architecture for learning nonlinear mappings	Advanced architecture evolved from ANN for spatial data
Input Handling	Treats all input features independently	Exploits local spatial correlations via convolution
Parameter Sharing	Each weight is unique to a connection	Filters (kernels) share parameters across spatial locations
Best Suited For	Tabular data, time-series, generic regression tasks	Image-like grids, spatially distributed data (e.g., thermal maps)
Computational Complexity	Typically lower for small networks	Higher due to convolution and pooling operations
BDT Applications	Surrogate modeling, load prediction, comfort estimation	Spatial pattern extraction, thermal field analysis, occupancy mapping

- thermal fields arranged on building floor grids,
- facade solar irradiation maps,
- occupancy heatmaps,
- multi-zone temperature snapshots from BDT simulations.

A convolutional layer performs

$$(h * k)(i, j) = \sum_{m,n} x(i - m, j - n)\, k(m, n) \tag{4.4}$$

where "k" is a learnable filter capturing spatial correlations in the input x. CNNs offer translation-insensitive feature extraction—critical when building conditions (e.g., hot/cold spots) shift due to occupancy or weather. Hybrid CNN–MPC architectures update local model parameters in real time, improving predictive models used inside MPC and RL agents described in Chap. 3.

Symbol Definitions

- $x(i, j)$: Input feature map at spatial location (i, j)
- $h * k$: Convolution operation between input x and kernel k
- $k(m, n)$: Learnable convolutional kernel (filter) weights
- (i, j): Spatial indices of the output feature map
- (m, n): Spatial indices within the kernel window

CNN layers are typically followed by nonlinear activations (e.g., ReLU) and pooling operations to reduce spatial dimensions while preserving salient features. In BDT pipelines, CNNs can

- extract spatial patterns from thermal or occupancy grids,
- compress high-dimensional simulation outputs into latent representations,
- support hybrid architectures combining CNN-based feature extraction with MPC or RL for control.

Highlight

Artificial Intelligence (AI) draws inspiration from human cognitive abilities such as perception, learning, and decision-making. Learning systems, including Artificial Neural Networks (ANNs) and Convolutional Neural Networks (CNNs), emulate aspects of biological neural processing by learning complex mappings directly from data. ANNs approximate nonlinear functions through layers of interconnected neurons, while CNNs exploit spatial hierarchies and local receptive fields, reflecting mechanisms observed in the human visual cortex. In contrast, model-based approaches such as Model Predictive Control (MPC) resemble deliberative reasoning, where predictions and constraints guide decision-making. Together, these paradigms[1] illustrate how AI systems operationalize cognitive principles—either through data-driven pattern recognition or through structured optimization leveraging explicit models.

[1] Model Predictive Control (MPC) is not inherently part of Artificial Intelligence (AI), but rather an advanced control theory paradigm which is increasingly integrated with AI and Machine Learning (ML) techniques.

Hybrid Learning Systems (Physics-Guided ML)

Buildings are complex, slow-dynamic, and partially observable (Chap. 3). Purely data-driven models may drift, while purely physics-driven models may be inaccurate. Hybrid systems integrate both approaches [20–22]:

- **Grey-Box Models**:

$$\dot{x} = f_{\text{phys}}(x, u, \phi) + f_{\text{ML}}(x, u; \theta) \tag{4.5}$$

  where f_{phys} encodes first-principles dynamics and f_{ML} provides data-driven corrections. ϕ denotes calibrated physics parameters, while "θ" represents learnable ML parameters.
- **Neural augmentation of EnergyPlus™/MATLAB™ models** for parameter re-identification and drift compensation.
- **Surrogate-Enhanced MPC**: MPC uses a fast ML surrogate instead of full simulation for real-time optimization.

Symbol Definitions

- x: System state vector (e.g., zone temperatures, humidity)
- u: Control input vector (e.g., HVAC setpoints)
- $\dot{x}$: Time derivative of the state vector
- $f_{\text{phys}}(x, u, \phi)$: Physics-based model component parameterized by ϕ
- $f_{\text{ML}}(x, u; \theta)$: Machine learning correction term parameterized by θ
- ϕ: Calibrated physical parameters (e.g., thermal capacitance, resistance)
- θ: Trainable ML parameters (e.g., neural network weights)

These architectures align with the BDT objective of real-time synchronization, continuous calibration, and adaptive prediction. By combining physics priors with data-driven flexibility, hybrid models could improve robustness under changing conditions such as occupancy shifts or weather variations.

Model-Based Decision-Making: Model Predictive Control (MPC)

Model Predictive Control (MPC) represents a canonical *model-based* decision-making paradigm within Computational Intelligence. Unlike model-free learning systems such as ANNs and CNNs, which directly approximate the input–output mapping, $u = \mathcal{F}(x, \theta)$ from data, MPC explicitly embeds a predictive system model within the optimization loop.

At each control step, MPC solves a finite-horizon optimization problem using a model of the form:

$$x_{k+1} = f(x_k, u_k, d_k) \tag{4.6}$$

where "x_k" denotes the system state at time step "k" (e.g., zone temperature), "u_k" is the control input (e.g., heating or cooling power), and "d_k" represents exogenous disturbances such as weather or occupancy. The function "$f(\cdot)$" captures the system dynamics, mapping the current state, control, and disturbance to the next state, x_{k+1}.

Future trajectories $\{x_k\}$ are predicted over a finite horizon and evaluated against a cost function which reflects control objectives (e.g., comfort tracking, energy efficiency) and operational constraints. The optimization yields an optimal control sequence $\{u_k\}$, of which only the first action is applied; the horizon then recedes and the process repeats.

Illustrative Example: Constrained MPC

To complement the conceptual MPC formulation in Chap. 3, consider a scalar RC thermal model with additive disturbances:

$$T_{k+1} = aT_k + bu_k + d_k$$

where T_k is the zone temperature, a is the thermal inertia coefficient ($0 < a < 1$), b is the actuation gain, u_k is the HVAC control signal (positive for heating, negative for cooling), and d_k represents an external disturbance (e.g., weather or occupancy effects). A constrained, finite-horizon MPC minimizes:

$$J = \sum_{k=0}^{N-1} \left[(T_k - T_{\text{set}})^2 + \lambda u_k^2 \right] + \mu (T_N - T_{\text{set}})^2$$

subject to actuator and comfort constraints:

$$u_{\min} \leq u_k \leq u_{\max}, \quad T_{\min} \leq T_k \leq T_{\max}, \quad k = 0, \ldots, N-1$$

and dynamics and initialization:

$$T_{k+1} = aT_k + bu_k + d_k, \quad T_0 = T_{\text{init}}$$

Here, T_{set} is the comfort setpoint. λ and μ are weights for control effort and terminal cost: $\lambda > 0$ penalizes control effort and $\mu \geq 0$ is a terminal weight enforcing horizon-end setpoint proximity. For operational robustness, comfort constraints can be softened via slack variables $\varepsilon_k \geq 0$:

$$T_{\min} - \varepsilon_k \leq T_k \leq T_{\max} + \varepsilon_k, \quad J \leftarrow J + \rho \sum_{k=0}^{N-1} \varepsilon_k$$

where $\rho \gg 0$ controls the penalty on constraint violation.

Symbol Definitions

- T_k: Zone temperature at step k
- u_k: Heating/cooling control input at step k
- d_k: Exogenous disturbance at step k

- a, b: Discrete-time RC model coefficients
- N: Prediction horizon length
- T_{set}: Desired temperature setpoint
- λ: Control-effort penalty weight
- μ: Terminal penalty weight
- u_{min}, u_{max}: Actuator bounds
- T_{min}, T_{max}: Comfort bounds
- ε_k: Slack variable for soft comfort constraints
- ρ: Slack penalty weight

Listing 4.1 presents a minimal, constrained MPC with optional disturbance preview and soft comfort constraints using `cvxpy`. It provides a pedagogical yet practically meaningful reference for the surrogate-assisted MPC formulations in Chap. 3.

```python


#!/usr/bin/env python3
# -*- coding: utf-8 -*-
"""
Constrained MPC for a first-order thermal RC model (scalar).
- Single-horizon solver (cvxpy + OSQP) with optional soft comfort constraints.
- Receding-horizon loop (apply first move, re-solve).
- Plot and CSV export for reproducibility.

Author: <Your Name>
Date: 2025-12-07
"""

import numpy as np
import cvxpy as cp


def solve_mpc_single_horizon(
    T0, a, b, N, Tset,
    lambda_u=0.5,              # higher effort penalty to reduce saturation
    mu_terminal=2.0,           # terminal cost weight
    u_min=-2.0, u_max=2.0,     # actuator bounds
    T_min=22.0, T_max=26.0,    # comfort bounds
    d=None,                    # disturbance preview of length N
    use_soft_comfort=True, rho_slack=1000.0,
    cooling_positive=False,    # if True, +u means cooling; else +u means heating
    solver=cp.OSQP, verbose=False
):
"""
    Solve one MPC horizon for:
        T_{k+1} = a*T_k + sgn*b*u_k + d_k
    where sgn = +1 if +u adds heat (heating positive),
          sgn = -1 if +u removes heat (cooling positive).
"""
    if d is None:
        d = np.zeros(N)
    sgn = -1.0 if cooling_positive else 1.0

```

```
    # Decision variables
    u = cp.Variable(N)
    T = cp.Variable(N + 1)
    eps = cp.Variable(N, nonneg=True) if use_soft_comfort else None

    # Constraints
    constraints = [T[0] == T0]
    for k in range(N):
        constraints += [T[k+1] == a * T[k] + sgn * b * u[k] + d[k],
                        u[k] >= u_min, u[k] <= u_max]
    for k in range(N):
        if use_soft_comfort:
            constraints += [T[k] >= T_min - eps[k],
                            T[k] <= T_max + eps[k]]
        else:
            constraints += [T[k] >= T_min, T[k] <= T_max]
    # Optional terminal comfort bound (comment if not desired)
    if use_soft_comfort:
        constraints += [T[-1] >= T_min - eps[-1],
                        T[-1] <= T_max + eps[-1]]
    else:
        constraints += [T[-1] >= T_min, T[-1] <= T_max]

    # Objective
    stage_cost = cp.sum((T[:-1] - Tset)**2 + lambda_u * cp.square(u))
    terminal_cost = mu_terminal * cp.square(T[-1] - Tset)
    slack_cost = rho_slack * cp.sum(eps) if use_soft_comfort else 0.0
    cost = stage_cost + terminal_cost + slack_cost

    # Solve
    prob = cp.Problem(cp.Minimize(cost), constraints)
    prob.solve(solver=solver, verbose=verbose)

    # Extract
    u_opt = u.value if u.value is not None else np.full(N, np.nan)
    T_opt = T.value if T.value is not None else np.full(N + 1, np.nan)
    eps_opt = eps.value if (use_soft_comfort and eps.value is not None) else
     None

    return prob.status, prob.value, u_opt, T_opt, eps_opt


def receding_horizon_controller(
    T_init, a, b, Tset,
    N=10, steps=20,
    lambda_u=0.5, mu_terminal=2.0,
    u_min=-2.0, u_max=2.0,
    T_min=22.0, T_max=26.0,
    forecast_fn=None,
    use_soft_comfort=True, rho_slack=1000.0,
    cooling_positive=False,
    solver=cp.OSQP, verbose=False
):
"""
    Apply receding horizon control for 'steps' iterations.
    At each step:
      - build preview d_t (length N),
      - solve horizon,
      - apply u[0],
      - update plant one step using measured disturbance if available (here:
     preview as proxy).
"""
    T_current = float(T_init)
    sgn = -1.0 if cooling_positive else 1.0

    u_applied = []
    T_log = [T_current]
```

```
    status_log = []

    for t in range(steps):
        d_preview = forecast_fn(t, N) if forecast_fn is not None else np.zeros(N)

        status, obj, u_opt, T_opt, eps_opt = solve_mpc_single_horizon(
            T0=T_current, a=a, b=b, N=N, Tset=Tset,
            lambda_u=lambda_u, mu_terminal=mu_terminal,
            u_min=u_min, u_max=u_max,
            T_min=T_min, T_max=T_max,
            d=d_preview,
            use_soft_comfort=use_soft_comfort, rho_slack=rho_slack,
            cooling_positive=cooling_positive,
            solver=solver, verbose=verbose
        )
        status_log.append(status)

        # Fallback if infeasible
        u0 = 0.0 if status not in ("optimal", "optimal_inaccurate") else float(u_opt[0])

        # Plant update (use preview as proxy for measured disturbance)
        T_next = a * T_current + sgn * b * u0 + d_preview[0]
        u_applied.append(u0)
        T_log.append(T_next)
        T_current = T_next

    return u_applied, T_log, status_log


def plot_and_save(u_applied, T_log, Tset, T_min, T_max, u_min, u_max, out_path="mpc_trajectory.png"):
    import matplotlib.pyplot as plt
    plt.rcParams.update({"font.size": 11})
    fig = plt.figure(figsize=(10, 4.5), dpi=160)

    # Temperature subplot
    ax1 = plt.subplot(1, 2, 1)
    ax1.plot(T_log, marker='o', label=r"$T_k$")
    ax1.axhline(Tset, color='r', linestyle='--', label='Setpoint')
    ax1.axhline(T_min, color='k', linestyle=':', linewidth=0.9, label=r"$T_{\min}$")
    ax1.axhline(T_max, color='k', linestyle=':', linewidth=0.9, label=r"$T_{\max}$")
    ax1.set_title(r"Zone Temperature $T_k$")
    ax1.set_xlabel("Step")
    ax1.set_ylabel("Temperature (°C)")
    ax1.grid(True, alpha=0.3)
    ax1.legend(loc='best')

    # Control subplot
    ax2 = plt.subplot(1, 2, 2)
    ax2.step(range(len(u_applied)), u_applied, where='post', label=r"$u_k$")
    ax2.axhline(u_min, color='k', linestyle='--', linewidth=0.8)
    ax2.axhline(u_max, color='k', linestyle='--', linewidth=0.8)
    ax2.set_title(r"Control Input $u_k$")
    ax2.set_xlabel("Step")
    ax2.set_ylabel("Control")
    ax2.grid(True, alpha=0.3)

    plt.tight_layout()
    fig.savefig(out_path, bbox_inches="tight")
    plt.close(fig)


def save_csv(u_applied, T_log, status_log, out_path="mpc_results.csv"):
```

```
    import csv
    with open(out_path, "w", newline="") as f:
        w = csv.writer(f)
        w.writerow(["step", "status", "T_k", "u_k"])
        # T_log has length steps+1; u_applied has length steps
        w.writerow([0, status_log[0] if status_log else "", T_log[0], ""])
        for k in range(len(u_applied)):
            w.writerow([k+1, status_log[k], T_log[k+1], u_applied[k]])


if __name__ == "__main__":
    # ------------------------------
    # Parameters
    # ------------------------------
    a, b = 0.9, 0.1
    N = 10
    steps = 20
    Tset = 24.0
    lambda_u = 0.5
    mu_terminal = 2.0
    u_min, u_max = -2.0, 2.0            # choose [-2,0] if cooling is negative-only
    T_min, T_max = 22.0, 26.0
    T0 = 26.0
    cooling_positive = False            # False: +u heats; True: +u cools

    # Optional disturbance forecaster (toy sinusoid)
    def example_forecast(t, horizon):
        k = np.arange(horizon)
        return 0.2 * np.sin(0.2 * (t + k))

    # Single-horizon diagnostic
    status, obj, u_opt, T_opt, eps_opt = solve_mpc_single_horizon(
        T0=T0, a=a, b=b, N=N, Tset=Tset,
        lambda_u=lambda_u, mu_terminal=mu_terminal,
        u_min=u_min, u_max=u_max,
        T_min=T_min, T_max=T_max,
        d=np.zeros(N),
        use_soft_comfort=True, rho_slack=1000.0,
        cooling_positive=cooling_positive,
        solver=cp.OSQP, verbose=False
    )
    print("Single-horizon status:", status)
    if status in ("optimal", "optimal_inaccurate"):
        print("Optimal first control:", float(u_opt[0]))
        print("Predicted temperatures (first 5):", np.round(T_opt[:5], 3))

    # Receding-horizon run
    u_applied, T_log, status_log = receding_horizon_controller(
        T_init=T0, a=a, b=b, Tset=Tset,
        N=N, steps=steps,
        lambda_u=lambda_u, mu_terminal=mu_terminal,
        u_min=u_min, u_max=u_max,
        T_min=T_min, T_max=T_max,
        forecast_fn=example_forecast,
        use_soft_comfort=True, rho_slack=1000.0,
        cooling_positive=cooling_positive,
        solver=cp.OSQP, verbose=False
    )

    print("RHC statuses (first 5):", status_log[:5])
    print("Applied controls (first 5):", np.round(u_applied[:5], 3))
    print("Temperature trajectory (first 6):", np.round(T_log[:6], 3))

    # Save outputs
    plot_and_save(u_applied, T_log, Tset, T_min, T_max, u_min, u_max, out_path="mpc_trajectory.png")
```

```
    save_csv(u_applied, T_log, status_log, out_path="mpc_results.csv")
    print("Saved figure: mpc_trajectory.png ; CSV: mpc_results.csv")
```

Listing 4.1 Receding-horizon MPC for a first-order thermal RC model with actuator and comfort constraints; solved as a convex QP in cvxpy (OSQP). Supports disturbance preview and optional soft comfort constraints via slack penalties.

Interpretation and Modeling Remark

It is important to emphasize that the illustrative MPC example above intentionally employs a minimal first-order thermal model in order to highlight the prediction–optimization loop rather than to provide a fully realistic building representation. The adopted dynamics:

$$T_{k+1} = aT_k + bu_k + d_k$$

do not explicitly include an ambient equilibrium temperature term. As a result, the open-loop system exhibits an inherent decay toward a low steady-state[2] value, see Fig. 4.4, unless sufficiently large control action is applied. Under actuator saturation, the optimizer therefore saturates the control input at its upper bound while still allowing comfort violations through the slack variables.

This behavior is consistent with the soft-constraint formulation introduced earlier and demonstrates an important practical concept: *when physical limitations prevent comfort maintenance, MPC trades constraint violations against control effort in an optimal manner.* In real building applications, the disturbance term, d_k, typically embeds the influence of outdoor temperature, solar gains, and internal loads, yielding dynamics of the form:

$$T_{k+1} = aT_k + (1-a)T_{\text{out},k} + bu_k + d_k$$

which naturally stabilize the zone temperature around the ambient conditions—here, designated by the outdoor (ambient) temperature, $T_{\text{out},k}$.

[2] Steady-state means:

$$T_{k+1} = T_k = T_\infty, \quad u_k = u = constant$$

For a constant control input $u_k = u$, the steady-state temperature satisfies

$$T_\infty = aT_\infty + bu \quad \Rightarrow \quad T_\infty = \frac{bu}{1-a}$$

With the selected parameters, $a = 0.1$ (strong thermal inertia); $b = 0.1$ (weak heating effectiveness) and $u = 2$ (max heating), even saturation of the actuator ($u = u_{\max}$) results in a steady-state temperature significantly below the comfort setpoint. Consequently, the MPC saturates the control input while tolerating comfort violations through slack variables, illustrating the behavior of soft constraints under physical limitations. This simplified example is intentionally chosen to motivate the use of enhanced predictive models—such as hybrid or learning-augmented formulations—where ambient effects and uncertain parameters are continuously refined within a Building Digital Twin.

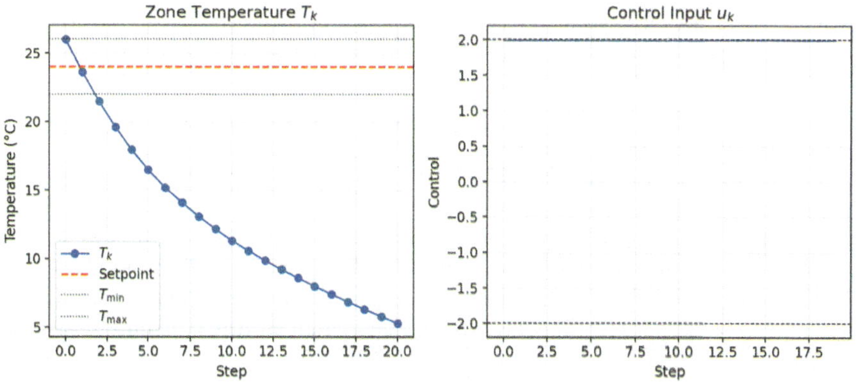

Fig. 4.4 Illustrative MPC response for a first-order thermal model without explicit ambient equilibrium coupling. Despite actuator saturation ($u_k = u_{\max}$), the zone temperature decays toward its steady-state value $T_\infty = \frac{bu}{1-a}$, highlighting the effect of soft comfort constraints and motivating hybrid and learning-enhanced model refinement

> The simplified formulation adopted here serves a didactic purpose and provides a transparent baseline which can be extended toward hybrid and learning-enhanced MPC architectures, where neural surrogate models or *online parameter estimation* could be used to continuously refine the predictive model within a Building Digital Twin.

Learning Systems in BDT Feedback Loops

Within the Building Digital Twin (BDT) operational pipeline, learning systems play a pivotal role in enhancing adaptability and robustness. Their contributions include

- **Dynamic Updates and Parameter Calibration**: Artificial Neural Networks (ANNs) estimate uncertain parameters (e.g., infiltration rates, internal gains from occupancy), enabling continuous calibration against real-time sensor data.
- **State Estimation**: CNN and ANN models complement classical estimators such as Kalman or particle filters when the building state is partially observable or sensor coverage is sparse.
- **Surrogate Modeling**: Learning systems approximate computationally expensive EnergyPlus™-based simulations, providing fast predictive models which accelerate optimization tasks (see Chap. 7).
- **Adaptive Control**: Model-free approaches (e.g., RL and data-driven ANNs) adjust control policy parameters "θ" to track evolving building conditions, complementing MPC's model-based predictions.

In essence, learning systems act as the "adaptive engine" beneath the predictive and optimization layers of the BDT architecture. They enable rapid correction of model drift, handle parametric uncertainty, and mitigate stochastic disturbances discussed in Chap. 3.

From Predictive Decision-Making to Evolutionary Optimization:

While MPC provides a principled framework for real-time, model-based decision-making, its performance and scalability are often influenced by the quality of model parameters, cost-function tuning, and constraint selection. These aspects motivate the use of population-based evolutionary computation methods, introduced in the following section, which enable global exploration and automated optimization beyond the local, gradient-driven nature of classical MPC.

4.3.2 Evolutionary Computation

Evolutionary computation is one of the most established pillars of CI. Inspired by natural selection and population dynamics, Evolutionary Algorithms (EAs) such as Genetic Algorithms (GA), Differential Evolution (DE), and Evolution Strategies (ES) operate on a population of candidate solutions and evolve them over generations.

Let $\mathcal{P}^{(t)} = \{x_i^{(t)}\}_{i=1}^{N}$ denote the population at generation t. A simple mutation operator in an ES-like scheme can be written as

$$x_i^{(t+1)} = x_i^{(t)} + \sigma^{(t)}\epsilon, \qquad \epsilon \sim \mathcal{N}(0, I) \tag{4.7}$$

where $\sigma^{(t)}$ is a step-size parameter and ϵ is a normally distributed random vector. After variation (mutation, crossover), selection chooses individuals for the next generation, often according to their fitness values $f(x)$:

$$\mathcal{P}^{(t+1)} \subset \mathcal{P}^{(t)} \cup \mathcal{O}^{(t)} \tag{4.8}$$

where $\mathcal{O}^{(t)}$ denotes the set of offspring produced from $\mathcal{P}^{(t)}$.

For BDTs, EAs are attractive because:

- they can explore highly non-convex, multi-modal search spaces where gradient information is unreliable or unavailable;
- they naturally handle mixed discrete–continuous decision variables (for example, equipment on/off states and continuous setpoints);
- they support multi-objective optimization, enabling simultaneous trade-offs between energy, cost, comfort, emissions, and other criteria.

The Python™ code[3] in Listing 4.2 implements a simplified Evolutionary Strategy (ES)-like algorithm to demonstrate the concepts introduced in Eqs. (4.7) and (4.8).

[3] The code requires the following Python™ libraries: `numpy` for numerical computations and `matplotlib` for plotting. These are freely available under Python™ Packages.

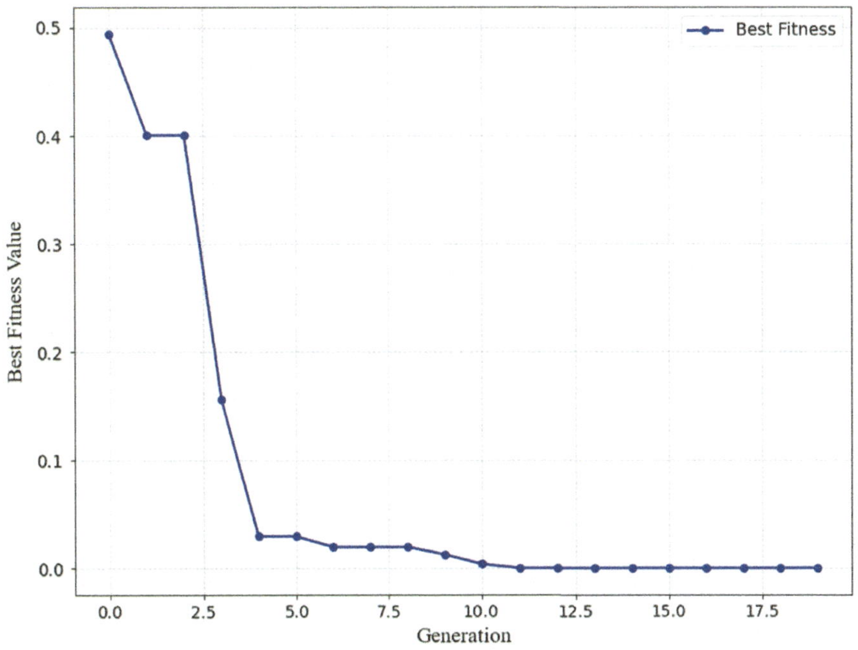

Fig. 4.5 Illustration of an Evolutionary Strategy (ES)-like algorithm applied to a simple Sphere function. The plot shows the best fitness value improving over generations, demonstrating mutation and selection dynamics

The algorithm operates on a population of candidate solutions, applying *Gaussian* mutation scaled by a step-size parameter "$\sigma^{(t)}$" and selecting the best individuals based on their fitness. For illustration, the *Sphere function*, $f(x) = \sum x_i^2$, serves as the optimization objective. Figure 4.5 depicts the improvement in best fitness values across generations, highlighting the evolutionary dynamics of variation and selection. Complementing this, Fig. 4.6 visualizes the spatial progression of the population in a two-dimensional search space. Snapshots at Generation 1, Generation 10, and Generation 20 reveal how candidate solutions converge toward the global optimum at the origin. Each point represents an individual, and the observed movement across generations illustrates the combined effect of Gaussian mutation and fitness-based selection. Together, these visualizations provide an intuitive understanding of how Evolutionary Algorithms (EAs) explore and exploit the search space.

Subsequent chapters, particularly Chaps. 5 and 7, build upon these concepts by employing evolutionary computation as one of several Computational Intelligence (CI) tools for optimization and decision support in building-related problems. In particular, Multi-Objective Evolutionary Algorithms (MOEAs) are later used to illustrate how competing performance criteria—such as energy use, comfort, emissions, and operational cost—can be explored and balanced within simulation-driven and data-assisted design workflows.

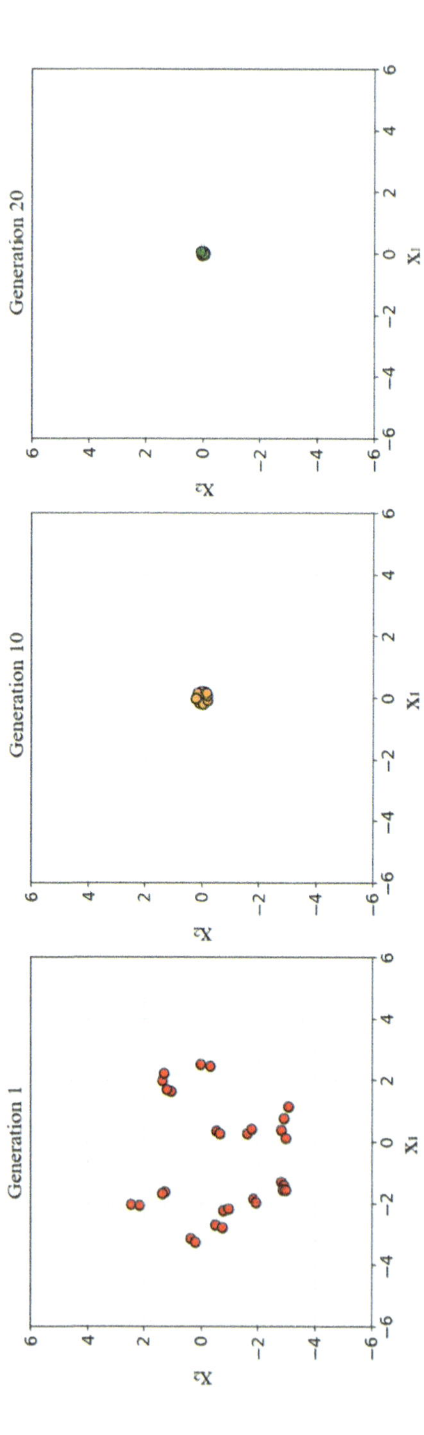

Fig. 4.6 Population evolution in 2D space for an ES-like algorithm. Snapshots at Generation 1, Generation 10, and Generation 20 show how individuals converge toward the optimum through mutation and selection

```python

import numpy as np
import matplotlib.pyplot as plt

# ------------------------------
# Parameters
# ------------------------------
num_generations = 20
population_size = 30
dimension = 2
sigma_initial = 0.2

# Fitness function (Sphere function)
def fitness(x):
    return np.sum(x**2)

# Initialize population randomly in [-5, 5]
population = np.random.uniform(-5, 5, (population_size, dimension))

# Track best fitness and snapshots
best_fitness = []
snapshots = {}

sigma = sigma_initial
for t in range(num_generations):
    # Evaluate fitness
    fitness_values = np.array([fitness(ind) for ind in population])
    best_fitness.append(np.min(fitness_values))

    # Mutation
    offspring = []
    for i in range(population_size):
        epsilon = np.random.normal(0, 1, dimension)
        child = population[i] + sigma * epsilon
        offspring.append(child)
    offspring = np.array(offspring)

    # Combine and select best
    combined = np.vstack((population, offspring))
    combined_fitness = np.array([fitness(ind) for ind in combined])
    selected_indices = np.argsort(combined_fitness)[:population_size]
    population = combined[selected_indices]

    # Save snapshots for first, middle, last generation
    if t in [0, num_generations // 2, num_generations - 1]:
        snapshots[t] = population.copy()

    sigma *= 0.99

# ------------------------------
# Figure 1: Best fitness over generations
# ------------------------------
plt.figure(figsize=(8, 6))
plt.plot(range(num_generations), best_fitness, marker='o', color='blue',
     linewidth=2, label='Best Fitness')
plt.xlabel('Generation', fontsize=16, fontname='Times New Roman')
plt.ylabel('Best Fitness Value', fontsize=16, fontname='Times New Roman')
plt.grid(True, linestyle='--', alpha=0.6)
plt.legend(fontsize=14)
plt.tight_layout()
plt.savefig('ea_progress_matplotlib.png', dpi=300)
plt.show()

# ------------------------------
# Figure 2: Population evolution snapshots
# ------------------------------
fig, axes = plt.subplots(1, 3, figsize=(15, 5))
```

```
gen_labels = ['Generation 1', f'Generation {num_generations // 2}', f'
    Generation {num_generations}']
colors = ['red', 'orange', 'green']

for ax, (gen, pop), label, color in zip(axes, snapshots.items(), gen_labels,
    colors):
    ax.scatter(pop[:, 0], pop[:, 1], c=color, edgecolors='black', s=60)
    ax.set_xlabel('x?', fontsize=18, fontname='Times New Roman')
    ax.set_ylabel('x?', fontsize=18, fontname='Times New Roman')
    ax.tick_params(axis='both', which='major', labelsize=14)
    ax.grid(True, linestyle='--', alpha=0.6)
    ax.set_xlim(-6, 6)
    ax.set_ylim(-6, 6)
    ax.set_title(label, fontsize=16, fontname='Times New Roman')

plt.tight_layout()
plt.savefig('ea_population_evolution_large_labels.png', dpi=300)
plt.show()
```

Listing 4.2 Python implementation of an ES-like Evolutionary Algorithm for illustration

While Evolutionary Algorithms (EAs) rely on principles of variation, selection, and survival to guide population-level optimization, other population-based paradigms emphasize collective intelligence emerging from decentralized agent interactions. Swarm Intelligence methods, introduced in the following section, share the population-based philosophy of evolutionary computation but differ fundamentally in how information is exchanged, how candidate solutions evolve, and how convergence is achieved (see Table 4.2 for a comparative overview).

4.3.3 Swarm Intelligence

Swarm intelligence refers to a family of optimization algorithms inspired by the collective behavior of decentralized, self-organizing populations such as bird flocks, fish schools, or colonies of ants. Unlike classical optimization methods, swarm-based algorithms do not rely on gradients, convexity assumptions, or explicit global models. Instead, they rely on distributed agent interactions and indirect communication mechanisms (for example, pheromone trails in ant colonies or velocity alignment in flocks), enabling highly adaptive and robust search capabilities. Representative methods include Particle Swarm Optimization (PSO) [23], Ant Colony Optimization (ACO) [24], and more recent nature-inspired variants such as the Grey Wolf Optimizer (GWO) [25] and the Whale Optimization Algorithm (WOA) [26].

Table 4.2 Comparison between evolutionary computation and Swarm intelligence paradigms

Aspect	Evolutionary Computation (EC)	Swarm Intelligence (SI)
Biological inspiration	Natural selection and genetics	Collective behavior of social organisms
Population structure	Explicit population with generational replacement	Set of interacting agents updated iteratively
Search operators	Mutation, crossover, and selection	Velocity updates, attraction, or local interaction rules
Information sharing	Indirect, via fitness-based selection	Direct or implicit, via neighborhood or global best solutions
Convergence behavior	Often slower but robust global exploration	Often faster convergence, risk of premature stagnation
Typical examples	Genetic Algorithms, Evolution Strategies, NSGA-II	Particle Swarm Optimization, Ant Colony Optimization
Strengths	Strong global search and multi-objective handling	Simple implementation and rapid adaptation
Typical applications in BDTs	Design-space exploration, calibration, Multi-Objective Optimization (MOO)	Control tuning, scheduling, real-time, or near-real-time optimization

Note

Although swarm intelligence algorithms derive their mechanisms from collective **biological behaviors (nature inspired)**—such as flocking, schooling, or pheromone-based trail formation—many other branches of Artificial Intelligence (AI) originate from human cognitive processes such as learning, reasoning, abstraction, and structured decision-making. This juxtaposition highlights the breadth of paradigms encompassed by Computational Intelligence (CI): *nature-inspired* heuristics (for example, PSO, ACO, GWO, WOA) and *human-inspired* methods (such as neural networks, neuro-fuzzy systems, and expert systems) operate side by side, often synergistically. Together they form a versatile, hybrid computational toolkit capable of tackling complex, uncertain, and high-dimensional optimization and decision-making tasks commonly encountered in Building Digital Twins.

Particle Swarm Optimization

PSO, introduced by Kennedy and Eberhart in 1995 [23], models the coordinated movement of a *flock of birds searching for food* in a high-dimensional landscape. Each particle "i" maintains a position vector, $\mathbf{x}_i^{(t)}$, and a velocity vector, $\mathbf{v}_i^{(t)}$. Its personal best location is denoted "$\mathbf{p}_i$", and the global best location discovered by the swarm is "$\mathbf{g}$". The velocity and position updates are given by

$$\mathbf{v}_i^{(t+1)} = \omega\,\mathbf{v}_i^{(t)} + c_1 r_1 \left(\mathbf{p}_i - \mathbf{x}_i^{(t)}\right) + c_2 r_2 \left(\mathbf{g} - \mathbf{x}_i^{(t)}\right) \tag{4.9}$$

$$\mathbf{x}_i^{(t+1)} = \mathbf{x}_i^{(t)} + \mathbf{v}_i^{(t+1)} \tag{4.10}$$

where "ω" is an inertia weight, "c_1" and "c_2" are cognitive and social acceleration coefficients, "t" can be *iteration index* or *time-step*, and $r_1, r_2 \sim \mathcal{U}(0, 1)$. These simple update rules enable PSO to perform fast, "*parallelizable*" global search with minimal parameter tuning. *In the context of BDTs, PSO has been used for multi-zone HVAC setpoint tuning, data-driven model calibration, and co-optimization of comfort, energy, and indoor air quality metrics under uncertainty* [27, 28].

Example: Particle Swarm Optimization on a 2D Quadratic Landscape

To make the update rules in Eq. (4.9)–(4.10) more concrete, consider minimizing a simple two-dimensional Sphere function:

$$f(\mathbf{x}) = x_1^2 + x_2^2$$

which has its global minimum at the origin. Each particle is a point $\mathbf{x}_i^{(t)} = [x_{i1}^{(t)}, x_{i2}^{(t)}]^\top$ in the (x_1, x_2) plane. At each iteration, the velocity update combines three effects:

- *Inertia* ($\omega \mathbf{v}_i^{(t)}$): tendency to keep moving in the same direction;
- *Cognitive Pull* ($c_1 r_1 (\mathbf{p}_i - \mathbf{x}_i^{(t)})$): attraction toward the particle's own best location;
- *Social Pull* ($c_2 r_2 (\mathbf{g} - \mathbf{x}_i^{(t)})$): attraction toward the global best location found by the swarm.

Listing 4.3 shows a compact Python™ implementation of PSO for this test problem. For visualization, the script also records the particle trajectories and plots the search process on top of a three-dimensional surface plot of the objective function; see Figs. 4.7 and 4.8. Further, to aid the understanding of PSO, Fig. 4.9 provides a conceptual view of particles flocking toward the optimum in the (x_1, x_2) plane.[4] Under the influence of attraction toward *personal* and

[4] The implementation in Listing 4.3 is intentionally simplified for pedagogical purposes. It demonstrates the core PSO update rules and does not include all practical enhancements such as velocity clamping, constriction coefficients, or constraint-handling strategies.

global best—the particle, $\mathbf{x}_i^{(t)}$, gets updated with a new velocity, $\mathbf{v}_i^{(t+1)}$, toward the next best position, $\mathbf{x}_i^{(t+1)}$.

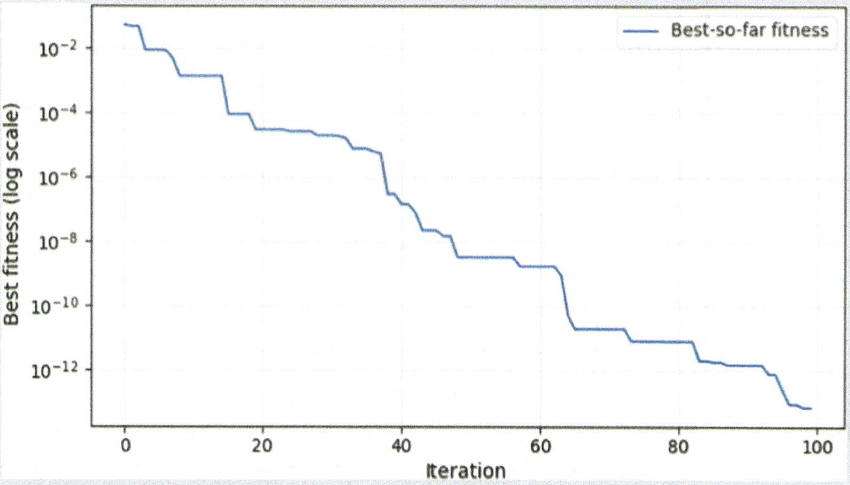

Fig. 4.7 Illustrative convergence behavior of Particle Swarm Optimization on a two-dimensional Sphere function using Listing 4.3. The best fitness value decreases rapidly as particles exploit information from their personal and global best locations

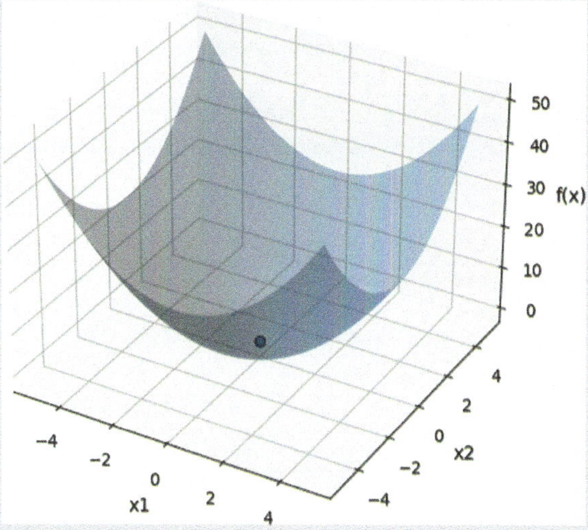

Fig. 4.8 Conceptual visualization of PSO particles moving on a quadratic energy landscape. The 3D surface represents the objective function $f(x_1, x_2)$, while the marker(s) indicate particle positions. Over iterations, the swarm contracts toward the global minimum near the origin

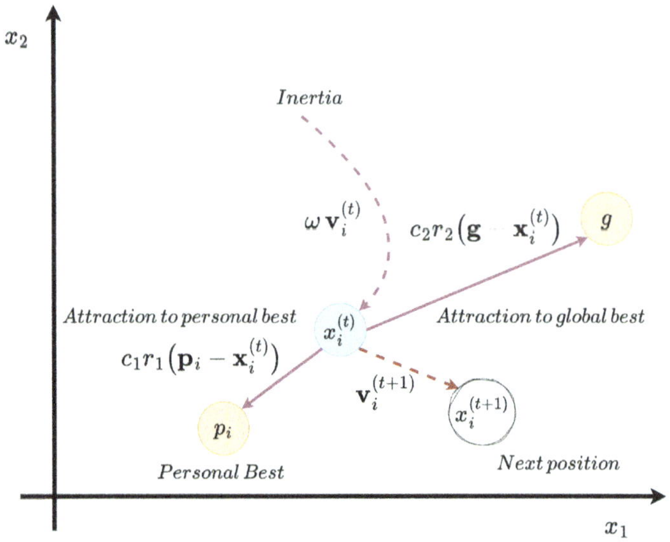

Fig. 4.9 Conceptual illustration of Particle Swarm Optimization in a two-dimensional search space. Each particle, $\mathbf{x}_i^{(t)}$, updates its velocity as a weighted combination of inertia, attraction to its personal best position $\mathbf{p}_i$, and attraction to the global best $\mathbf{g}$, resulting in flock-like trajectories converging toward the optimum

```python
import numpy as np
import matplotlib.pyplot as plt
from mpl_toolkits.mplot3d import Axes3D  # noqa: F401

# ------------------------------
# 1. Objective function
# ------------------------------
def sphere(x):
    """2D Sphere function: f(x1, x2) = x1^2 + x2^2."""
    return np.sum(x**2)

# ------------------------------
# 2. Particle Swarm Optimizer
# ------------------------------
def pso(objective_func, dim=2, bounds=(-5.0, 5.0),
        num_particles=30, max_iter=100,
        w=0.7, c1=1.5, c2=1.5):
    lower, upper = bounds

    # Initialize particle positions and velocities
    positions = np.random.uniform(lower, upper, size=(num_particles, dim))
    velocities = np.zeros_like(positions)

    # Personal bests
    p_best_pos = positions.copy()
    p_best_val = np.array([objective_func(p) for p in positions])

    # Global best
    g_idx = np.argmin(p_best_val)
    g_best_pos = p_best_pos[g_idx].copy()
    g_best_val = p_best_val[g_idx]
```

```
    convergence = []
    trajectories = [positions.copy()]

    for t in range(max_iter):
        r1 = np.random.rand(num_particles, dim)
        r2 = np.random.rand(num_particles, dim)

        # Velocity update
        velocities = (
            w * velocities
            + c1 * r1 * (p_best_pos - positions)
            + c2 * r2 * (g_best_pos - positions)
        )

        # Position update
        positions = positions + velocities

        # Enforce bounds
        positions = np.clip(positions, lower, upper)

        # Evaluate
        fitness = np.array([objective_func(p) for p in positions])

        # Update personal bests
        better_mask = fitness < p_best_val
        p_best_pos[better_mask] = positions[better_mask]
        p_best_val[better_mask] = fitness[better_mask]

        # Update global best
        g_idx = np.argmin(p_best_val)
        if p_best_val[g_idx] < g_best_val:
            g_best_val = p_best_val[g_idx]
            g_best_pos = p_best_pos[g_idx].copy()

        convergence.append(g_best_val)
        trajectories.append(positions.copy())

    return g_best_pos, g_best_val, np.array(convergence), trajectories

# ------------------------------
# 3. Run PSO and visualize
# ------------------------------
if __name__ == "__main__":
    best_pos, best_val, curve, trajectories = pso(sphere)

    print("Best position:", best_pos)
    print("Best value:", best_val)

    # Convergence curve
    plt.figure(figsize=(7, 4))
    plt.semilogy(curve, label="Best-so-far fitness")
    plt.xlabel("Iteration")
    plt.ylabel("Best fitness (log scale)")
    plt.title("PSO Convergence on 2D Sphere Function")
    plt.grid(alpha=0.3)
    plt.legend()
    plt.tight_layout()
    plt.show()

    # 3D surface + final particle positions
    fig = plt.figure(figsize=(7, 5))
    ax = fig.add_subplot(111, projection="3d")

    x = np.linspace(-5, 5, 80)
    y = np.linspace(-5, 5, 80)
    X, Y = np.meshgrid(x, y)
    Z = X**2 + Y**2
```

```
    ax.plot_surface(X, Y, Z, alpha=0.4, linewidth=0, antialiased=True)

    final_positions = trajectories[-1]
    Zp = np.array([sphere(p) for p in final_positions])
    ax.scatter(final_positions[:, 0], final_positions[:, 1], Zp,
               s=30, edgecolor="k")

    ax.set_xlabel("x1")
    ax.set_ylabel("x2")
    ax.set_zlabel("f(x)")
    plt.tight_layout()
    plt.show()
```

Listing 4.3 Minimal PSO implementation for the 2D Sphere function with trajectory and convergence plotting.

Ant Colony Optimization

Ant Colony Optimization originates from the behavioral observation that ants collectively discover the shortest paths between their nest and food sources through indirect communication (stigmergy) mediated by *pheromones*. As documented by Dorigo et al. [29, 30], real ants deposit pheromone along the paths they traverse; the probability of an ant choosing a given path increases with the amount of pheromone present, enabling the colony to converge to near-optimal routes. *At the algorithmic level, this behavior is transformed into a probabilistic decision rule.* For an artificial ant at node "i", the probability of moving to node "j" is given by

$$P_{ij}^{(t)} = \frac{\left[\tau_{ij}^{(t)}\right]^{\alpha} \left[\eta_{ij}\right]^{\beta}}{\sum_{k \in \mathcal{N}_i} \left[\tau_{ik}^{(t)}\right]^{\alpha} \left[\eta_{ik}\right]^{\beta}} \tag{4.11}$$

where "k" is the index over all possible neighbor nodes which the ant can move to from its current node "i". "t" is the iteration index or time step,[5] $\tau_{ij}^{(t)}$ is the pheromone intensity, η_{ij} is a visibility term (often the reciprocal of edge length or cost), "α" and "β" control the relative influence of pheromone and visibility, and "$\mathcal{N}_i$" is the neighborhood of node "i". Fig. 4.10 illustrates the concept at a higher level. After constructing solutions, pheromone trails are updated as

$$\tau_{ij}^{(t+1)} = (1-\rho)\,\tau_{ij}^{(t)} + \sum_{m=1}^{M} \Delta\tau_{ij}^{(m)} \tag{4.12}$$

where "ρ" is the evaporation rate and $\Delta\tau_{ij}^{(m)}$ is the pheromone deposited by ant "m". These equations allow ACO to address routing, scheduling, allocation, and resource balancing problems, such as those originally demonstrated in telecommunications (for example, routing and load balancing in dynamic networks).

[5] Pheromone intensity, $\tau_{ij}^{(t)}$, changes over time because pheromone trails are updated after each iteration based on the quality of solutions found.

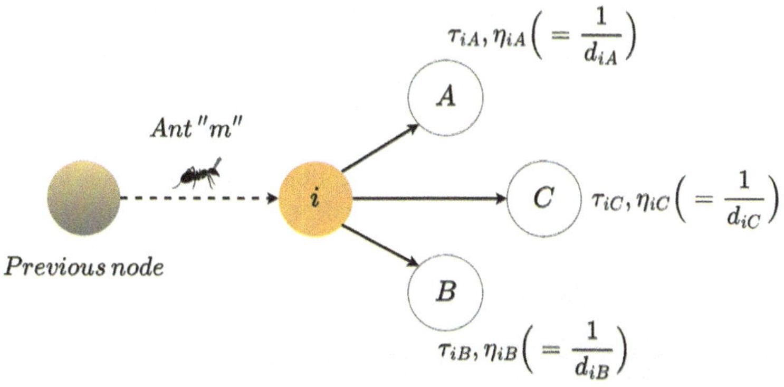

Fig. 4.10 Illustration of the probabilistic decision process in Ant Colony Optimization: Ant "m" arrives at node "i" from a previous node and selects the next node (A, B, or C) based on pheromone intensity "τ_{ij}" and visibility, $\eta_{ij} = \frac{1}{d_{ij}}$; where "$d_{ij}$" denotes the *edge cost* or *distance*

In ACO, several artificial ants construct solutions to an optimization problem and share information about their quality through a communication mechanism which resembles the one used by real ants [29].

In the context of BDTs, this analogy carries over naturally. A building can be interpreted as a multi-agent environment where airflow, heat transfer, sensor communication, and even occupant movement involve routing-like phenomena. *This opens the door to swarm-based strategies for sensor network routing, demand aggregation, or distributed scheduling in smart buildings.*

Example: ACO-Style Routing and Load Balancing Analogy for Building Systems

The following simplified illustration shows how the classical ACO decision rule in Eq. (4.11) can be applied to a building-routing analogy: imagine a set of communication nodes (sensors, actuators, controllers) which must choose optimal paths to transmit information with minimal congestion and packet loss. If the "edge cost" is defined as a function of delay or load, and the pheromone represents recent path quality, then each agent chooses the next hop according to

$$P_{ij} = \frac{[\tau_{ij}]^{\alpha}\,[1/\mathrm{d}_{ij}]^{\beta}}{\sum_k [\tau_{ik}]^{\alpha}\,[1/\mathrm{d}_{ik}]^{\beta}}$$

A minimal Python™ snippet demonstrating this decision rule is provided in Listing[6] 4.4. The snippet computes the probabilistic routing choice for an agent at node i, given current pheromone intensities and communication costs. While very simple, this example mirrors the core principles of ACO routing algorithms originally used in telecommunications and demonstrates how similar ideas can be repurposed for distributed decision-making in BDTs.

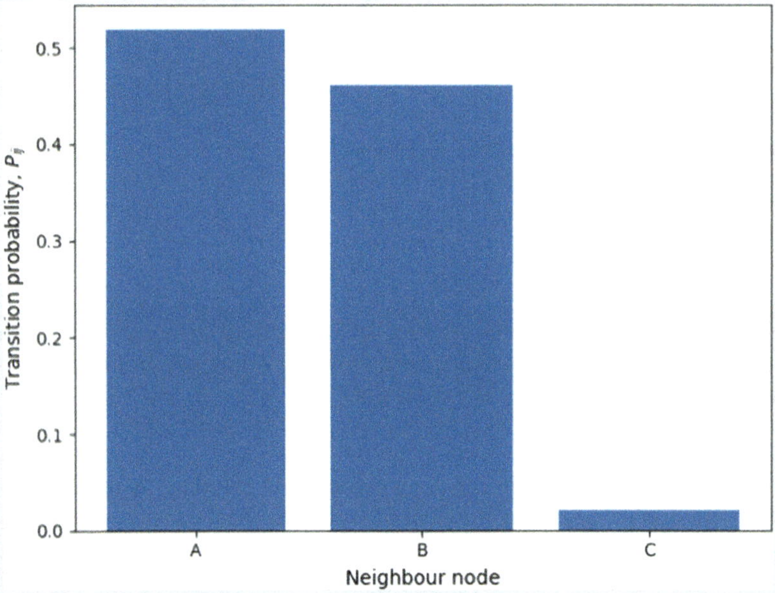

Fig. 4.11 Example ACO transition probabilities from a node with three outgoing links, computed using Listing 4.4. The bar heights indicate the relative probability of selecting each link based on pheromone intensity and communication cost

Figure 4.11 illustrates the transition probabilities computed by Listing 4.4 for a node with three outgoing links to neighbors A, B, and C. Link B has slightly lower cost (for example, lower latency or congestion) than link A, while link C is both costly and poorly reinforced. As a result, the ACO rule assigns the highest selection probability to link A, closely followed by link B, and a very small probability to link C. In a building context, these links could represent alternative communication paths between zones or controllers: links with historically good performance (high pheromone) and low current cost become more likely to be chosen, enabling adaptive routing and load balancing without any central coordinator.

[6] This code snippet illustrates only the ACO transition probability rule and is not a complete implementation of the Ant Colony Optimization algorithm. Full ACO includes pheromone updating, evaporation, solution construction by multiple ants, and iterative global search, which are omitted here for clarity.

```python
import numpy as np
import matplotlib.pyplot as plt


def compute_transition_probabilities(pheromone, cost, alpha=1.0, beta=2.0):
    """
    Compute ACO transition probabilities from a node i to its neighbours j.

    Parameters
    ----------
    pheromone : array-like
        Pheromone intensities tau_ij on outgoing edges from node i.
    cost : array-like
        Positive edge costs cost_ij (e.g., delay, congestion, or energy).
    alpha : float
        Exponent controlling the influence of pheromone.
    beta : float
        Exponent controlling the influence of visibility (1/cost).

    Returns
    -------
    probabilities : np.ndarray
        Normalized transition probabilities P_ij summing to 1.
    """
    pheromone = np.asarray(pheromone, dtype=float)
    cost = np.asarray(cost, dtype=float)

    if pheromone.shape != cost.shape:
        raise ValueError("Pheromone and cost arrays must have the same shape.")
    if np.any(cost <= 0):
        raise ValueError("Costs must be strictly positive.")

    visibility = 1.0 / cost              # eta_ij
    numerator = (pheromone ** alpha) * (visibility ** beta)

    total = numerator.sum()
    if total <= 0.0:
        raise ValueError("Numerator sum is zero; check pheromone and cost values.")

    probabilities = numerator / total
    return probabilities


# Example usage: node i with three outgoing links to neighbours A, B, C
pheromone = np.array([0.8, 0.4, 0.2])
cost      = np.array([2.0, 1.5, 5.0])
neighbours = ["A", "B", "C"]

P = compute_transition_probabilities(pheromone, cost)

print("Transition probabilities:", P)
print("Sum of probabilities:", P.sum())

# Visualization
plt.bar(range(len(P)), P)
plt.xticks(range(len(P)), neighbours)
plt.xlabel("Neighbour node")
plt.ylabel(r'Transition probability, $P_{ij}$')
plt.tight_layout()
plt.show()
```

Listing 4.4 Minimal Python snippet illustrating the ACO transition probability rule for a routing-style decision.

Grey Wolf Optimizer (GWO)

Inspired by the social hierarchy and cooperative hunting strategies of grey wolves, the Grey Wolf Optimizer (GWO), proposed by Mirjalili et al. [25], models encircling, pursuing, and attacking prey through simple yet powerful position-update mechanisms. A pack consists of four dominant roles—α, β, δ, and ω—which guide the search. The best three candidate solutions found so far are regarded as α, β, and δ, and all other wolves follow their lead.

The "encircling" mechanism is expressed as[7]:

$$\mathbf{D} = \left|\mathbf{C}\,\mathbf{x}_{\text{prey}}(t) - \mathbf{x}(t)\right| \tag{4.13}$$

$$\mathbf{x}(t+1) = \mathbf{x}_{\text{prey}}(t) - \mathbf{A} \cdot \mathbf{D} \tag{4.14}$$

Here, "t" denotes the iteration index (i.e., the current time-step or iteration). "$\mathbf{x}_{\text{prey}}$" and "$\mathbf{x}(t)$" denote the positions of prey and the wolves, respectively[8]. **A** and **C** are coefficient vectors defined as

$$\mathbf{A} = 2a\,\mathbf{r}_1 - a \tag{4.15}$$

$$\mathbf{C} = 2\,\mathbf{r}_2 \tag{4.16}$$

where $\mathbf{r}_1$ and $\mathbf{r}_2$ are random vectors in [0, 1], and "a" decreases linearly from 2 to 0 over the iterations—driving the algorithm from exploration to exploitation.

To estimate the true "*prey*" location,[9] see Fig. 4.12, each wolf updates its position relative to the estimated positions from α, β, and δ:

$$\mathbf{x}(t+1) = \frac{1}{3}\left(\mathbf{x}_\alpha(t) + \mathbf{x}_\beta(t) + \mathbf{x}_\delta(t)\right) - \frac{1}{3}\left(\mathbf{A}_\alpha \cdot \mathbf{D}_\alpha + \mathbf{A}_\beta \cdot \mathbf{D}_\beta + \mathbf{A}_\delta \cdot \mathbf{D}_\delta\right) \tag{4.17}$$

Owing to its low parameter count, stable convergence behavior, and balanced exploration-exploitation, GWO has been widely applied in building energy and comfort optimization—including model calibration, HVAC control tuning, and hybrid ML approaches such as SVR–GWO for load prediction. Recent applications demon-

[7] "**D**" is an auxiliary "distance-like" vector which measures the (scaled) absolute difference between the prey's position and the wolf's current position.

[8] The leading wolves (α, β, δ) define the estimated prey location, and the coefficients **A** and **C** are computed based on the iteration-dependent parameter "a".

[9] In GWO, the true prey location (the global optimum) is unknown. Instead, the three best solutions found so far—denoted α, β, and δ—are used to approximate the prey's position. Each wolf computes three candidate updates relative to these leaders, and the final position is obtained by averaging them. Thus, the leaders collectively encode the current estimate of the prey's location.

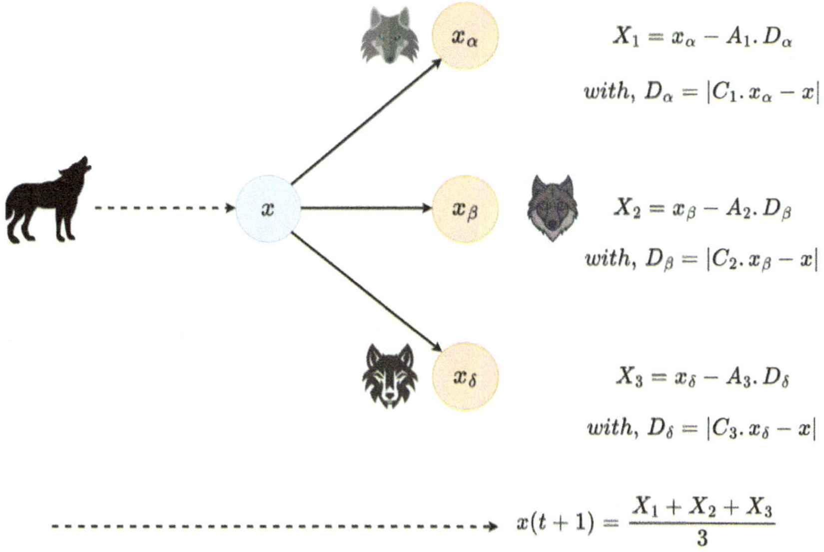

Fig. 4.12 Each wolf computes three candidate positions relative to the leaders ((α), (β), (δ)) using the encircling equations, and the next position ($x(t+1)$) is obtained by averaging (X_1), (X_2), and (X_3). This cooperative behavior models the hunting dynamics of grey wolves

strate strong performance for predicting heating and cooling loads, optimizing daylighting layouts, and solving complex multi-objective energy-comfort problems [31–33].

Example: Grey Wolf Optimizer on a Simple Benchmark

To illustrate how the update rules in Eqs. (4.13)–(4.17) operate in practice, consider a Grey Wolf Optimizer applied to the n-dimensional *Sphere function* $f(\mathbf{x}) = \sum_{i=1}^{n} x_i^2$. At each iteration "$t$", the wolves update their positions according to the guidance of the current best three leaders (α, β, δ), while the parameter "a" decreases linearly from 2 to 0, gradually shifting the search from exploration to exploitation.

Listing 4.5 shows a minimal yet complete Python™ implementation of GWO for this test problem. Figure 4.13 shows the corresponding convergence curve which records the best fitness value found at each iteration. Although simple, this example clearly demonstrates the role of the hierarchical leadership structure and the iteration index "t" in driving the pack toward the *global optimum*.

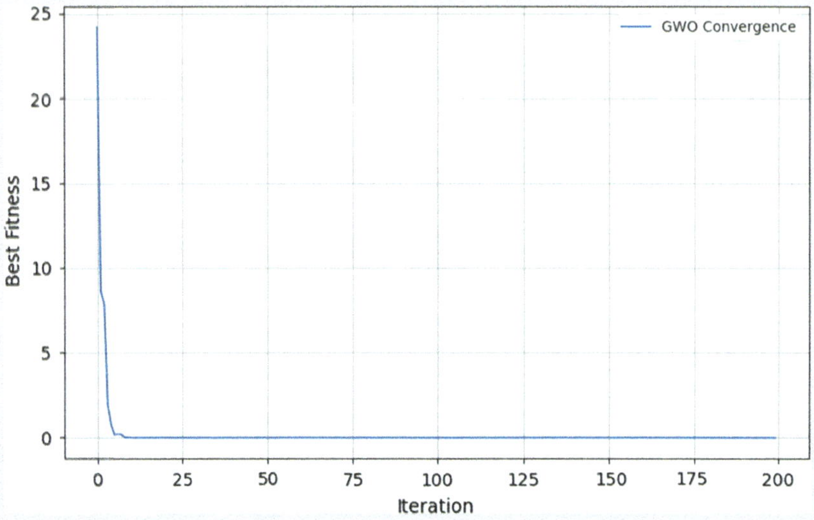

Fig. 4.13 Illustrative convergence behavior of the Grey Wolf Optimizer applied to a 5-dimensional Sphere test function using Listing 4.5. The curve shows the best fitness value (alpha wolf) at each iteration, highlighting the transition from exploration to exploitation as "a" decreases linearly

```
import numpy as np
import matplotlib.pyplot as plt

def gwo(objective_func, dim, bounds, num_wolves=20, max_iter=100):
    """
    Basic Grey Wolf Optimizer (GWO).

    Parameters
    ----------
    objective_func : callable
        Function f(x) to minimize. Expects a 1D NumPy array.
    dim : int
        Problem dimension.
    bounds : tuple (lower, upper)
        Lower and upper bounds for each decision variable.
    num_wolves : int
        Number of wolves in the pack.
    max_iter : int
        Maximum number of iterations.

    Returns
    -------
    best_position : np.ndarray
        Position of the alpha wolf (best solution found).
    best_fitness : float
        Best objective value.
    convergence_curve : list
        Best fitness value at each iteration.
```

```python
    """
    lower, upper = bounds

    # Initialize wolves randomly within bounds
    wolves = np.random.uniform(lower, upper, (num_wolves, dim))
    fitness = np.array([objective_func(w) for w in wolves])

    # Identify alpha, beta, delta wolves
    alpha_idx, beta_idx, delta_idx = np.argsort(fitness)[:3]
    alpha, beta, delta = wolves[alpha_idx], wolves[beta_idx], wolves[delta_idx]

    convergence_curve = []

    for t in range(max_iter):
        # Linearly decrease a from 2 to 0
        a = 2.0 - t * (2.0 / max_iter)

        for i in range(num_wolves):
            # --- alpha guidance ---
            r1, r2 = np.random.rand(dim), np.random.rand(dim)
            A1, C1 = 2 * a * r1 - a, 2 * r2
            D_alpha = np.abs(C1 * alpha - wolves[i])
            X1 = alpha - A1 * D_alpha

            # --- beta guidance ---
            r1, r2 = np.random.rand(dim), np.random.rand(dim)
            A2, C2 = 2 * a * r1 - a, 2 * r2
            D_beta = np.abs(C2 * beta - wolves[i])
            X2 = beta - A2 * D_beta

            # --- delta guidance ---
            r1, r2 = np.random.rand(dim), np.random.rand(dim)
            A3, C3 = 2 * a * r1 - a, 2 * r2
            D_delta = np.abs(C3 * delta - wolves[i])
            X3 = delta - A3 * D_delta

            # New position of wolf i
            wolves[i] = (X1 + X2 + X3) / 3.0

        # Clip wolves to feasible bounds
        wolves = np.clip(wolves, lower, upper)

        # Evaluate fitness and update alpha, beta, delta
        fitness = np.array([objective_func(w) for w in wolves])
        alpha_idx, beta_idx, delta_idx = np.argsort(fitness)[:3]
        alpha, beta, delta = wolves[alpha_idx], wolves[beta_idx], wolves[delta_idx]

        convergence_curve.append(fitness[alpha_idx])

    return alpha, fitness[alpha_idx], convergence_curve


# Objective function: Sphere function
def sphere(x):
    return np.sum(x**2)


# Run GWO on a 5D Sphere
best_solution, best_score, convergence_curve = gwo(
    objective_func=sphere,
    dim=5,
    bounds=(-10.0, 10.0),
    num_wolves=30,
    max_iter=200
)
```

```python

print("Best solution:", best_solution)
print("Best score:", best_score)

# Plot convergence curve
plt.figure(figsize=(8, 5))
plt.plot(convergence_curve, label="GWO convergence")
plt.xlabel("Iteration")
plt.ylabel("Best fitness")
plt.legend()
plt.grid(True)
plt.tight_layout()
plt.show()
```

Listing 4.5 Minimal Grey Wolf Optimizer applied to the n-dimensional Sphere function.

Whale Optimization Algorithm (WOA)

The Whale Optimization Algorithm (WOA), introduced by Mirjalili and Lewis [26], mimics the bubble-net hunting strategy of *humpback whales*. In WOA, each candidate solution (search agent) represents a whale, and the current best solution $\mathbf{x}^*(t)$ plays the role of the prey. The algorithm alternates between three behaviors: (*i*) encircling the prey, (*ii*) bubble-net attacking along a logarithmic spiral, and (*iii*) random search for prey. This structure gives WOA a relatively clean exploration-exploitation balance with only a few control parameters. Extensive tests on 29 benchmark functions and several structural design problems show that WOA is competitive with, or superior to, many state-of-the-art meta-heuristics while remaining easy to tune [26].

Encircling behavior. Once a provisional best solution $\mathbf{x}^*(t)$ is known, the other whales update their positions by "encircling" it. This is modeled by

$$\mathbf{D}(t) = \left|\mathbf{C}(t) \circ \mathbf{x}^*(t) - \mathbf{x}(t)\right| \tag{4.18}$$

$$\mathbf{x}(t+1) = \mathbf{x}^*(t) - \mathbf{A}(t) \circ \mathbf{D}(t) \tag{4.19}$$

where $\mathbf{x}(t)$ represents the position vector of a whale (i.e., a candidate solution) at iteration "t", "$\circ$" denotes element-wise multiplication, and $\mathbf{A}(t)$ and $\mathbf{C}(t)$ are coefficient vectors defined as

$$\mathbf{A}(t) = 2a(t)\,\mathbf{r}_1 - a(t) \tag{4.20}$$

$$\mathbf{C}(t) = 2\,\mathbf{r}_2 \tag{4.21}$$

Here $\mathbf{r}_1, \mathbf{r}_2 \sim \mathcal{U}(0, 1)$ are random vectors, while $a(t)$ decreases linearly from 2 to 0 over the course of the iterations. For $|\mathbf{A}| < 1$, whales move toward $\mathbf{x}^*$ (exploitation); for $|\mathbf{A}| > 1$, they are pushed away from reference positions, encouraging exploration.

Spiral bubble–net attack. The bubble-net maneuver combines a shrinking circle around the prey with a spiral path. WOA captures the spiral component by

$$\mathbf{x}(t+1) = \mathbf{D}'(t)\,e^{bl}\cos(2\pi l) + \mathbf{x}^*(t) \tag{4.22}$$

where $\mathbf{D}'(t) = \left|\mathbf{x}^*(t) - \mathbf{x}(t)\right|$, $b > 0$ controls the shape of the logarithmic spiral, and $l \sim \mathcal{U}[-1, 1]$ is a random scalar or vector. At each iteration, WOA chooses between the encircling step, Equation (4.19), and the spiral update, Eq. (4.22), with probability $p = 0.5$, mimicking the stochastic alternation between circular and spiral bubble-net attacks.

Search for prey (exploration). To escape local optima, WOA also allows whales to move with respect to randomly selected peers instead of $\mathbf{x}^*$ when $|\mathbf{A}| \geq 1$:

$$\mathbf{D}(t) = \left|\mathbf{C}(t) \circ \mathbf{x}_{\text{rand}}(t) - \mathbf{x}(t)\right| \tag{4.23}$$

$$\mathbf{x}(t+1) = \mathbf{x}_{\text{rand}}(t) - \mathbf{A}(t) \circ \mathbf{D}(t) \tag{4.24}$$

where "$\mathbf{x}_{\text{rand}}$" is a whale chosen uniformly at random from the population. This mechanism enables wide exploratory moves early in the run, while the decreasing "$a(t)$" gradually biases the search toward exploitation near $\mathbf{x}^*$.

In building applications, WOA has been used to tune HVAC schedules, calibrate surrogate models, and optimize multi-objective energy-comfort metrics [34, 35]. Its few control parameters, adaptive spiral/encircling behavior, and robust convergence properties make it a suitable candidate for high-dimensional BDT calibration and controller tuning problems.

Whale Optimization Algorithm: Spiral Encircling on a Simple Test Function

The following illustration shows how the mathematical operators in Eqs. (4.18)–(4.22) translate into an actual search process. We use WOA to minimize a simple quadratic objective function (the Sphere function):

$$f(\mathbf{x}) = \sum_{j=1}^{d} x_j^2$$

which has its global minimum at, $\mathbf{x} = \mathbf{0}$. Each whale is a candidate vector $\mathbf{x}_i(t)$ in $\mathbb{R}^d$.

Key Points for the Reader

- The *encircling* Eqs. (4.18)–(4.19) pull whales toward the current best solution $\mathbf{x}^*(t)$.
- The *spiral* Eq. (4.22) makes the best-tracking whales follow a contracting spiral around $\mathbf{x}^*(t)$, similar to the bubble-net maneuver observed in humpback whales.
- The exploration step, Eq. (4.24), briefly allows whales to chase randomly chosen peers instead of $\mathbf{x}^*(t)$, helping the swarm to escape poor local minima. These points are depicted visually in Fig. 4.14.

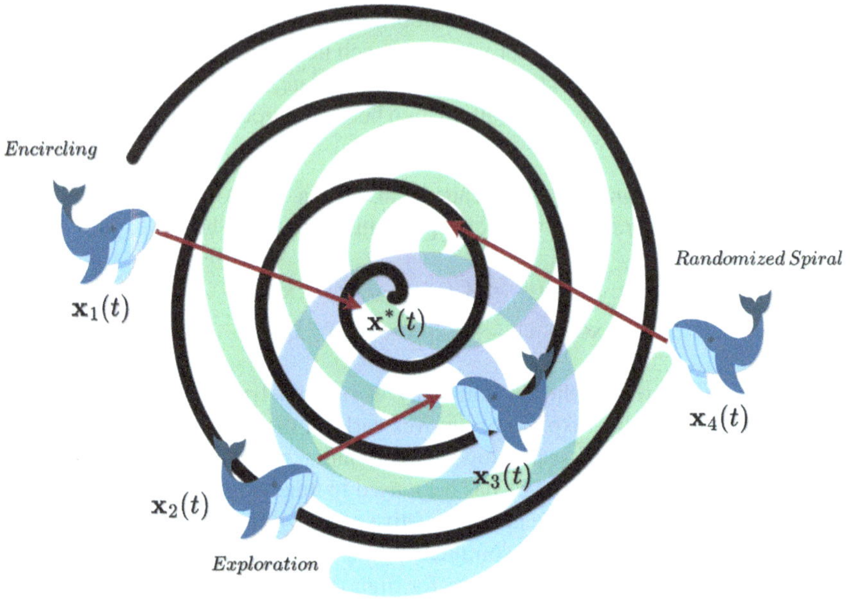

Fig. 4.14 Conceptual illustration of the Whale Optimization Algorithm (WOA). Each whale is a candidate solution. Depending on the random choice of operators, whales either *(i)* encircle the current best solution $x^*(t)$, *(ii)* follow a logarithmic spiral trajectory toward $x^*(t)$ (bubble-net attack), or *(iii)* move relative to other randomly selected whales to explore new regions of the search space

Listing 4.6 provides a compact Python™ implementation of WOA for the Sphere function. Figure 4.15 then plots the best fitness value $f(\mathbf{x}^*(t))$ versus iteration, illustrating how the whales progressively tighten their spiral around the optimum.[10]

```

import numpy as np
import matplotlib.pyplot as plt

# ------------------------------------------------------------
# 1. Objective function (Sphere)
# ------------------------------------------------------------
def sphere(x):
    """Simple convex test function: f(x) = sum(x_j^2)."""
    return np.sum(x**2)

# ------------------------------------------------------------
# 2. Whale Optimization Algorithm (WOA)
```

[10] The code in Listing 4.6 is intentionally simplified for pedagogical purposes. It demonstrates the core WOA update rules and is not a full-featured implementation with constraint handling or problem-specific enhancements.

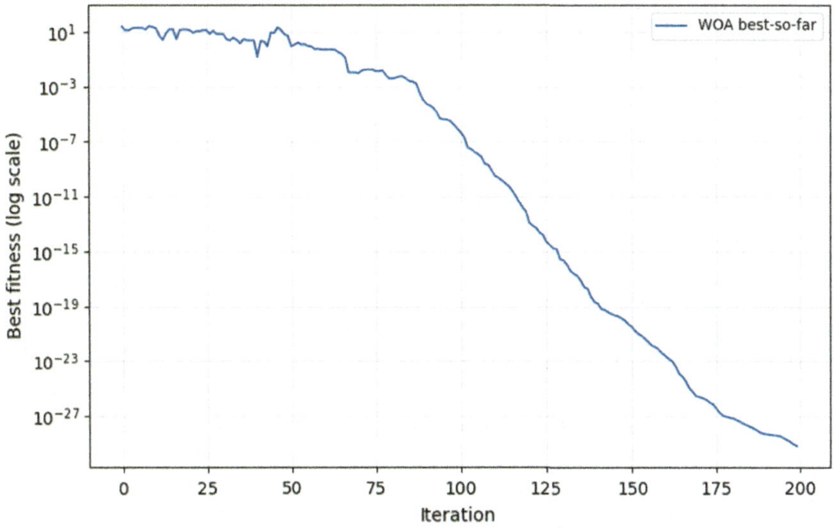

Fig. 4.15 Convergence of WOA on a 5-dimensional Sphere test function using the code from Listing 4.6. The best fitness value decreases rapidly during early iterations as whales explore the search space, and then gradually stabilizes as the swarm spirals around the global optimum

```
# ------------------------------------------------------------
def whale_optimization(objective_func, dim, bounds, num_whales=20, max_iter
    =100):
    lower, upper = bounds
    whales = np.random.uniform(lower, upper, size=(num_whales, dim))
    fitness = np.array([objective_func(w) for w in whales])

    alpha_idx, beta_idx, delta_idx = np.argsort(fitness)[:3]
    alpha = whales[alpha_idx].copy()
    beta  = whales[beta_idx].copy()
    delta = whales[delta_idx].copy()

    convergence_curve = []

    for t in range(max_iter):
        a = 2.0 - 2.0 * t / float(max_iter - 1)
        for i in range(num_whales):
            r1 = np.random.rand(dim)
            r2 = np.random.rand(dim)
            A = 2.0 * a * r1 - a
            C = 2.0 * r2
            p = np.random.rand()

            if p < 0.5:
                if np.all(np.abs(A) < 1.0):
                    D = np.abs(C * alpha - whales[i])
                    whales[i] = alpha - A * D
                else:
                    rand_idx = np.random.randint(num_whales)
                    X_rand = whales[rand_idx]
                    D = np.abs(C * X_rand - whales[i])
                    whales[i] = X_rand - A * D
            else:
                D_prime = np.abs(alpha - whales[i])
```

```
                b = 1.0
                l = np.random.uniform(-1.0, 1.0, size=dim)
                whales[i] = D_prime * np.exp(b * l) * np.cos(2.0 * np.pi * l)
     + alpha

        whales = np.clip(whales, lower, upper)
        fitness = np.array([objective_func(w) for w in whales])
        alpha_idx, beta_idx, delta_idx = np.argsort(fitness)[:3]
        alpha = whales[alpha_idx].copy()
        beta  = whales[beta_idx].copy()
        delta = whales[delta_idx].copy()
        convergence_curve.append(fitness[alpha_idx])

    return alpha, np.min(fitness), np.array(convergence_curve)

if __name__ == "__main__":
    dim = 5
    bounds = (-10.0, 10.0)
    best_solution, best_score, curve = whale_optimization(sphere, dim=dim,
     bounds=bounds, num_whales=30, max_iter=200)
    print("Best solution:", best_solution)
    print("Best score:", best_score)

    plt.figure(figsize=(8, 5))
    plt.semilogy(curve, label="WOA best-so-far")
    plt.xlabel("Iteration")
    plt.ylabel("Best fitness (log scale)")
    plt.grid(True, alpha=0.3)
    plt.legend()
    plt.tight_layout()
    plt.show()
```

Listing 4.6 Minimal Python implementation of WOA on the Sphere function.

Compared to earlier population-based methods such as PSO, ACO, and GWO, the Whale Optimization Algorithm introduces several novel mechanisms. First, WOA uses a mathematically explicit logarithmic-spiral operator which enables nonlinear, oscillatory contraction toward the prey—a behavior absent in prior meta-heuristics. Second, WOA *probabilistically alternates between shrinking encirclement and spiral bubble-net attack*, yielding two complementary exploitation modes. Third, exploration is directly controlled by the magnitude of $\mathbf{A}(t)$, offering a simple, low-parameter way to regulate search dynamics. These features give WOA strong local exploitation, robust performance in high-dimensional landscapes, and low parameter sensitivity, supporting its effectiveness in building-energy and BEMS optimization.

Common Structure Across Swarm Algorithms

Despite their differing biological inspirations, swarm algorithms exhibit several unifying properties:

- **Decentralized Search**: Solutions emerge from simple agent-level interactions rather than global models.
- **Stochasticity**: Randomness is integral to exploration (for example, random coefficients in PSO or stochastic path selection in ACO).
- **Positive Feedback**: Good solutions reinforce agent behavior (for example, pheromone accumulation in ACO or attraction toward best solutions in PSO, GWO, and WOA).
- **Adaptation**: Parameters or behaviors evolve based on ongoing search performance.
- **Few Control Parameters**: Many swarm algorithms (PSO, GWO) are simple to configure and thus suitable for real-world BDT applications.

These properties make swarm intelligence highly suited to the Digital Twin setting, where search landscapes are often non-convex, models are uncertain, and fast global optimization is required for tasks such as setpoint tuning, sensor routing, and multi-objective control.

4.3.4 Fuzzy Systems and Approximate Reasoning

While the evolutionary and swarm-based methods in the previous sections focus on *search* in high-dimensional spaces, fuzzy logic control provides an interpretable way of encoding expert knowledge in the form of linguistic IF- - THEN rules. Within Computational Intelligence, fuzzy systems occupy a complementary role to neural networks and Evolutionary Algorithms (EAs): they offer a flexible nonlinear mapping between inputs and outputs, but retain a rule-based structure which is easier to interpret and tune in practice. Since the early 1990s, fuzzy controllers have been widely deployed in buildings for thermal comfort, indoor air quality, and lighting control, often using indices such as Predicted Mean Vote (PMV) as high-level comfort targets [36–39].[11]

A standard Mamdani-type fuzzy controller [41] maps a vector of crisp inputs $\mathbf{u} = [u_1, \ldots, u_m]^\top$ to a scalar crisp control signal "y" using three steps:

1. **Fuzzification**: Each input, u_j, is mapped into one or more fuzzy sets via membership functions $\mu_{A_{j\ell}}(u_j) \in [0, 1]$, where $A_{j\ell}$ denotes the ℓ-th linguistic term (for example, "low", "medium", "high") associated with input, u_j.
2. **Inference**: A rule base of the general form

$$R_k : \quad \text{IF } u_1 \text{ is } A_{1k} \text{ AND } u_2 \text{ is } A_{2k} \text{ THEN } y \text{ is } B_k \tag{4.25}$$

[11] Beyond buildings, similar fuzzy-control architectures have been applied to safety-critical systems such as high-speed railways, where they encode expert driving strategies and constraints directly in the rule base [40].

is evaluated using fuzzy logical operators. Here, "k" indexes the rule, A_{jk} is the antecedent fuzzy set for input u_j in rule R_k, and B_k is the consequent fuzzy set for the output variable "y". For Mamdani inference with the minimum "t-norm", the firing strength, w_k, of rule, R_k, is

$$w_k = \min_j \mu_{A_{jk}}(u_j) \tag{4.26}$$

where $\mu_{A_{jk}}(u_j)$ is the membership degree of u_j in fuzzy set A_{jk}. The contribution of that rule to the output membership function $\mu_{B_k}(y)$ is then truncated at height w_k.

3. **Aggregation and Defuzzification**: All rule contributions are aggregated into a single output membership function:

$$\mu_{\text{out}}(y) = \max_k \left[w_k \wedge \mu_{B_k}(y) \right] \tag{4.27}$$

where $\mu_{\text{out}}(y)$ denotes the combined fuzzy description of the output and "$\wedge$" represents the chosen "t-norm" (for min-max inference, "$\wedge$" reduces to the minimum operator and the outer max is the usual maximum over rules). A crisp output is finally obtained via a defuzzification operator. For the widely used centroid (center-of-area) method, the control signal, $y^\star$, is

$$y^\star = \frac{\displaystyle\int y\, \mu_{\text{out}}(y)\, \mathrm{d}y}{\displaystyle\int \mu_{\text{out}}(y)\, \mathrm{d}y} \tag{4.28}$$

Here $y^\star$ is the scalar command sent to the actuator (for example, a valve position or fan-speed setpoint).

In BEMSs, fuzzy controllers have been used to regulate thermal comfort, ventilation rates, and lighting while explicitly embedding linguistic concepts such as "slightly warm" or "too cold" [37, 38]. Several field studies report energy savings of 20–30% for heating and cooling and even higher reductions for lighting, compared to conventional on-off or fixed-gain PID schemes, while maintaining or improving comfort and indoor air quality [36, 39]. At the same time, these works also highlight important *limitations*: PD-type fuzzy controllers can be sensitive to measurement noise and disturbances, and PID-type fuzzy controllers are rarely deployed in practice due to the difficulty of constructing and tuning a large, consistent rule base [37, 38]. *As the number of inputs grows (for example, temperature, humidity, CO_2, illumination, occupancy), rule bases can suffer from "rule explosion" unless combined with learning or evolutionary tuning techniques, an issue which motivates hybrid CI schemes and their integration with Building Digital Twins (BDTs) discussed later in this chapter.*

The next example shows how the structure in Eqs. (4.25)–(4.28) can be realized for a simple thermal-comfort controller using `scikit-fuzzy`.

Example: Fuzzy-Logic Thermal Comfort Controller

Consider a single-zone space where we want to adjust a generic heating/cooling command, "`adjust`", based on two inputs:

- indoor air temperature, T [°C];
- temperature error, $e = T_{set} - T$ [°C].

We define three fuzzy sets for each input: *low/comfortable/high* for "T", and *negative/zero/positive* for "e", as well as three fuzzy sets for the output: *cool/none/heat*. A typical rule base is

$$R_1 : \text{IF } T \text{ is low OR } e \text{ is negative THEN adjust is heat,}$$
$$R_2 : \text{IF } T \text{ is comfortable AND } e \text{ is zero THEN adjust is none,}$$
$$R_3 : \text{IF } T \text{ is high OR } e \text{ is positive THEN adjust is cool.}$$

Rule R_1 captures the intuition that if the space is cold, or the setpoint is above the measured temperature, the controller should increase heating. Conversely, R_3 reduces the control signal when the space is hot or the measured temperature overshoots the setpoint. When both temperature and error are in comfortable/zero ranges, R_2 keeps the actuator near its current level. Similar rule-based comfort regulators have been implemented in real buildings, showing tangible energy savings and improved PMV-based comfort when compared with fixed-gain controllers [36, 37, 39].

Listing 4.7 implements this fuzzy controller in Python™ using `scikit-fuzzy`, and generates two diagnostic plots: (i) the membership functions of all variables, and (ii) the resulting control surface $u^\star(T, e)$ obtained via Mamdani inference and centroid defuzzification. Figures 4.16 and 4.17 illustrate these outputs, while Fig. 4.18 provides a high-level block diagram of the fuzzy control pipeline.[12]

```
"""
Example: Fuzzy Logic Controller for Thermal Comfort

This script implements a simple fuzzy logic controller that adjusts
heating/cooling based on:
    - indoor air temperature (degC)
    - temperature error (setpoint - measured)
```

[12] The implementation in Listing 4.7 is deliberately simple and serves as an illustrative example. In practice, additional inputs (such as humidity, CO_2, or daylight levels), more refined linguistic sets, and data-driven tuning of membership functions and rules [e.g., neuro-fuzzy or GA-assisted approaches; 37, 38] are often required—particularly when robustness is a design objective for BDT-based controllers.

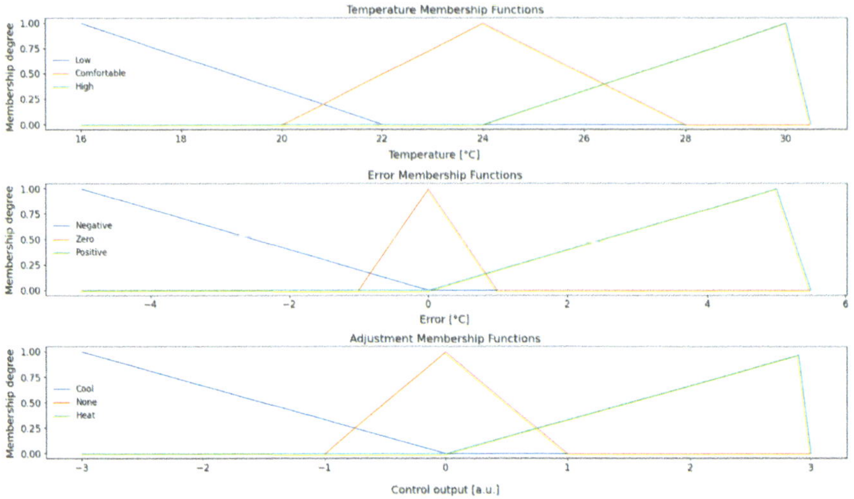

Fig. 4.16 Membership functions used in the fuzzy thermal comfort controller of Listing 4.7. Temperature and error are partitioned into three fuzzy sets each (*low/comfortable/high* and *negative/zero/positive*), while the control output is described by *cool/none/heat*

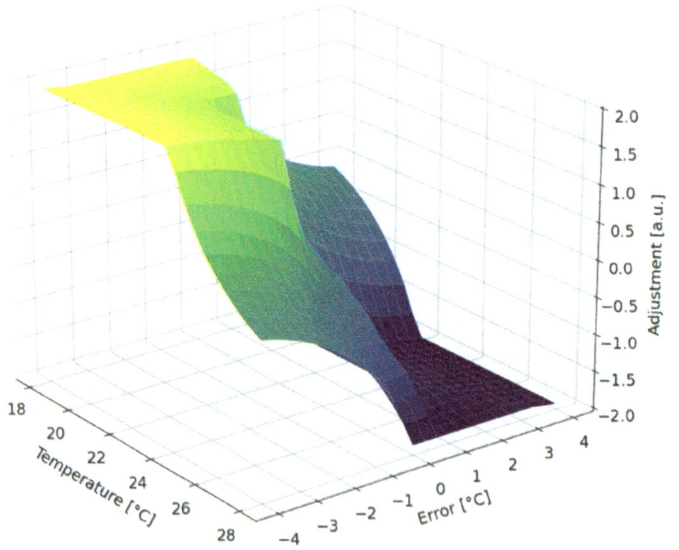

Fig. 4.17 Control surface of the fuzzy thermal comfort controller, showing the crisp adjustment command as a function of temperature and temperature error. The surface reflects the combined effect of the rule base and centroid defuzzification: heating is increased in cold or underheated conditions, reduced near comfort, and reversed toward cooling when the space becomes too warm

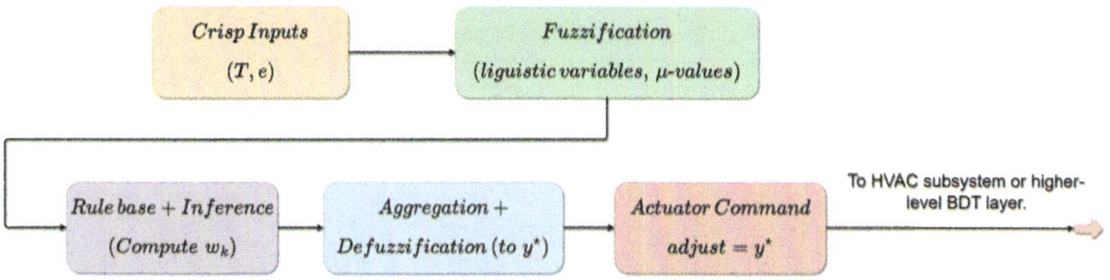

Fig. 4.18 Conceptual structure of the fuzzy thermal comfort controller. Crisp inputs (temperature and error) are fuzzified into linguistic variables, processed by the rule base and inference engine, and then defuzzified into a single actuator command that can be passed to a building HVAC subsystem or a higher-level BDT control layer

```
Outputs:
    - 'adjust' command in arbitrary units
      (negative = cool, positive = heat)

It also plots:
    1) Membership functions for all fuzzy variables
    2) The resulting control surface (adjust vs. temperature & error)
"""

import numpy as np
import skfuzzy as fuzz
from skfuzzy import control as ctrl
import matplotlib.pyplot as plt
from mpl_toolkits.mplot3d import Axes3D  # noqa: F401

# ----------------------------------------------------------
# 1. Define fuzzy variables
# ----------------------------------------------------------
temp = ctrl.Antecedent(np.arange(16, 31, 0.5), 'temperature')
error = ctrl.Antecedent(np.arange(-5, 6, 0.5), 'error')
adjust = ctrl.Consequent(np.arange(-3, 3.1, 0.1), 'adjust')
# (default defuzzification method in scikit-fuzzy is 'centroid')

# ----------------------------------------------------------
# 2. Membership functions
# ----------------------------------------------------------
temp['low']           = fuzz.trimf(temp.universe, [16, 16, 22])
temp['comfortable']   = fuzz.trimf(temp.universe, [20, 24, 28])
temp['high']          = fuzz.trimf(temp.universe, [24, 30, 30])

error['negative']     = fuzz.trimf(error.universe, [-5, -5, 0])
error['zero']         = fuzz.trimf(error.universe, [-1, 0, 1])
error['positive']     = fuzz.trimf(error.universe, [0, 5, 5])

adjust['cool']        = fuzz.trimf(adjust.universe, [-3, -3, 0])
adjust['none']        = fuzz.trimf(adjust.universe, [-1, 0, 1])
adjust['heat']        = fuzz.trimf(adjust.universe, [0, 3, 3])

# ----------------------------------------------------------
# 3. Plot membership functions
# ----------------------------------------------------------
plt.rcParams.update({'font.size': 14})

fig, axes = plt.subplots(3, 1, figsize=(8, 10))

# Temperature
```

```
axes[0].plot(temp.universe, temp['low'].mf, label='Low')
axes[0].plot(temp.universe, temp['comfortable'].mf, label='Comfortable')
axes[0].plot(temp.universe, temp['high'].mf, label='High')
axes[0].set_title('Temperature Membership Functions', fontsize=16)
axes[0].set_xlabel('Temperature [°C]', fontsize=14)
axes[0].set_ylabel('Membership degree', fontsize=14)
axes[0].legend(fontsize=12)
axes[0].grid(alpha=0.3)

# Error
axes[1].plot(error.universe, error['negative'].mf, label='Negative')
axes[1].plot(error.universe, error['zero'].mf, label='Zero')
axes[1].plot(error.universe, error['positive'].mf, label='Positive')
axes[1].set_title('Error Membership Functions', fontsize=16)
axes[1].set_xlabel('Error [°C]', fontsize=14)
axes[1].set_ylabel('Membership degree', fontsize=14)
axes[1].legend(fontsize=12)
axes[1].grid(alpha=0.3)

# Adjustment
axes[2].plot(adjust.universe, adjust['cool'].mf, label='Cool')
axes[2].plot(adjust.universe, adjust['none'].mf, label='None')
axes[2].plot(adjust.universe, adjust['heat'].mf, label='Heat')
axes[2].set_title('Adjustment Membership Functions', fontsize=16)
axes[2].set_xlabel('Control output [a.u.]', fontsize=14)
axes[2].set_ylabel('Membership degree', fontsize=14)
axes[2].legend(fontsize=12)
axes[2].grid(alpha=0.3)

plt.tight_layout()
plt.show()

# ------------------------------------------------------------
# 4. Rule base
# ------------------------------------------------------------
rule1 = ctrl.Rule((temp['low'] | error['negative']), adjust['heat'])
rule2 = ctrl.Rule((temp['comfortable'] & error['zero']), adjust['none'])
rule3 = ctrl.Rule((temp['high'] | error['positive']), adjust['cool'])

comfort_ctrl = ctrl.ControlSystem([rule1, rule2, rule3])

# Example: evaluate controller at a single operating point
setpoint = 24.0
measured = 26.0
err_val = setpoint - measured  # -2.0 °C

sim = ctrl.ControlSystemSimulation(comfort_ctrl)
sim.input['temperature'] = measured
sim.input['error'] = err_val
sim.compute()

print(f"Temperature = {measured:.1f} °C, Error = {err_val:.1f} °C")
print(f"Fuzzy controller output: {sim.output['adjust']:.3f}")

# ------------------------------------------------------------
# 5. Control surface over (T, e)
# ------------------------------------------------------------
T_vals = np.linspace(18, 28, 25)
E_vals = np.linspace(-4, 4, 25)
T_grid, E_grid = np.meshgrid(T_vals, E_vals)
A_grid = np.zeros_like(T_grid)

for i in range(T_grid.shape[0]):
    for j in range(T_grid.shape[1]):
        sim = ctrl.ControlSystemSimulation(comfort_ctrl)  # reset state
        sim.input['temperature'] = T_grid[i, j]
        sim.input['error'] = E_grid[i, j]
```

```
            sim.compute()
            A_grid[i, j] = sim.output['adjust']

fig = plt.figure(figsize=(9, 7))
ax = fig.add_subplot(111, projection='3d')
surf = ax.plot_surface(T_grid, E_grid, A_grid,
                       cmap='viridis', linewidth=0,
                       antialiased=True)

ax.set_xlabel('Temperature [°C]', fontsize=14)
ax.set_ylabel('Error [°C]', fontsize=14)
ax.set_zlabel('Adjustment [a.u.]', fontsize=14)

plt.tight_layout()
plt.show()
```

Listing 4.7 Fuzzy logic controller for thermal comfort with membership plots and control surface.

Interpreting the Membership Functions: A Worked Numerical Example

To illustrate how the membership functions in Fig. 4.16 are used within the fuzzy controller of Listing 4.7, consider the evaluated operating point:

$$T = 26°\text{C}, \qquad e = -2°\text{C}$$

1. Temperature Memberships
The membership functions for temperature were defined as

$$\begin{aligned} \text{low} &: [16, 16, 22] \\ \text{comfortable} &: [20, 24, 28] \\ \text{high} &: [24, 30, 30] \end{aligned}$$

Low Temperature: Since $T = 26$ lies entirely outside the support [16, 22]:

$$\mu_{\text{low}}(26) = 0$$

Comfortable Temperature: This triangular set peaks at 24°C and decreases linearly toward 28°C. The `descending slope` is defined by

$$\mu_{\text{comfortable}}(T) = \frac{28 - T}{28 - 24}, \qquad T \in [24, 28]$$

Thus,

$$\mu_{\text{comfortable}}(26) = \frac{28 - 26}{4} = \frac{2}{4} = 0.5$$

High Temperature: The `rising portion` of the triangle is defined on [24, 30] by

$$\mu_{\text{high}}(T) = \frac{T - 24}{30 - 24}$$

Therefore,

$$\mu_{\text{high}}(26) = \frac{26 - 24}{6} = \frac{2}{6} \approx 0.33$$

2. Error Memberships
The error sets are

$$\text{negative} : [-5, -5, 0], \qquad \text{zero} : [-1, 0, 1], \qquad \text{positive} : [0, 5, 5]$$

Negative Error: On the `rising portion` from −5 to 0:

$$\mu_{\text{negative}}(e) = \frac{0 - e}{0 - (-5)}$$

Thus, for $e = -2$,

$$\mu_{\text{negative}}(-2) = \frac{0 - (-2)}{5} = \frac{2}{5} = 0.4$$

Zero Error: Since $e = -2$ is outside $[-1, 1]$:

$$\mu_{\text{zero}}(-2) = 0$$

Positive error: All negative inputs yield

$$\mu_{\text{positive}}(-2) = 0$$

3. Rule Activation
The rule base is

$$\begin{aligned}
R_1 &: \text{IF } T \text{ is low OR } e \text{ is negative THEN adjust is heat,}\\
R_2 &: \text{IF } T \text{ is comfortable AND } e \text{ is zero THEN adjust is none,}\\
R_3 &: \text{IF } T \text{ is high OR } e \text{ is positive THEN adjust is cool.}
\end{aligned}$$

Rule R1 (heat):

$$w_1 = \max(\mu_{\text{low}}, \mu_{\text{negative}}) = \max(0,\ 0.4) = 0.4$$

Rule R2 (none):

$$w_2 = \min(\mu_{\text{comfortable}}, \mu_{\text{zero}}) = \min(0.5,\ 0) = 0$$

Rule R3 (cool):

$$w_3 = \max(\mu_{\text{high}}, \mu_{\text{positive}}) = \max(0.33, 0) = 0.33$$

4. Defuzzified Outcome
The heat and cool membership functions are clipped at heights 0.4 and 0.33, respectively, aggregated, and defuzzified using the centroid operator, Eq. (4.28). For this operating point, the resulting crisp output is $y^\star \approx 0.13$ (as printed by Listing 4.7), that is, a near-neutral adjustment with a slight bias toward the "heat" consequent because rule R_1 fires more strongly than R_3. This illustrates how fuzzy inference blends partially conflicting linguistic rules into a single smooth control command.

4.4 Optimization and Search in Computational Intelligence

The preceding sections established Computational Intelligence (CI) as a unifying paradigm for learning, adaptation, and decision-making under uncertainty. A central theme underlying all CI techniques—whether learning-based, evolutionary, or swarm-based—is `search`. At its core, CI reframes control and decision-making problems as structured searches over complex, uncertain, and high-dimensional spaces.

In conventional control theory, optimization is often formulated as a well-posed mathematical problem: objective functions are smooth, constraints are convex, and gradients are either known or efficiently approximated. Building systems, however, rarely conform to these assumptions. Real buildings exhibit nonlinear dynamics, discrete operational modes, stochastic disturbances, delayed responses, and incomplete observability. As a result, the optimization problems encountered in Building Digital Twins (BDTs) are fundamentally *search problems*, rather than classical optimization problems.

Computational Intelligence embraces this reality by prioritizing exploration, robustness, and adaptability over analytical tractability. Instead of relying on local gradients or exact models, CI methods operate directly on objective evaluations—often obtained from simulations, measurements, or surrogate models—and iteratively improve candidate solutions through heuristic search. This shift from "solving equations" to "exploring landscapes" is essential for Digital Twins which must operate under uncertainty, evolve over time, and balance competing objectives.

Conceptual Insight!

In building systems, optimization does not begin with equations—it begins with search.

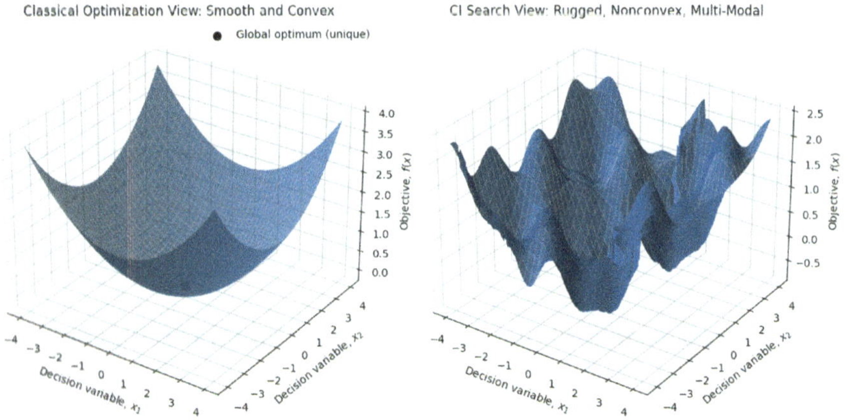

Fig. 4.19 Conceptual comparison between classical optimization and Computational Intelligence (CI) search landscapes. The left panel illustrates a smooth, convex, unimodal objective surface typical of problems where gradient-based methods are reliable and a unique optimum is readily identifiable. The right panel depicts a rugged, non-convex, multi-modal landscape representative of Building Digital Twin optimization, where discontinuities, uncertainty, and black-box simulation effects can obscure gradients and create many competing basins, motivating CI-based search strategies

4.4.1 Search Landscapes in Buildings and Digital Twins

A useful way to understand the role of CI in Building Digital Twins is through the notion of a *search landscape*. In this abstraction, each point in the landscape corresponds to a candidate solution—such as a set of control parameters, calibration coefficients, or design variables—while the landscape height represents one or more performance metrics, including energy consumption, comfort deviation, emissions, or operational cost.

Unlike textbook optimization problems, building performance landscapes are rarely smooth or unimodal. They are shaped by several intrinsic characteristics:

- strong nonlinearities in thermal, electrical, and occupant interactions,
- multi-modal behavior, where several locally optimal solutions coexist,
- discontinuities introduced by logical control rules, equipment constraints, and mode switching,

- noise and stochasticity arising from weather variability, occupancy uncertainty, and sensor imperfections,
- simulation artifacts, including solver tolerances, convergence failures, and discretization effects.

From a Digital Twin perspective, these landscapes are further complicated by model mismatch and temporal drift: the "true" landscape evolves as building usage patterns, system degradation, and external conditions change [42, 43]. Traditional gradient-based optimization[13] methods struggle in such environments, as gradients may be ill-defined, misleading, or entirely unavailable—particularly when objective evaluations rely on black-box simulations or measured data.

For example, a simple energy cost minimization might assume a smooth quadratic objective such as

$$f(x) = (x-5)^2, \quad \text{where the gradient:} \tag{4.29}$$

$$\nabla f(x) = 2(x-5), \quad \text{clearly points toward the optimum at:} \tag{4.30}$$

$$x^* = 5 \tag{4.31}$$

In reality, Building Digital Twin objectives often resemble rugged, noisy surfaces with multiple local minima, making such gradient cues unreliable, see Figs. 4.19 and 4.20 for a conceptual illustration.

CI-based search methods, by contrast, are explicitly designed to navigate rugged and uncertain landscapes. They do not assume convexity, smoothness, or perfect models, making them naturally aligned with the realities of simulation-driven and data-assisted building optimization.

4.4.2 Deterministic Versus Stochastic Optimization

Another defining distinction in CI-based optimization is between *deterministic* and *stochastic* search processes [44]. Deterministic optimization methods follow fixed rules: given identical initial conditions and data, they produce identical solution trajectories. While efficient for well-behaved problems, such methods are highly sensitive to initialization and prone to entrapment in local optima when the underlying landscape is complex.

Stochastic optimization, in contrast, deliberately incorporates randomness into the search process. Random perturbations, probabilistic decisions, or population diversity allow the search to escape local optima and explore broader regions of the solution space. In the context of CI, stochasticity is not a weakness but a mechanism for exploration, robustness, and resilience.

For Building Digital Twins, stochastic optimization offers several advantages:

[13] Here, "gradient-based" refers to optimization methods which use `gradients` (derivatives) of the objective function to decide how to change the variables to improve the objective.

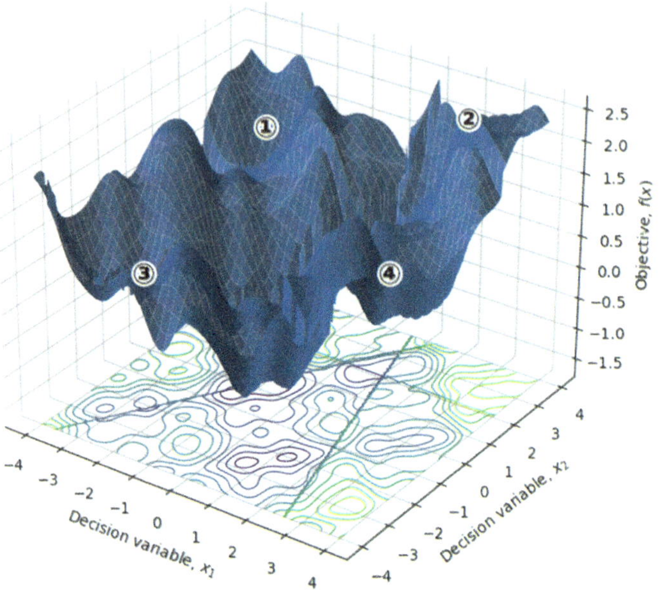

Fig. 4.20 Conceptual illustration of a rugged optimization landscape representative of building and Building Digital Twin applications. The surface exhibits sharp ridges and cliff-like transitions (①) reflecting equipment mode switching and rule-based logic, plateau-like regions (②) associated with comfort deadbands and hysteresis, rough multi-modal basins (③) influenced by stochastic disturbances and measurement uncertainty, and step-like discontinuities (④) that can arise from black-box simulation artifacts such as discretization and solver tolerances. These landscape characteristics undermine gradient reliability and motivate search-based Computational Intelligence approaches

- robustness to noisy and uncertain objective evaluations,
- improved exploration when gradient information is unreliable or unavailable,
- adaptability under evolving building dynamics and operational conditions,
- natural accommodation of uncertainty in model parameters and sensor data.

Importantly, stochastic optimization does not imply uncontrolled randomness. Instead, randomness is structured and guided by performance feedback, enabling an effective balance between exploration and exploitation. Table 4.3 draws a high level comparison between the deterministic and stochastic optimization in the context of buildings.

Key Principle!

In Computational Intelligence (CI), stochasticity is a design feature which enables resilience and global exploration.

Table 4.3 Conceptual comparison between deterministic and stochastic optimization in building applications

Aspect	Deterministic optimization	Stochastic optimization
Search behavior	Single trajectory	Multiple exploratory paths
Sensitivity to initialization	High	Reduced
Robustness to noise	Limited	High
Suitability for BDTs	Restricted	Well suited

4.4.3 Global Versus Local Search

Closely related to the deterministic–stochastic distinction is the concept of *global versus local search*. Local search methods refine a single candidate solution by making incremental improvements within its neighborhood. While computationally efficient, they rely heavily on good initial guesses and are vulnerable to local optima in multi-modal landscapes.

Global search strategies, by contrast, explore multiple regions of the solution space simultaneously, see Fig. 4.21. Rather than committing early to a single trajectory, global search maintains diversity among candidate solutions and uses performance feedback to guide exploration [45]. This paradigm is particularly valuable in building systems, where optimal solutions may be widely separated in parameter space and prior knowledge of good initial conditions is limited.

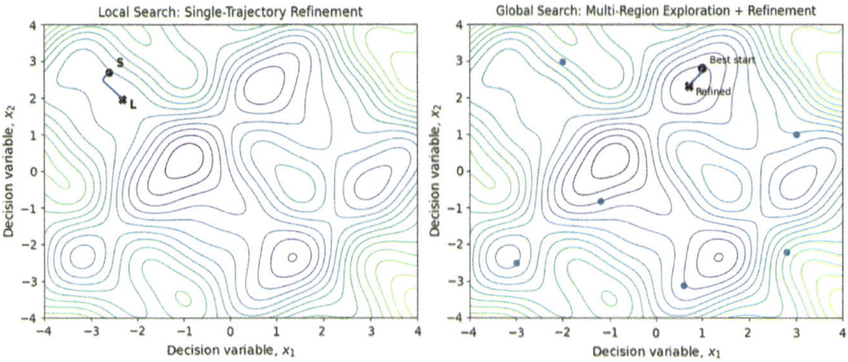

Fig. 4.21 Conceptual illustration of local versus global search on a multi-modal optimization landscape. Local search refines a single trajectory from an initial point and converges to the nearest basin of attraction, often yielding a local optimum ($S \rightarrow L$). Global search explores multiple regions of the decision space (multi-start candidates) before refinement, reducing sensitivity to initialization and improving the likelihood of identifying higher-quality basins prior to local improvement ("Best start" → "Refined"). *The blue markers denote candidate points sampled during global exploration*

In CI-based Digital Twins, global and local search are often complementary rather than competing concepts. Global search identifies promising regions of the landscape, while local refinement improves solution quality within those regions. This layered perspective prepares the reader for the population-based and multi-objective frameworks which are introduced and applied in Chap. 5.

4.4.4 Multi-objective Optimization Fundamentals

Building optimization problems are inherently multi-objective. energy efficiency, thermal comfort, indoor air quality, emissions, and operational cost are often conflicting objectives, and no single solution simultaneously optimizes all of them [46–48]. Consequently, the goal of optimization in Building Digital Twin (BDT) frameworks is not to identify a single "best" solution, but to characterize and explore the trade-offs among competing objectives.

Multi-Objective Optimization (MOO) formalizes this perspective through the concept of *Pareto optimality*. A solution is Pareto-optimal if no other solution improves one objective without degrading at least one other. The collection of such solutions forms the *Pareto front*, see Fig. 4.22, representing the feasible trade-off surface available to the decision-maker.

This viewpoint aligns naturally with the envisioned role of Digital Twins as decision-support systems. Rather than prescribing a single operating point, a CI-enabled BDT framework can present a spectrum of alternatives, allowing stakeholders to select solutions based on contextual priorities, regulatory constraints, or evolving operational considerations.

Decision-Making Insight!

In Building Digital Twin frameworks, optimization is better viewed not as the search for a single solution, but as the systematic exploration of trade-offs among competing objectives.

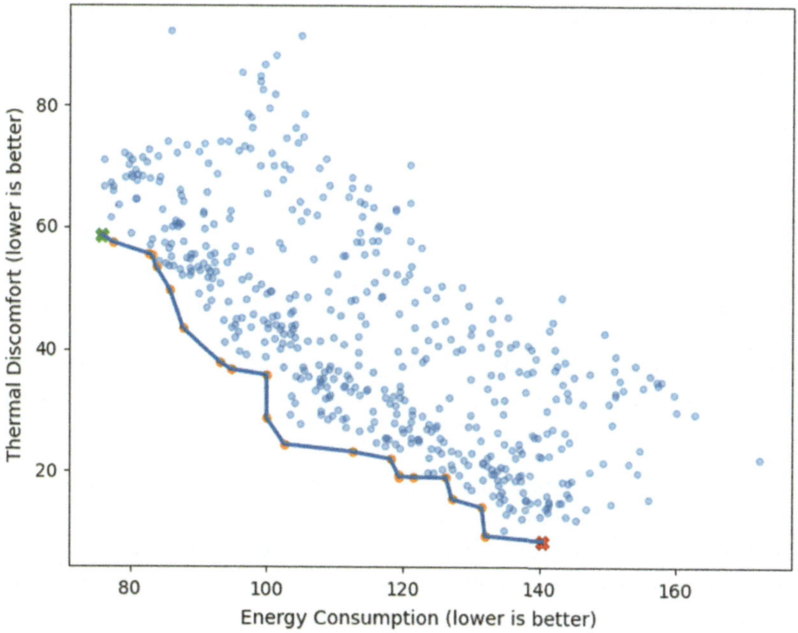

Fig. 4.22 Illustrative Pareto front depicting the trade-off between energy consumption and thermal discomfort in building optimization. Each marker represents a feasible operating configuration. The emphasized curve denotes the set of *non-dominated* trade-offs (Pareto-optimal solutions), where improving one objective inevitably degrades another. The endpoints reflect operating regimes that prioritize minimizing energy (green) or minimizing discomfort (red), respectively. *Such representations highlight how Building Digital Twin frameworks can be envisioned as decision-support systems, enabling stakeholders to explore and select from a portfolio of optimal compromises rather than relying on a single operating point*

This multi-objective perspective forms a central conceptual pillar for the Computational Intelligence (CI) frameworks developed in the subsequent chapters. Chapter 5 consolidates core CI and Multi-Objective Optimization (MOO) principles—*drawing on a coherent body of research developed over the course of the author's doctoral research/studies*—and examines their application across Building Energy Management Systems (BEMSs), parameter estimation, and HVAC-related subsystems. Chapter 6 then establishes the simulation backbone required for Building Digital Twin development, introducing EnergyPlus™ as a physics-based building performance modeling environment and detailing the modeling and simulation workflows which enable subsequent integration with CI techniques. Building on these foundations, Chap. 7 illustrates how these elements can be brought together within a simulation-driven, surrogate-assisted optimization framework oriented toward Building Digital Twins (BDTs).

4.5 Why Computational Intelligence for Digital Twins?

The preceding sections collectively establish why Computational Intelligence (CI) is not merely an optional enhancement, but a foundational enabler for Building Digital Twin frameworks. As discussed in Chap. 3, BDTs operate at the intersection of physics-based modeling, sensing, prediction, and decision-making. However, their effectiveness is constrained by several fundamental challenges:

- nonlinearity arising from coupled thermal, electrical, and occupant dynamics,
- stochasticity due to weather variability, occupancy uncertainty, and measurement noise,
- partial observability, as not all relevant states can be directly measured,
- long-memory and multi-timescale behavior, where slow thermal dynamics interact with fast control actions.

Classical control and optimization methods, while powerful, are often ill-equipped to address all these challenges simultaneously—particularly when models are imperfect and objectives conflict. Computational Intelligence addresses this gap by offering search-based, adaptive, and uncertainty-aware mechanisms that complement physics-based Digital Twins rather than replace them.

CI can be envisioned to augment Digital Twin frameworks with capabilities for global exploration, adaptive learning, and multi-objective reasoning [49, 50]. In doing so, CI enables Digital Twins to evolve beyond static simulation models toward flexible decision-support systems capable of learning from data, adapting to change, and navigating complex trade-offs under uncertainty.

Bridge Statement

Where physics-based models describe how buildings behave, Computational Intelligence determines how they can be operated under competing objectives and uncertainty.

This synthesis serves as the bridge between Part I and Part II. Having established the foundations of CI, search techniques, and multi-objective reasoning, the subsequent chapters move toward practical implementations—introducing operational frameworks, simulation-driven optimization, and surrogate-assisted methodologies. These approaches, including Parameter Estimation (PE), Surrogate-Assisted (SA) NSGA-II to accelerate EnergyPlus™ simulations, and Multi-Objective Optimization (MOO) of building HVAC systems, collectively advance the development of Building Digital Twin-oriented decision-support systems.

Table 4.4 CI Strategies for building Digital Twins

Strategy	Techniques	Purpose
Surrogate Modeling and Meta-Models	Gaussian Processes, Neural Networks, Polynomial Chaos expansions	Approximate high-fidelity simulations and reduce computational cost
Hybrid Learning and Physics (Grey-Box)	Physics-informed surrogate models, data assimilation	Combine first-principles understanding with measured data
Evolutionary Hyperparameter Optimization	Multi-Objective Evolutionary Algorithms (MOEAs)	Tune RL policies, predictive models, and MPC parameters
Robust and Safe Reinforcement Learning	Reward shaping, constraint handling, runtime assurance	Ensure safety-critical building operation
CI for Model Drift Detection	Kalman filtering, particle filtering, evolutionary re-identification	Maintain model relevance over time
Fast Multi-Objective Optimization	Surrogate-assisted CI (e.g., SA-NSGA-II)	Enable tractable exploration of complex trade-offs
Human-Centric and Occupant Behavior Modeling	Data-driven occupancy inference, comfort prediction models	Support human-in-the-loop decision-making

4.6 Computational Intelligence for Building Digital Twins

The challenges associated with Building Digital Twins, as identified in the challenge taxonomy of Sect. 3.9, motivate the use of Computational Intelligence (CI) strategies which extend beyond traditional modeling and control approaches. Rather than addressing individual limitations in isolation, CI offers a collection of complementary techniques capable of jointly handling uncertainty, nonlinearity, scalability, and model mismatch within emerging Digital Twin frameworks.

In particular, CI-based methods enable Digital Twins to operate under imperfect knowledge, evolve as new data becomes available, and reason explicitly about competing performance objectives. Table 4.4 summarizes key strategies illustrating how CI can be systematically leveraged to mitigate these challenges.

Collectively, these strategies demonstrate how CI provides the analytical flexibility required to address the fundamental limitations of static models and classical optimization, thereby enabling more adaptive and resilient Building Digital Twin frameworks.

4.7 CI-Driven Surrogate Modeling for Building Digital Twins

High-fidelity building performance simulation is a critical component of Building Digital Twin development, yet its computational cost often limits direct use within optimization, control, or real-time decision-support loops. Surrogate modeling offers a practical mechanism to alleviate this bottleneck by learning computationally efficient approximations of complex simulation models while preserving essential system behavior.

Within a CI-enabled Digital Twin context, surrogate models can support rapid evaluation of candidate solutions, facilitate uncertainty-aware optimization, and enable scalable multi-objective search across large design and operational spaces. Importantly, surrogate modeling does not replace physics-based simulation, but rather complements it by enabling efficient interaction between simulation and optimization layers. These components are brought together in Chap. 7, where surrogate-assisted optimization is examined as a central mechanism for achieving scalable and adaptive decision-support frameworks centered on Building Digital Twins.

Computational Intelligence for Adaptive Digital-Twin Decision Support

Computational Intelligence (CI) transforms Building Digital Twins (BDTs) from passive simulation artifacts into adaptive decision-support frameworks. By enabling surrogate-assisted evaluation, multi-objective reasoning, and learning under uncertainty, CI provides the mechanisms through which Digital Twins can evolve from static models toward responsive, data-informed operational systems.

4.8 Summary

This chapter has positioned Computational Intelligence (CI) as the conceptual bridge between physics-based building simulation and the emerging paradigm of Building Digital Twins. Rather than treating optimization as a purely numerical exercise, the chapter has reframed it as a process of structured exploration within complex, uncertain, and multi-objective search spaces—a perspective which aligns naturally with the decision-support role envisioned for Digital Twins.

Through the discussion of search landscapes, global versus local optimization behavior, and Pareto-based trade-offs, the chapter has clarified why classical optimization and control techniques are insufficient in isolation for building-scale problems. Computational Intelligence (CI) has been shown to provide the methodological tools required to reason under nonlinearity, uncertainty, partial observability, and competing objectives, while remaining complementary to physics-based modeling approaches.

The chapters which follow develop this foundation in a deliberately staged manner. Chapter 5 consolidates core CI methodologies as they have been applied to building energy systems, including Building Energy Management Systems (BEMSs), Parameter Estimation (PE), and the optimization of HVAC and related subsystems. These topics are treated not as isolated case studies, but as transferable analytical frameworks which establish how building behavior can be modeled, calibrated, and optimized using CI principles. Chapter 6 then focuses on the simulation backbone required for such analyses, introducing EnergyPlus™ as a high-fidelity building performance modeling environment and detailing the modeling, parameterization, and simulation workflows upon which CI-based methods can later be integrated.

Taken together, this chapter prepares the reader for the transition from conceptual foundations to methodological and computational building blocks. By clearly delineating the roles of CI theory, simulation infrastructure, and optimization workflows, it sets the stage for subsequent chapters, specifically in this part of the book, where these elements are progressively combined toward simulation-driven, CI-enabled frameworks—key enablers for advancing the Building Digital Twin paradigm.

References

1. Wani MH (2022) Intelligent control strategies for efficient building energy management systems (BEMS). Department of Electrical and Electronic Engineering, University of Auckland, PhD thesis
2. Bezdek JC (1998) Computational intelligence: defined-by everyone! In: Computational intelligence: soft computing and fuzzy-neuro integration with applications. Springer, pp 10–37
3. Pedrycz W (1997) Computational intelligence: an introduction. CRC Press.
4. Duch W (2007) What is computational intelligence and where is it going? In: Challenges for computational intelligence. Springer, pp 1–13
5. Hanne T, Dornberger R (2016) Computational intelligence. In: Computational intelligence in logistics and supply chain management. Springer, pp 13–41
6. Zhang L, Chen Z, Zhang X, Pertzborn A, Jin X (2023) Challenges and opportunities of machine learning control in building operations. Build Simul Springer 16:831–852
7. Garces-Jimenez A, Gomez-Pulido JM, Gallego-Salvador N, Garcia-Tejedor AJ (2021) Genetic and Swarm algorithms for optimizing the control of building HVAC systems using real data: a comparative study. Mathematics 9(18):2181
8. Domingos LMM, Sousa MJ, et al (2025) Swarm intelligence applications for the cities of the future. CRC Press, Taylor & Francis Group
9. Chojecki A, Ambroziak A, Borkowski P (2023) Fuzzy controllers instead of classical PIDs in HVAC equipment: dusting off a well-known technology and today's implementation for better energy efficiency and user comfort. Energies 16(7):2967
10. Arsecularatne B, Rodrigo N, Chang R (2024) Digital Twins for reducing energy consumption in buildings: a review. Sustainability 16(21):9275
11. Agostinelli S, Cumo F, Guidi G, Tomazzoli C (2021) Cyber-physical systems improving building energy management: Digital Twin and artificial intelligence. Energies 14(8):2338
12. Li Y, Bae Y, Im P (2021) Surrogate model of flexible research platform energyplus models to enable sensitivity analysis. ORNL/LTR-2021/1923 technical report, Oak Ridge National

Laboratory, Oak Ridge, TN, USA. https://doi.org/10.2172/1817464, https://doi.org/10.2172/1817464

13. Alghamdi S, Tang W, Sittimont K, Alterman D (2024) Optimising building energy and comfort predictions with intelligent computational model. Sustainability 16(8):343. https://doi.org/10.3390/su16083432
14. Yang X, Lian X, Zhang Z, Wang P, Ma S (2025) Hybrid physics-neural-network MPC with stochastic disturbance forecasting for high-inertia thermally activated building systems. Build Simul 18:2677–2695. https://doi.org/10.1007/s12273-025-1338-3
15. Jagait RK, Fekri MN, Grolinger K, Mir S (2021) Load forecasting under concept drift: online ensemble learning with recurrent neural network and ARIMA. IEEE Access 9:98992–9900. https://doi.org/10.1109/ACCESS.2021.3095420, https://ieeexplore.ieee.org/document/9476011
16. LeCun Y, Bottou L, Bengio Y, Haffner P (1998) Gradient-based learning applied to document recognition. Proc IEEE 86(11):2278–2324. https://doi.org/10.1109/5.726791
17. Han JM, Choi ES, Malkawi A (2022) CoolVox: advanced 3D convolutional neural network models for predicting solar radiation on building facades. Build Simul 15(5):755–776. https://doi.org/10.1007/s12273-021-0837-0, https://link.springer.com/article/10.1007/s12273-021-0837-0
18. Almadhor A, Ghazouani N, Bouallegue B et al (2025) Digital Twin based deep learning framework for personalized thermal comfort prediction and energy efficient operation in smart buildings. Sci Rep 15:24654. https://doi.org/10.1038/s41598-025-10086-y
19. Tekler ZD, Chong A (2022) Occupancy prediction using deep learning approaches across multiple space types: a minimum sensing strategy. Build Environ 226:10968. https://doi.org/10.1016/j.buildenv.2022.109689, https://www.sciencedirect.com/science/article/pii/S0360132322009192
20. Mao R, Lan Y, Liang L, Yu T, Mu M, Leng W, Long Z (2025) Rapid CFD prediction based on machine learning surrogate model in built environment: a review. Fluids 10(8):19. https://doi.org/10.3390/fluids10080193
21. Zhang L, Kaufman Z, Leach M (2024) Physics-informed hybrid modeling methodology for building infiltration. Energy Build. https://doi.org/10.1016/j.enbuild.2024.114580, https://doi.org/10.1016/j.enbuild.2024.114580. (United States: N. P., Web)
22. Bhavsar S, Pitchumani R, Reynolds M, Merket N, Reyna J (2023) Machine learning surrogate of physics-based building-stock simulator for end-use load forecasting. Energy Build. https://doi.org/10.1016/j.enbuild.2023.113395. (United States: N. P., Web)
23. Kennedy J, Eberhart R (1995) Particle Swarm optimization. In: Proceedings of ICNN'95-international conference on neural networks, vol 4. IEEE, pp 1942–1948
24. Blum C (2005) Ant colony optimization: introduction and recent trends. Phys Life Rev 2(4):353–373
25. Mirjalili S, Mirjalili SM, Lewis A (2014) Grey wolf optimizer. Adv Eng Softw 69:46–61
26. Mirjalili S, Lewis A (2016) The whale optimization algorithm. Adv Eng Softw 95:51–67
27. Ambroziak A, Chojecki A (2023) The PID controller optimisation module using fuzzy self-tuning PSO for air handling unit in continuous operation. Eng Appl Artif Intell 117:105485
28. Wani M, Hafiz F, Swain A et al (2025) Balancing energy consumption and thermal comfort in buildings: a multi-criteria framework. Ann Oper Res 346:1841–1867. https://doi.org/10.1007/s10479-023-05747-y
29. Dorigo M, Birattari M, Stutzle T (2006) Ant colony optimization: artificial ants as a computational intelligence technique. IEEE Comput Intell Mag 1(4):28
30. Dorigo M, Stützle T (2018) Ant colony optimization: overview and recent advances. In: Handbook of metaheuristics, pp 311–351
31. Ghalambaz M, Yengejeh RJ, Davami AH (2021) Building energy optimization using grey wolf optimizer (GWO). Case Stud Thermal Eng 27:101250
32. Sakib M, Ahmad S, Anwar K, Saqib M (2025) Optimizing support vector regression using grey wolf optimizer for enhancing energy efficiency and building prototype architecture. Clust Comput 28(1):60

33. Tuğal I (2025) Energy efficiency in building: entropy-based grey wolf optimization for improved MLP performance. Energy Rep 13:4247–4260
34. Feng Z, An J, Han M, Ji X, Zhang X, Wang C, Liu X, Kang L (2024) Office building energy consumption forecast: adaptive long short term memory networks driven by improved beluga whale optimization algorithm. J Build Eng 91:109612
35. Liu Z, Ge H, Song T, Ma S (2025) Research on building energy-saving based on GA-BP coupled improved multi-objective whale optimization algorithm. Energy Build 328:115141
36. Kolokotsa D, Tsiavos D, Stavrakakis G, Kalaitzakis K, Antonidakis E (2001) Advanced fuzzy logic controllers design and evaluation for buildings' occupants thermal-visual comfort and indoor air quality satisfaction. Energy Build 33(6):531–543
37. Kolokotsa D (2007) Artificial intelligence in buildings: a review of the application of fuzzy logic. Adv Build Energy Res 1(1):29–54
38. Ghadi YY, Rasul M, Khan M (2014) Recent developments of advanced fuzzy logic controllers used in smart buildings in subtropical climate. Energy Proc 61:1021–1024
39. Ghadi YY, Rasul M, Khan MMK (2016) Design and development of advanced fuzzy logic controllers in smart buildings for institutional buildings in subtropical Queensland. Renew Sustain Energy Rev 54:738–744
40. Liu W, Han J, Lu X (2013) A high speed railway control system based on the fuzzy control method. Expert Syst Appl 40(15):6115–6124
41. Abadi DNM, Khooban MH (2015) Design of optimal Mamdani-type fuzzy controller for nonholonomic wheeled mobile robots. J King Saud Univ Eng Sci 27(1):92–100
42. Li H, Hong T (2025) A Digital Twin platform for building performance monitoring and optimization: performance simulation and case studies. Build Simul 18:1561–1579. https://doi.org/10.1007/s12273-025-1290-2
43. Zhang H, Thilker CA, Xiao F et al (2025) Stochastic occupancy-integrated MPC for multi-objective optimal built environment control. Build Simul 18:1963–1999. https://doi.org/10.1007/s12273-025-1300-4
44. Das D, Sadiq AS, Mirjalili S (2025) Optimization methods: deterministic versus stochastic. Springer Nature Singapore, Singapore, pp 13–19. https://doi.org/10.1007/978-981-96-3849-9_2
45. Pellegrini R, Serani A, Liuzzi G, Rinaldi F, Lucidi S, Campana EF, Iemma U, Diez M (2018) Hybrid global/local derivative-free multi-objective optimization via deterministic particle swarm with local linesearch. In: Nicosia G, Pardalos P, Giuffrida G, Umeton R (eds) Machine learning, optimization, and big data. Springer International Publishing, Cham, pp 198–209
46. Duan Z, Li B, Zi Y et al (2025) Building retrofit multi-objective optimization using neural networks and genetic algorithm three for energy carbon and comfort. Sci Rep 15:38076. https://doi.org/10.1038/s41598-025-21871-0
47. Nasouri M, Delgarm N (2024) Efficiency-based pareto optimization of building energy consumption and thermal comfort: a case study of a residential building in Bushehr. Iran J Thermal Sci 33:1037–1054. https://doi.org/10.1007/s11630-023-1933-5
48. Dehghan F, Porras Amores C (2025) Simulation-based multi-objective optimization for building retrofits in Iran: addressing energy consumption, emissions, comfort, and indoor air quality considering climate change. Sustainability 17(5):205. https://doi.org/10.3390/su17052056
49. Nikula RP, Paavola M, Ruusunen M, Keski-Rahkonen J (2020) Towards online adaptation of Digital Twins. Open Eng 10(1):776–783. https://doi.org/10.1515/eng-2020-0088
50. Lee H, Yang H (2025) Cognitive Digital Twin for multiobjective production scheduling. J Manuf Sci Eng 147(9):n. pag. https://doi.org/10.1115/1.4068980, https://asmedigitalcollection.asme.org/manufacturingscience/article-abstract/147/9/091004/1219140/Cognitive-Digital-Twin-for-Multiobjective?redirectedFrom=fulltext

Chapter 5
Intelligent Control Strategies for Efficient Building Energy Management Systems (BEMS)

Abstract This chapter presents a comprehensive, multi-phase exploration into intelligent optimization strategies for modern Building Energy Management Systems (BEMS). Structured around three progressive phases, the chapter advances from foundational thermal modeling to real-time, preference-sensitive control mechanisms tailored for HVAC systems. In Phase-I, a physics-informed methodology is employed to estimate thermal parameters of a real commercial building, forming a reliable basis for subsequent modeling and control. Phase-II extends this work through the application of multi-objective optimization via NSGA-II to a forced air extraction system, capturing trade-offs between energy performance and thermal comfort. Finally, Phase-III introduces a preference-aware optimization framework where decision-maker priorities are directly embedded into the control architecture using scalarization and ϵ-constraint formulations. The proposed framework integrates a computationally efficient thermal approximation scheme with a Particle Swarm Optimization (PSO) engine, enabling near real-time set-point optimization. Simulation results using real-world weather data confirm the framework's ability to deliver adaptive, robust, and efficient control outcomes across varying seasonal and operational contexts. This chapter not only demonstrates the feasibility of integrating Artificial Intelligence (AI) and control theory into practical BEMS design, but also highlights the transformative role of preference embedding in achieving context-aware, human-centric energy management. It lays a conceptual and methodological foundation for scalable, intelligent building systems.

Keywords Co-simulation · CLPSO · Computational fluid dynamics · EnergyPlus™ · MATLAB™ · MOO · MCDM · NSGA-II · Parameter estimation · Preference integration · System identification · Thermal parameters

5.1 Overall Structure of the Author's Research Project

The culmination of author's Ph.D. research project is encapsulated in the thesis titled: "*Intelligent Control Strategies for Efficient Building Energy Management Systems (BEMSs)*" [1]. This research explored novel control methodologies that

M. H. Wani et al., *Digital Twins for Building Optimization: Enhancing Energy and Comfort*, Studies in Infrastructure and Control,
https://doi.org/10.1007/978-3-032-23274-8_5

enhance building energy efficiency while ensuring occupant comfort, leveraging advanced meta-heuristics, AI-driven optimization, and multi-objective frameworks. The insights gained from this work directly contribute to the emerging paradigm of smart, adaptive, and self-learning building systems.

The digital transformation of the built environment has introduced AI-driven control strategies that optimize both energy efficiency and occupant comfort in smart buildings. However, traditional BEMS solutions struggle with real-time adaptability. This chapter explores how intelligent control methods, powered by meta-heuristics and multi-objective optimization, can bridge this gap by accurately estimating thermal parameters, optimizing HVAC operations, and integrating Decision-Maker preferences into energy management frameworks.

The chapter is structured in three phases:

- Phase-I presents a meta-heuristic approach for estimating key thermal parameters of buildings, which are crucial for accurate energy modeling.
- Phase-II introduces a Multi-Objective Optimization (MOO) framework, applied to a dynamic kitchen environment, to balance energy efficiency and thermal comfort.
- Phase-III explores how Decision-Maker preferences can be incorporated into BEMS, leveraging AI-driven control methods for adaptive, real-time optimization.

5.2 Phase-I: Estimation of Thermal Parameters of a Building

Building energy consumption is highly dependent on thermal parameters such as heat transfer coefficients, solar absorptance, and emissivity. However, these parameters are often unknown or difficult to measure directly. To overcome this, a meta-heuristic optimization framework, leveraging Comprehensive Learning Particle Swarm Optimization (CLPSO), is employed for accurate thermal parameter estimation. This section introduces the methodology, experimental setup, and validation using real-world data.

5.2.1 Introduction

Buildings are responsible for a considerable portion of global energy use. According to recent data from the U.S. Energy Information Administration (EIA), over 40% of the energy generated worldwide is used by buildings, encompassing both residential and commercial sectors. Notably, more than one-third of this energy consumption is dedicated to space heating and cooling [2]. These figures clearly suggest that reducing energy consumption in Heating, Ventilation, and Air Conditioning (HVAC) systems in buildings is crucial for a sustainable future. Such reductions would directly affect the thermal comfort of occupants.

One of the important concerns in the area of Building Energy Management Systems (BEMS) is, therefore, to create a balance between two *objectives*: *occupants' comfort* and *energy consumption* [3]. To this end, devising intelligent control methods for building HVAC systems has been recognized as an important objective in the research area of Artificial Intelligence (AI) and Smart Grids. One of such control methods, which exploits the full potential of the HVAC system, is the Model Predictive Control (MPC) [4–13].

However, the success of the MPC for building HVAC system is critically dependent on the availability of an accurate *building thermal model*, which can successfully imitate the *thermal dynamics* of the actual building. The accuracy of the building thermal model, is in-turn, dependent upon the accuracy of the *thermal parameters* of the model; particularly the *heat transfer coefficient*, *solar absorptance*, and *emissivity*; as these are known to have significant effect on the thermal dynamics and, thereby, the energy consumption of a building [14–18]. Since, these parameters are strictly determined by the characteristics of the building materials, their values vary from one building structure/type to another. Furthermore, the values are not readily available [19]. This makes the development of a reliable *building thermal model* a challenging task. An easy but dependable approach for estimating proper values of such parameters is thereby needed; which is the focus of this study.

Devising reliable methodologies for the estimation of thermal parameters has been a major research issue since late 80's [20]; as these facilitate the development of optimal strategies for achieving *building energy efficiency*. In the past, many researchers have proposed various methods for building *thermal parameter* estimation [21–25]. Most of such researches focus on estimation of building thermal parameters using lumped-element Resistive-Capacitive (RC) models and have been applied to mono-zoned buildings. Additionally, these methods demand measurements, either on a full-scale building (hardly controllable, as buildings are always in non-steady state) or on a reduced model of a building (easier to control however less realistic). In either case, particularly the former, the tests are costly and burdensome, requiring a very significant time investment and the information collected on-site often lacks 100% correctness. For instance, in [21], a *least squares parameter estimation* method is used to estimate the *thermal parameters* of a simple unheated single-zone test cell; where it has been shown that this method often fails to give consistent estimate and may be difficult to be applied for multi-zone conditioned buildings. Bouache et al. [22] use a reduced laboratory model to validate the estimated thermal parameter values of a simplified lumped RC model for an actual building. Their approach is based on *least squares* technique and is capable of performing rapid estimation. However, setting up of laboratory model and taking measurements is expensive and time consuming. Moreover, the difference between the estimated thermal parameter values and those measured is around 21%. Some further practical issues in implementing *least squares estimation* techniques for building *thermal parameter* estimation are investigated in [26].

In [25] a *regression analysis* method based on averages is applied to estimate the *thermal parameters* of a room in a fully monitored office building. Although their estimation results are quite accurate, the methodology demands expensive and comprehensive experimental set-up (measurement devices) to be used for detailed monitoring of each room in the building. Further, the method requires excessively long test period to obtain reliable parameter estimates, which limits its practicality for commercial applications. An interesting approach is presented in [24] for *thermal parameter* estimation of a multi-zone thermal network, using an Unscented Kalman Filter (UKF). However, the study only demonstrates the efficacy of the technique via simulated data; the validation of the technique with *actual data* from *real buildings* is not considered.

In recent years, *meta-heuristic* based algorithms as well as novel *multi-objective frame work*, have been proposed to identify certain class of *non-linear* systems. Such algorithms have also found a well demonstrated application in identifying the parameters of highly *non-linear* systems; such as in estimating *thermal parameters* of single-zone or multi-zone buildings [27]. Yang et al. [19], for example, propose an *evolutionary algorithm (EA)*, SaNSDE+ (an improved variant of SaNSDE), to estimate *thermal parameters* of a building model. Even though, the results do suggest that their proposed *evolutionary algorithm*, is very effective in estimating *thermal parameters* of the building model; the *generalization* capability of the model has not been shown.

The main objective of the present study is to investigate if *evolutionary algorithms* can be utilized to successfully estimate the *thermal parameters* of a building. To this end, we present a methodology for estimating the *thermal parameters* of a building; using a *meta-heuristic* algorithm, based on a well-known variant of Particle Swarm Optimization (PSO), referred to as Comprehensive Learning Particle Swarm Optimization (CLPSO) [28]. Practicality of the method is demonstrated by considering *actual energy consumption data* from a *real building*, located in Auckland, New Zealand. The estimation of the building *thermal parameters* is formulated as an *optimization task*; in which the *building thermal model*, developed in EnergyPlus™ [29], is regarded as a *black-box*. Performance of the algorithm is further demonstrated by considering *simulated energy consumption data*; generated after subjecting the building thermal model to a simulated geographical environment of a different city (Christchurch) in New Zealand; having significantly different weather profile compared to that of Auckland.

The remainder of this phase of the chapter is organized as follows: Sect. 5.2.2 summarizes the proposed problem formulation and model description. Section 5.2.3 outlines the optimization strategy used in our work for thermal parameter estimation. Section 5.2.4 provides a concise overview of the simulation configuration and the accompanying case study. Results are described in Sect. 5.2.5, finally followed by key insights in Sect. 5.2.6, respectively.

5.2.2 Problem Formulation and Model Description

The limitations of traditional parameter estimation techniques—particularly when applied to complex, black-box building energy models—underscore the need for an alternative approach that does not rely on gradient information or direct measurement of internal properties. To address these challenges, this work formulates the building thermal parameter estimation task as a numerical optimization problem. The subsequent sections outline the proposed framework, detailing the problem formulation, optimization strategy, and modeling environment using EnergyPlus™.

The goal of this study is to identify key thermal parameters of a building by minimizing the discrepancy between measured and simulated energy performance. This inverse modeling task is framed as a single-objective numerical optimization problem, in which the parameter values to be estimated form the decision vector. The search aims to find a vector that yields simulated HVAC energy consumption closely matching the measured (target) values.

Formulation of the Objective Function

Let $\Theta_j = \{\theta_1^j, \theta_2^j, \ldots, \theta_{10}^j\}$ denote the jth candidate solution, where each θ_i^j represents a thermal parameter to be estimated. The simulation output for each time step is a function of the parameter vector and the building model, expressed as

$$\hat{y}_k = f(X, \Theta_j) \tag{5.1}$$

Here, X denotes the known structure of the EnergyPlus™ simulation model, and $f(\cdot)$ encapsulates the model response at each simulation step.

The cost function $J(\Theta_j)$, used to evaluate the fitness of each candidate, is defined as the mean squared error (MSE) between measured (y_k) and simulated ($\hat{y}_k$) HVAC energy consumption, normalized to mitigate scaling effects:

$$J(\Theta_j) = \frac{1}{N} \sum_{k=1}^{N} \left(\frac{y_k - \hat{y}_k}{y_{\max} - y_{\min}} \right)^2 \tag{5.2}$$

where

- N is the number of simulation steps,
- y_{max} and y_{min} are the maximum and minimum observed values of measured energy consumption.

This objective penalizes deviations while ensuring numerical stability.

Table 5.1 Building thermal parameters considered

Parameter	Description [Unit]
θ_1	Interior surface convection coefficient (external wall) $[WK^{-1}m^{-2}]$
θ_2	Exterior surface convection coefficient (external wall), $[WK^{-1}m^{-2}]$
θ_3	Solar absorptance (external wall), $[-]$
θ_4	Thermal absorptance/emissivity (external wall) $[-]$
θ_5	Interior surface convection coefficient (internal wall), $[WK^{-1}m^{-2}]$
θ_6	Exterior surface convection coefficient (internal wall), $[WK^{-1}m^{-2}]$
θ_7	Interior surface convection coefficient (roof), $[WK^{-1}m^{-2}]$
θ_8	Exterior surface convection coefficient (roof), $[WK^{-1}m^{-2}]$
θ_9	Solar absorptance (roof membrane) $[-]$
θ_{10}	Thermal absorptance/emissivity (roof membrane) $[-]$

Decision Variables: Target Parameters

The parameters selected for estimation are thermophysical quantities that typically exhibit uncertainty or are hard to measure directly in real buildings. These include convection coefficients, absorptance properties, and emissivity values for external walls, internal partitions, and roofing layers. The list of parameters is summarized in Table 5.1.

Each parameter is bounded within a physically reasonable search interval, based on ASHRAE guidelines and material datasheets. These bounds are enforced during optimization.

Building Simulation Environment

The building model used in this study is developed using EnergyPlus™, a whole-building energy simulation engine. A detailed $3D$ envelope was created using SketchUp™ and imported via OpenStudio®. The model represents a multi-zone commercial hotel building located in Auckland, New Zealand. Key elements of the model include

- two primary thermal zones,
- HVAC system based on Constant Air Volume (CAV) with packaged rooftop units,
- building envelope details including windows, roof membrane, and internal partitions.

To replicate real-world operational conditions, simulations were performed using measured weather data for Auckland and Christchurch. These were chosen to evaluate the robustness of the model in two different climatic profiles. The measured HVAC energy consumption data for 30 days (sampled hourly) serves as a reference for model calibration. This data is assumed to represent the "ground truth" against which each candidate solution is compared.

A detailed description of the building layout, zoning, envelope construction, and HVAC configuration can be found in our earlier work [30].

5.2.3 Optimization Strategy Using CLPSO

Once the thermal parameter estimation problem is formulated as a single-objective optimization task, an appropriate optimization algorithm must be selected to efficiently explore the solution space. Given the high dimensionality and black-box nature of the EnergyPlus™ model, traditional gradient-based optimization methods are unsuitable. Instead, this study adopts a population-based meta-heuristic approach—Comprehensive Learning Particle Swarm Optimization (CLPSO)—to solve the estimation problem effectively.

CLPSO is an advanced variant of the canonical Particle Swarm Optimization (PSO) algorithm. It is particularly well-suited for problems involving complex, non-differentiable objective functions and expensive simulation calls, such as those found in energy modeling applications. The algorithm maintains diversity in the population by allowing each decision variable (i.e., each thermal parameter) to learn from different exemplars in the swarm, thereby avoiding premature convergence and improving exploration of the solution space.

CLPSO-Based Parameter Search Mechanism

Each particle in the swarm represents a candidate solution vector:

$$\Theta_j = \{\theta_1^j, \theta_2^j, \ldots, \theta_{10}^j\} \tag{5.3}$$

During each iteration, the particle updates its velocity and position based on the best experience of other particles in the swarm, according to the following update rule:

$$\begin{aligned} V_j^d &\leftarrow \omega V_j^d + c \cdot r_j^d \cdot (pbest_{f_j(d)}^d - \theta_j^d) \\ \theta_j^d &\leftarrow \theta_j^d + V_j^d \end{aligned} \tag{5.4}$$

Here:

- θ_j^d and V_j^d denote the position and velocity of the dth dimension of the jth particle, respectively.
- $pbest_{f_j(d)}^d$ is the best value of the dth dimension among peers of the particle j.
- ω is the inertia weight.
- c is the learning coefficient.

- r_j^d is random number in [0, 1].
- $f_j(d)$ determines from which peer the particle learns in the dth dimension.

The use of distinct exemplars for each dimension allows CLPSO to maintain diversity and robustness across complex search landscapes.

Integration with EnergyPlus™

Since the objective function requires evaluation through a simulation-based black-box model, the CLPSO algorithm is integrated with EnergyPlus™ via a co-simulation interface. The building model resides in EnergyPlus™, while the optimizer runs in MATLAB®. A middleware (MLE+) facilitates this interaction, see Fig. 5.1, allowing for iterative exchange of parameter vectors and simulation outputs at runtime.

Each time a particle's fitness is to be evaluated:

i. Its parameter vector Θ_j is passed to EnergyPlus™.
ii. EnergyPlus™ simulates the building's energy performance for the specified thermal parameter set.
iii. The resulting HVAC energy profile $\hat{y}_k$ is returned to MATLAB®.
iv. The fitness value, $J(\Theta_j)$, is computed using Eq. 5.2.

This process is repeated for all particles across multiple generations until convergence or until a stopping criterion (e.g., maximum Function Evaluations) is met.

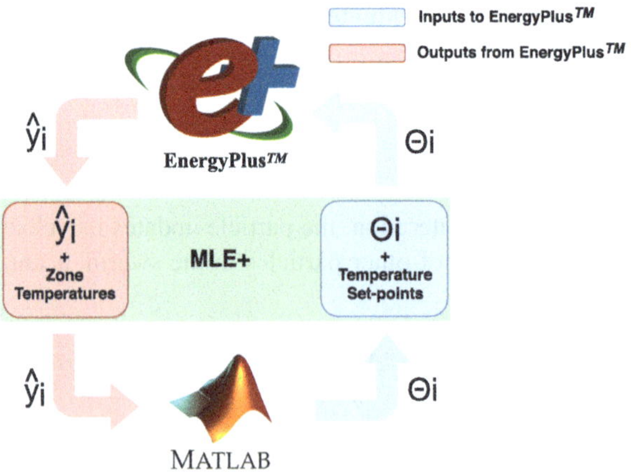

Fig. 5.1 Co-simulation framework used in our study

Table 5.2 Algorithm parameters and values

Parameter	Value
Swarm size	30
Max. Function Evaluations (FEs)	65,000
Inertia weight (ω)	0.9 $\rightarrow$ 0.7 (linearly decreasing)
Learning coefficient (c)	1.49445
Refreshing gap	5

CLPSO Parameter Configuration

To balance exploration and exploitation, the configuration summarized in Table 5.2 is adopted based on empirical studies. The algorithm is executed over 10 independent runs to account for stochasticity, and the best solution from each run is retained for analysis.

The use of CLPSO provides a robust search mechanism for estimating thermal parameters in a data-driven, simulation-based setting. Its ability to navigate complex search spaces without requiring gradient information makes it ideal for energy modeling problems where traditional identification techniques struggle. The next section presents the simulation setup and case study used to validate the approach.

5.2.4 *Simulation Setup and Case Study*

The simulation framework outlined earlier is now applied to validate the proposed CLPSO-based thermal parameter estimation method. Simulations are conducted using the EnergyPlus™ building model and reference data introduced in Sect. 5.2.2.

The optimizer operates by iteratively modifying candidate thermal parameter sets and invoking EnergyPlus™ to simulate the building's HVAC performance over a 30-day period. For each candidate, the normalized mean squared error between simulated and measured HVAC energy consumption is computed as the fitness score.

To assess the generalizability of the estimated parameters, the process is repeated for two distinct climate profiles—*Scenario-I*: Auckland (original location of the building) and *Scenario-II*: Christchurch (with notably different weather conditions, see Fig. 5.2). This cross-location validation serves to confirm the robustness of the approach. The CLPSO search is executed over 10 independent runs, with a limit of 65, 000 Function Evaluations (FEs) per run. The best parameter set from each run is recorded for statistical analysis.

This setup enables a comprehensive evaluation of estimation accuracy, convergence behavior, and model generalizability. The results are discussed in the next section.

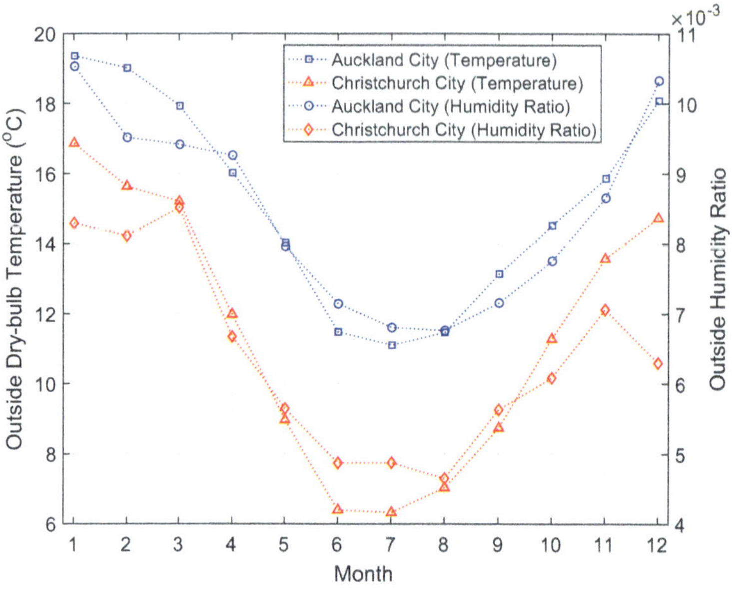

Fig. 5.2 Weather comparison of two cities [30]

5.2.5 *Results and Discussion*

The performance of the proposed CLPSO-based estimation framework is evaluated using the simulation setup described previously (Sect. 5.2.3). Results are presented in terms of estimation accuracy, convergence characteristics, and robustness across climatic profiles.

Parameter Estimation Accuracy

Table 5.3 summarizes the estimated values of the ten thermal parameters over ten independent runs of CLPSO, along with the corresponding minimum, maximum, and mean squared error (MSE) for each run. The best parameter set is selected based on the lowest fitness value.

These results show that the optimizer converges to consistent parameter values across independent trials, confirming the repeatability and reliability of the framework. The low MSE values indicate a high degree of accuracy in matching simulated HVAC energy with measured data.

Table 5.3 Estimated thermal parameters and corresponding error values across 10 CLPSO runs [30]

Parameters	Estimated Values for Scenario-I ("*Known Parameters*")			Estimated Values for Scenario-II		
	Best	Mean	SD	Best	Mean	SD
θ_1	7.22	7.22	6.59*e*-08	7.26	7.26	9.36*e*-16
θ_2	28.51	28.51	2.92*e*-08	28.49	28.49	3.74*e*-15
θ_3	0.55	0.55	8.42*e*-16	0.37	0.37	0
θ_4	0.31	0.31	4.96*e*-10	0.36	0.36	5.85*e*-17
θ_5	5.23	5.23	2.93*e*-10	5.20	5.20	9.36*e*-16
θ_6	8.14	8.14	0.0077	6.94	6.94	9.36*e*-16
θ_7	7.63	7.63	0	7.64	7.64	0
θ_8	20.94	20.94	2.27*e*-11	20.14	20.14	3.74*e*-15
θ_9	0.81	0.81	2.38*e*-04	0.78	0.78	0
θ_{10}	0.12	0.12	5.28*e*-05	0.11	0.11	1.46*e*-17
Fitness	Mean	4.12*e*-06		4.56*e*-06		
	SD	1.47*e*-08		8.92*e*-22		

Convergence Behavior

The convergence trends of the CLPSO algorithm are illustrated in Fig. 5.3, which shows the fitness value across iterations for a typical run.

A rapid decline in the objective function during the initial phase of optimization is observed, followed by a smooth tapering as the algorithm approaches the global minimum. This behavior confirms that the algorithm maintains a good balance between exploration and exploitation.

Prediction Accuracy: Simulated vs. Model Predicted Data

Figure 5.4 depicts the difference between the measured and simulated HVAC energy consumption profiles for *Scenario-I*, using the best *estimated parameter* set.

The close alignment between the simulated and model predicted profiles indicates that the estimated parameters accurately capture the thermal dynamics of the building.

Cross-Validation Under Different Weather Conditions

To evaluate generalization, the same estimated parameters are applied to simulate energy performance under Christchurch weather conditions, i.e., *Scenario-II*, see

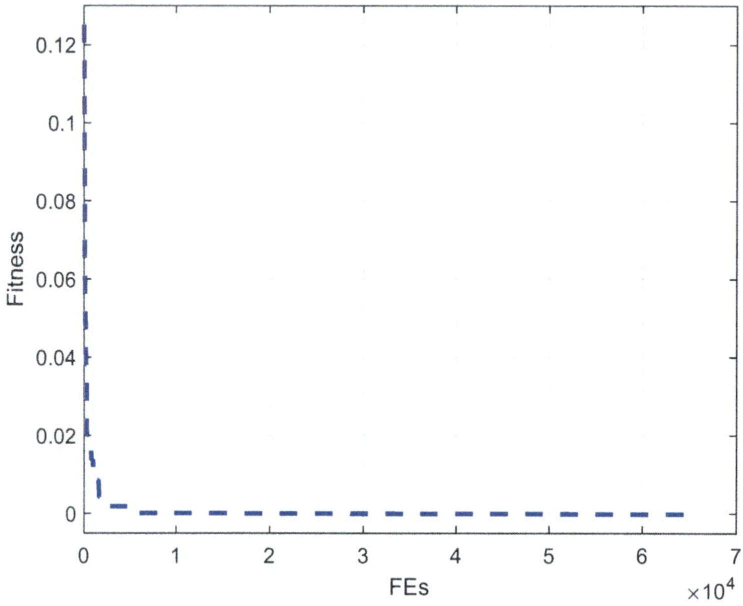

Fig. 5.3 Convergence curve of CLPSO optimizer, for *Scenario-I*, showing reduction in objective function over iterations [30]

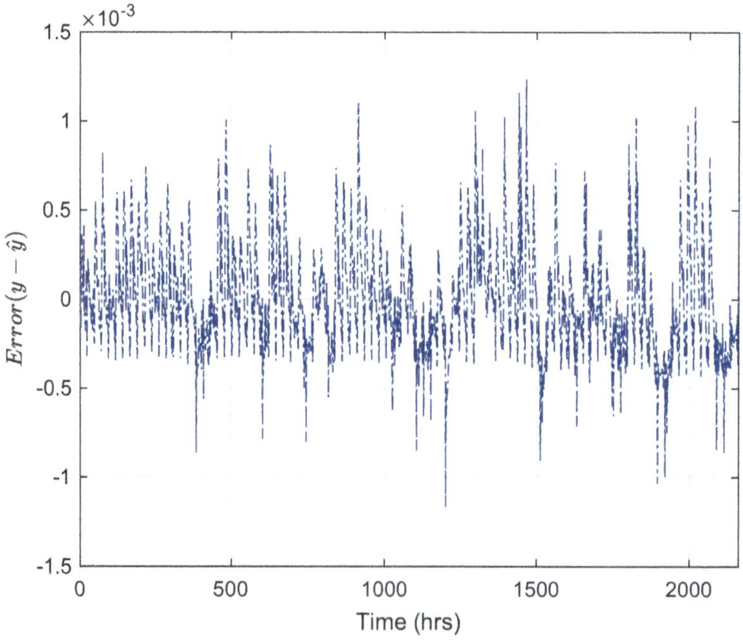

Fig. 5.4 Error between measured ("y") and model predicted ("$\hat{y}$") energy consumption, for *Scenario-I*, using *estimated* parameters [30]

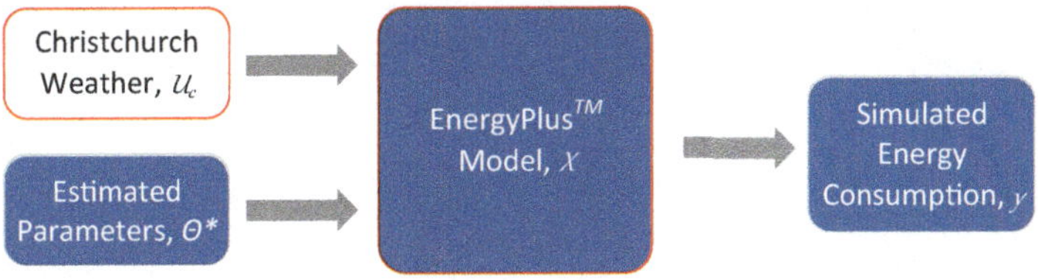

Fig. 5.5 Illustration of generation of *simulated energy consumption*, y, in Scenario-II [30]

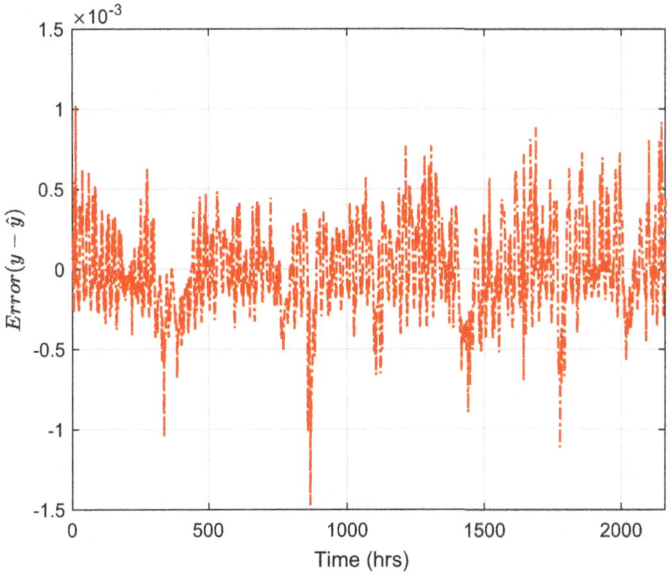

Fig. 5.6 Error between simulated ("y") and model predicted ("$\hat{y}$") energy consumption, for *Scenario-II*, using "Auckland-derived" parameters (from *Scenario-I*) [30]

Fig. 5.5. Figure 5.6 presents the comparison between simulated and model predicted energy in this alternate climate.

While a slight increase in error is observed due to climatic variance, the overall trend remains well-aligned, demonstrating the robustness and transferability of the parameter set.

Statistical Error Trends

Figure 5.7 shows a box plot illustrating the statistical distribution of the MSE values across all 10 independent runs, for *Scenario-I* and *Scenario-II*. The small interquartile range and low median values reinforce the stability of the optimization framework and the low sensitivity to initial conditions.

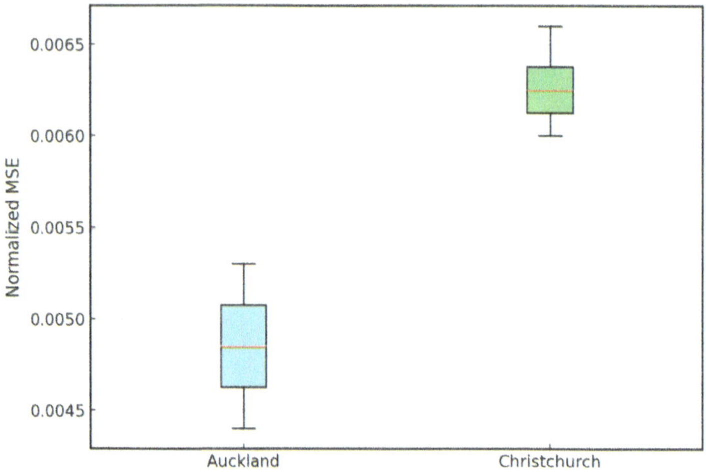

Fig. 5.7 Box plot of normalized MSE values over 10 CLPSO runs

5.2.6 Summary and Key Insights

This section concludes Phase-I of the research by summarizing the core insights drawn from the implementation and evaluation of the proposed CLPSO-based thermal parameter estimation framework. The methodology was designed to enable the reliable estimation of building thermal parameters—such as surface convection coefficients, absorptance, and emissivity—using simulation-driven optimization. It bridges the gap between traditional identification techniques and the practical requirements of high-fidelity building energy models.

The results clearly demonstrated that meta-heuristic optimization, when integrated with a robust simulation engine like EnergyPlus™, can offer accurate and repeatable parameter estimation even in complex multi-zone environments. The optimizer consistently converged across ten independent runs, exhibiting low variability and high convergence fidelity, as evidenced by the narrow interquartile range of MSE values.

Furthermore, the cross-validation experiments, where parameters estimated for Auckland (Scenario-I) were applied to simulations under Christchurch weather conditions (Scenario-II), confirmed the robustness and generalizability of the approach. Despite the climatic differences, the parameter set maintained predictive accuracy, indicating the potential of such estimations to scale across varying operational profiles. This reinforces the concept of transferable parameter sets in simulation-based Digital Twins.

Assumptions and Justifications

A few key assumptions underpinned this framework, each of which was considered reasonable within the scope of the study:

- **Constant Thermal Parameters Over Time:** The simulations span a 30-day window during which material properties are unlikely to change appreciably. Long-term degradation effects are outside the current scope.
- **Accurate Geometric Representation:** The model developed using SketchUp™ and OpenStudio® was assumed to faithfully replicate the building's physical layout, which is justifiable due to the modeling workflow rigor discussed in Chap. 6.
- **Simulation Model as a Black-box:** EnergyPlus™ was treated as a validated black-box simulator. This abstraction is widely accepted in the literature and facilitates flexible integration with optimization engines.
- **Randomness in Optimizer Smoothed Statistically:** Multiple optimization runs were used to mitigate the effect of randomness, and convergence trends were analyzed statistically.
- **EPW Files as Representative Weather Data:** The use of standard EPW files for Auckland and Christchurch ensured climatic realism while averaging out short-term stochastic disturbances.

Relevance to the Broader Framework

Within the context of this book's broader goal—developing Digital Twins for building energy optimization—this chapter represents the foundational phase. Accurate thermal parameter identification is essential for constructing simulation-backed digital twins that can later be used in real-time control, predictive analysis, and optimization.

The framework presented here contributes to a scalable and automated solution for building model calibration. Rather than relying solely on sensor-rich environments or intrusive measurements, this approach leverages optimization intelligence to identify model parameters using available consumption data.

Looking Ahead

While thermal parameter estimation establishes the basis of a reliable simulation environment, it is only the first step toward intelligent BEMS. The next research phase, explored in Sect. 5.3, builds upon this foundation by formulating a multi-objective optimization (MOO) problem aimed at balancing energy efficiency with occupant comfort in dynamic indoor environments such as commercial kitchens.

This transition reflects a progression from static estimation to adaptive control, highlighting the growing complexity and intelligence required for next-generation BEMS solutions.

Highlight

While accurately estimating thermal parameters is crucial, it represents only the first step toward intelligent BEMS. The next challenge is to integrate these parameters into a multi-objective optimization (MOO) framework, balancing energy efficiency with occupant comfort. This is particularly relevant in dynamic environments like commercial kitchens, office spaces, or retail stores, where HVAC loads fluctuate throughout the day. The following section explores the application of MOO to real-world HVAC optimization problems.

5.3 Phase-II: A Novel MOO Framework Using a Kitchen Environment as Case Study

Building upon the simulation-based calibration framework developed in Phase-I—serving as a foundational step toward digital twin development—this section transitions from passive parameter estimation to active multi-objective control. A kitchen environment presents unique challenges due to fluctuating occupancy, high heat loads, and demanding ventilation requirements. These conditions necessitate a more responsive and robust building energy management approach.

The proposed Multi-Objective Optimization (MOO) framework integrates thermal comfort constraints, air quality considerations, and energy efficiency targets into a unified, real-time HVAC control strategy. By leveraging a validated Computational Fluid Dynamics (CFD) model of a commercial kitchen and advanced Pareto Front optimization using NSGA-II, we systematically explore trade-offs among competing objectives–namely occupant comfort, spatial uniformity, and extraction energy use. The framework lays the groundwork for intelligent control regimes in energy-intensive indoor environments and continues the trajectory set forth in Phase-I.

5.3.1 Introduction and Literature Review

In daily indoor routines, the kitchen stands out as a highly occupied and health-critical zone. A cooking in-charge typically spends around 3.6 h each day on cooking and associated tasks [31–33], underscoring the importance of a healthy and thermally comfortable kitchen environment. However, cooking fumes have been found to contain more than 300 Volatile Organic Compounds (VOCs), many of which become carcinogenic and mutagenic at elevated temperatures [34]. In modern homes and commercial buildings—such as restaurants and hotels—kitchens are often located in confined areas with limited ventilation, preventing the timely discharge of cooking fumes and rapid dissipation of indoor heat.

Surveys indicate that over 80% of respondents have experienced unfavorable kitchen conditions, primarily due to poor thermal and air quality environments [35]. The first major concern is the thermal environment. For instance, Wei et al. [32] reported vertical temperature differences as high as 6.3 °C during cooking—more than double the acceptable 3 °C limit set by thermal comfort standards [36]. The second issue pertains to indoor air quality (IAQ). During cooking, significant amounts of CO, CO_2, NO_2, and particulate matter are generated, often exceeding national health standards by up to tenfold [37–42].

Given these health risks, improving the performance of forced extraction systems—commonly used in kitchens—is of vital importance. These systems must be designed to maintain both air quality and thermal comfort while minimizing energy consumption [33, 34].

To assess thermal and air quality conditions in kitchens, researchers have employed both experimental and simulation-based approaches. While experimental methods offer reliability and physical fidelity [43–45], they face limitations in terms of flexibility, cost, and repeatability [34]. In contrast, advances in computational modeling—particularly Computational Fluid Dynamics (CFD)—have enabled scalable and detailed analysis of airflow, heat transfer, and pollutant dispersion in kitchen environments.

CFD has a long-standing history in HVAC research, starting with foundational work by Nielsen et al. [46], who validated simulated airflow fields against Laser Doppler Anemometry (LDA) measurements. Subsequent studies expanded the modeling to range hood efficiency [47], comparative ventilation strategies [48], and broader air quality studies in kitchen settings. On the thermal comfort front, research has explored model development [49], human variability [50, 51], and comfort indices [52]. Yet, most of these studies focus on general indoor environments rather than kitchens. For instance, CFD-based analysis of bedrooms [53], classrooms [54], and showers [55] dominate the literature, leaving a gap in kitchen-specific optimization.

Although a few studies have attempted comprehensive evaluations—combining IAQ and thermal comfort in kitchen contexts [33, 34]—they still largely ignore energy consumption aspects. For example, Chen et al. [34] established CFD-based evaluation criteria for comfort and air quality, while Liu et al. [33] proposed an integrated HVAC-air curtain system for Chinese kitchens. Yet, the energy implications of extraction systems under dynamic conditions remain understudied.

Given the strong temporal and spatial interactions between energy consumption, air quality, and thermal comfort [56], there is a critical need for a unified framework that balances all three. This is especially important considering the significant time humans spend in kitchens and the energy-intensive nature of forced extraction systems.

To address this gap, this chapter proposes a novel framework for optimizing the performance of forced extraction systems in kitchens. The framework targets three primary objectives:

(a) minimization of occupant discomfort (thermal and IAQ-related),
(b) minimization of energy consumption by the extraction fan, and
(c) minimization of spatial gradients in temperature and CO_2 concentration.

These objectives are inherently conflicting, and thus the problem is posed as a Multi-Objective Optimization Problem (MOOP). The objective functions are formulated based on controllable variables (e.g., reference set-points) and ambient conditions. A suitable multi-objective optimization technique is then used to compute optimal set-points that strike an effective balance between comfort, energy, and spatial uniformity.

To evaluate this framework, the optimal set-points—specifically the knee-point solution—are fed into a CFD model of a kitchen environment to simulate the resulting performance. The contributions of this phase can be summarized as follows:

I. A CFD model is developed using FloVENT™ [57, 58], and validated against experimental data collected from a standardized test room (kitchen) built in compliance with BS EN IEC 61591 [59].
II. CFD simulation data is used to generate a multivariate polynomial, which serves as one of the objective functions in the optimization framework. A stepwise approach for its formulation is detailed in this chapter.
III. The Non-dominated Sorting Genetic Algorithm II (NSGA-II) is employed to perform the multi-objective optimization [60–62].
IV. An "a posteriori" decision-making is applied using the Minimum Manhattan Distance (MMD) method [60] to extract the knee-point solution from the Pareto front, ensuring a well-balanced trade-off among the objectives.
V. A comparative analysis is conducted between the optimal fan operation derived from the framework and conventional baseline operation, using the validated CFD model as a benchmark.
VI. This comparison is extended across nine different simulated ambient conditions to assess the robustness and generalizability of the proposed framework.

The results presented in this phase fall into two categories: (1) *Experimental*, involving CFD validation; and (2) *Simulated*, evaluating the multi-objective framework. In this study, CO_2 concentration is selected as the proxy for air quality, given its dominance among cooking-emitted gases. Although it is not the only pollutant, maintaining acceptable CO_2 levels also supports broader pollutant reduction [56].

The rest of this phase is organized as follows: Sect. 5.3.2 presents the CFD test room setup and modeling approach. Sections 5.3.3 and 5.3.4 detail the proposed optimization methodology, objective formulation and framework design. Section 5.3.5 discusses the results, and Sect. 5.3.6 concludes the phase.

5.3.2 CFD Test Room Setup and Model Development

To simulate the airflow and thermal dynamics within a typical kitchen environment, a physical test room was constructed and configured in accordance with the *BS EN IEC 61591* standard [59]. The test space includes a standardized cooktop, a forced extraction system (range hood), and environmental sensors positioned at multiple vertical locations to measure temperature and gas concentrations such as CO_2. Table 5.4 summarizes further details about the kitchen test room [63].

A high-fidelity Computational Fluid Dynamics (CFD) model of this test room was developed using $FloVENT^{\text{TM}}$ [57, 58]. The model incorporates realistic boundary conditions, internal heat sources, air inlet and outlet flows, and human-equivalent thermal profiles. The geometry is consistent with the physical layout of the test room, and the mesh is refined in regions near the heat sources and extraction zones to ensure numerical accuracy in resolving velocity and temperature fields. The geometric configuration of the model is illustrated in Fig. 5.8.

The CFD model was validated using experimental data collected from the test room under steady-state cooking conditions. Measured data for temperature and

Table 5.4 Material and structural details of basic elements of the room [63]

Element(s)	Structural details	Material specifications
Walls, Ceiling, Door & Floor	Walls/Ceiling/Door thickness: 16mm Floor thickness: 2mm Door dimensions: 750mm (X-axis), 2000mm (Y-axis)	–
Extraction System (ES)	Canopy dimensions‡: 520mm, 80mm & 895mm Flue dimensions‡: 320mm, 910mm & 375mm Duct radius: 75mm	Main body (stainless steel) Fan (stainless steel)
Cooking Area (CA)	Cook-top dimensions‡: 600mm, 60mm & 900mm Radius of the Hob(s): 100mm	Main body (stainless steel) Hob(s) (iron)
Ventilation Grid	Air-dispersion screen dimensions‡: 16mm, 700mm & 1000mm Air inlet dimensions: 350mm (Y-axis), 700mm (Z-axis)	Air-dispersion screen (plywood)
Cabinets	C1 & C2 dimensions‡: 300mm, 600mm & 600mm C3 & C4 dimensions‡: 600mm, 900mm & 600mm	Main body (chipboard)

‡Note that the dimensions are given along X, Y & Z-axis, respectively

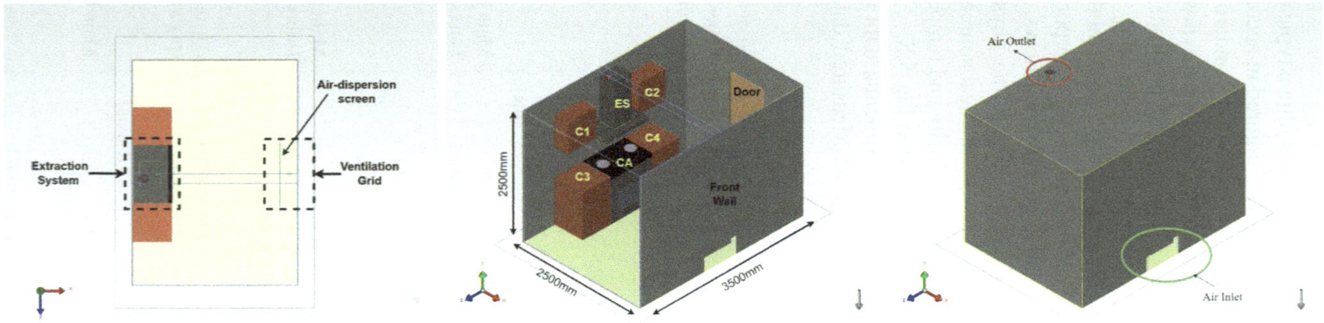

(a) Top view (b) Isometric view (c) Air flow direction

Fig. 5.8 3D CAD model of the kitchen test room used for CFD modeling in FloVENT™ [63]

CO_2 concentration at multiple sensor heights were compared with simulation results. Figure 5.9 shows the comparison of measured and simulated temperature profiles, while Fig. 5.10 presents the corresponding validation results for CO_2 concentration. The validation results demonstrate strong agreement, confirming the model's fidelity in replicating real-world indoor environmental behavior.

Upon validation, the CFD model was used to simulate various operational scenarios representing different ambient environmental conditions. These simulations produced performance data across multiple comfort and air quality dimensions. The resulting dataset was used to construct multivariate polynomial models that capture the system's response to varying control inputs. These surrogate models thus serve as objective functions in the multi-objective optimization framework presented in the next section.

This approach eliminates the need for repeated physical trials and allows extensive exploration of design space under constrained energy, comfort, and air quality objectives.

For a more detailed account of the experimental setup, CFD modeling workflow, and validation methodology, the reader is referred to our original work in [63].

5.3.3 Multi-objective Problem Formulation

Following the CFD-based simulation and validation described in the previous section, this part of the study focuses on formulating the optimization problem as a multi-objective framework. The goal is to simultaneously minimize occupant discomfort, energy consumed by the forced extraction system, and spatial gradients in environmental variables—namely temperature and CO_2 concentration.

Three criterion functions are defined, each representing one of the optimization objectives:

(a) **Discomfort Index** (f_1)—quantifies deviation of the indoor environment from a predefined reference state of thermal comfort and air quality;
(b) **Energy Consumption Index** (f_2)—estimates the fan power required by the extraction system to maintain target ventilation levels;
(c) **Gradient Index** (f_3)—represents non-uniformity within the room by measuring spatial variance in temperature and CO_2 concentration fields.

Each objective is treated as a function of controllable variables called *reference signals*, which serve as set-points for maintaining desired comfort and air quality. These include

- air temperature set-point (T_{ref}),
- CO_2 concentration set-point (C_{ref}),
- velocity set-point (v_{ref}) at the air supply or extraction point.

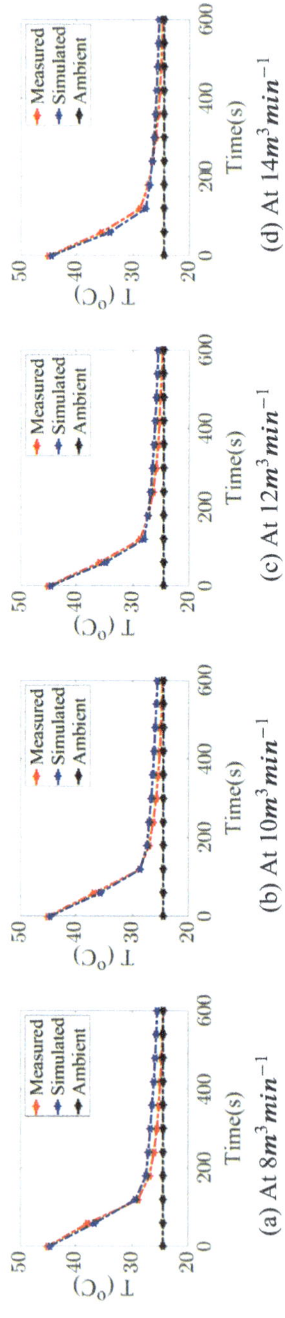

Fig. 5.9 Validation of simulated versus measured temperature profiles across vertical sensor locations [63]

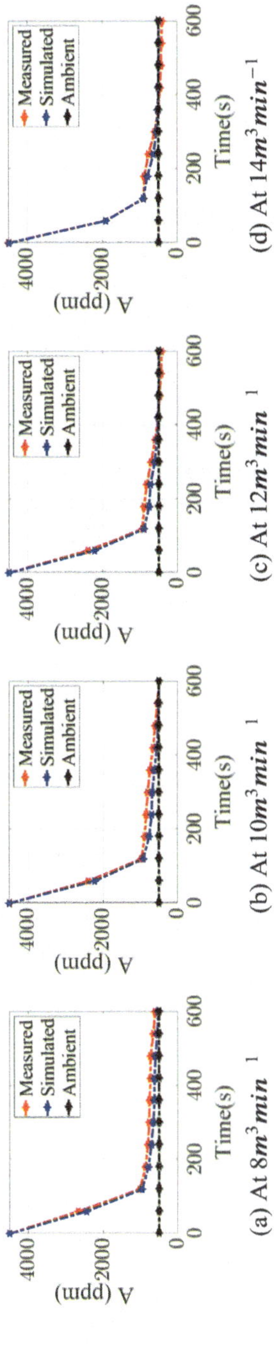

Fig. 5.10 Validation of simulated versus measured CO_2 concentration profiles [63]

The general form of the multi-objective optimization problem is written as

$$\min_{\mathbf{x}} \mathbf{f}(\mathbf{x}) = [f_1(\mathbf{x}),\ f_2(\mathbf{x}),\ f_3(\mathbf{x})]^T \quad (5.5)$$

subject to bounds on the decision variables:

$$\mathbf{x} = [T_{\text{ref}},\ C_{\text{ref}},\ v_{\text{ref}}] \in \mathcal{X} \quad (5.6)$$

The feasible decision space $\mathcal{X}$ is defined by upper and lower bounds derived from practical HVAC control limits and validated CFD scenarios. The reader is advised to refer to [63] for a detailed understanding on objectives and problem formulation.

Each objective function $f_i(\cdot)$ is modeled using surrogate-based regression derived from CFD simulations, ensuring computational efficiency during optimization. These functions were approximated using multivariate second-order polynomials fitted to the CFD response data over nine ambient condition scenarios.

Given the conflicting nature of the objectives—for instance, improving comfort often requires higher fan speeds and energy use—the problem cannot be solved through scalarization. Instead, a Pareto-optimal solution set is desired, from which a balanced (knee-point) solution is chosen using an "a posteriori" decision technique described in the next section.

This formulation allows simultaneous control of indoor air quality, energy consumption, and spatial comfort in a way that is flexible, generalizable, and grounded in physical simulations.

5.3.4 Optimization Framework Design and Implementation

Following the formulation of the multi-objective problem, this section outlines the computational framework developed to solve it effectively. Given the conflicting nature of objectives described earlier, a classical single-objective optimization method is insufficient. Thus, we employ the widely used meta-heuristic known as the Non-dominated Sorting Genetic Algorithm II (NSGA-II), renowned for effectively handling complex multi-objective problems.

NSGA-II Algorithm Overview

NSGA-II operates by generating a population of candidate solutions, progressively evolving this population through selection, crossover, and mutation operators. Each generation is evaluated against multiple criteria, ensuring convergence toward a diverse set of Pareto-optimal solutions.

The algorithm's evolution mechanism is guided by two primary processes:

1. **Non-dominated Sorting:** Candidates are sorted into Pareto-based fronts based on domination criteria.
2. **Crowding Distance Computation:** Ensures diversity along the Pareto front by preserving solutions spread across objective space.

Mathematically, for two solution vectors $\mathbf{x}_a$ and $\mathbf{x}_b$, $\mathbf{x}_a$ dominates $\mathbf{x}_b$ if and only if:

$$f_i(\mathbf{x}_a) \leq f_i(\mathbf{x}_b) \quad \forall i \in \{1, 2, 3\} \quad \text{and} \quad f_j(\mathbf{x}_a) < f_j(\mathbf{x}_b) \quad \text{for at least one } j \tag{5.7}$$

Implementation and Pseudocode

The main steps of NSGA-II tailored for our multi-objective kitchen environment problem are outlined clearly in the Algorithm 5.1. (This pseudocode is directly adapted from [63].)

Algorithm 5.1 NSGA-II Multi-Objective Optimization Framework

Initialize population P_0 randomly within variable bounds.
Evaluate objective functions f_1, f_2, and f_3 for all solutions in P_0.
Set generation counter $g = 0$.
while stopping criteria not met **do**
    Perform non-dominated sorting on population P_g.
    Calculate crowding distances within each Pareto front.
    Select parents via tournament selection based on rank and crowding distance.
    Apply crossover and mutation operators to generate offspring Q_g.
    Evaluate objectives for all offspring solutions.
    Combine parent and offspring populations: $R_g = P_g \cup Q_g$.
    Perform non-dominated sorting and select best N solutions for the next generation P_{g+1} based on rank and crowding distance.
    $g = g + 1$
end while
Return final Pareto-optimal population.

The implementation involves the following detailed operations:

- **Initialization:** Each candidate solution ($\mathbf{x}$) comprises three reference signals—temperature set-point (T_{ref}), CO_2 concentration set-point (C_{ref}), and velocity set-point (v_{ref}).
- **Evaluation:** For each candidate solution, the surrogate polynomial functions derived from CFD simulations (Sect. 5.3.3) are evaluated to compute $f_1(\mathbf{x})$, $f_2(\mathbf{x})$, and $f_3(\mathbf{x})$.
- **Non-Dominated Sorting:** Solutions are ranked based on dominance criteria into successive Pareto fronts (F_1, F_2, ...), where F_1 contains the non-dominated solutions.

- **Crowding Distance:** Calculated to preserve diversity. For a solution $\mathbf{x}_i$ within a front, the crowding distance ($CD(\mathbf{x}_i)$) is computed as:

$$CD(\mathbf{x}_i) = \sum_{k=1}^{3} \frac{f_k(\mathbf{x}_{i+1}) - f_k(\mathbf{x}_{i-1})}{f_k^{\max} - f_k^{\min}} \tag{5.8}$$

- **Selection, Crossover, and Mutation:** Solutions are selected based on their rank and crowding distance. Crossover and mutation generate new candidate solutions to maintain exploration of the decision space.

Selecting a Balanced Optimal Solution: Knee-Point Identification

After running the optimization algorithm, a set of Pareto-optimal solutions is generated. To select the most balanced (or "best compromise") solution, an "a posteriori" decision-making technique known as the *Minimum Manhattan Distance (MMD)* method is employed. Mathematically, MMD is calculated for each solution as

$$\text{MMD}(\mathbf{x}) = \sum_{i=1}^{3} \frac{\left| f_i(\mathbf{x}) - f_i^{\text{ideal}} \right|}{f_i^{\max} - f_i^{\min}} \tag{5.9}$$

Here, f_i^{ideal} represents the ideal value for each objective, and the solution with the smallest MMD value is identified as the knee-point, reflecting the best trade-off among all objectives.

Integration with CFD Simulations

To validate the optimization results, the knee-point reference signals are implemented in the CFD model. The simulated outcomes under optimal conditions are then compared against conventional or baseline operations across diverse ambient scenarios, discussed comprehensively in the next section (Sect. 5.3.5).

5.3.5 Results and Performance Comparison

The evaluation of the proposed multi-objective optimization (MOO) framework is performed using a simulation-based performance assessment, underpinned by a previously validated CFD model (see Sect. 5.3.2).

Table 5.5 Optimal knee-point solutions (set-points) derived using NSGA-II for the nine simulated weather scenarios (ambient temperature, CO concentration, and wind velocities) typical of Auckland conditions [63]

Knee-point solution	Scenario: 1A/1B/1C	Scenario: 2A/2B/2C	Scenario: 3A/3B/3C
x_1^*	20.04/20.24/20.03	20.29/20.52/20.47	23.05/23.21/23.46
x_2^*	457.14/460.79/451.65	456.11/467.88/461.28	450.81/452.24/455.90
Q^*	9.93/15.39/17.61	9.83/15.07/16.18	10.05/16.55/19.84

Optimal Solutions and Pareto Front Analysis

The MOO framework leverages the NSGA-II algorithm to identify optimal solutions (knee-points) across nine distinct weather scenarios representative of typical Auckland conditions (Scenario-1: July; Scenario-2: October; Scenario-3: January), with sub-scenarios reflecting variations in wind velocities (1, 2.5, and 5.8 m/s). Table 5.5 summarizes these optimal solutions.

Figure 5.11 illustrates the three-dimensional Pareto Front for Scenario-2A, highlighting the complexity of balancing comfort, energy consumption, and environmental uniformity. The selected knee-point represents the optimal trade-off among conflicting objectives.

Comparative Performance Analysis

The comparative performance of the optimal fan operation against the conventional baseline is assessed across three primary metrics: occupant comfort, energy consumption, and spatial gradients in temperature and CO_2 concentration. Figures 5.12 and 5.13 show the superiority of the optimal fan operation in maintaining desired comfort levels while significantly reducing energy consumption compared to the conventional operation at high extraction rates.

Gradient analysis (Figs. 5.14 and 5.15) further emphasizes the optimal operation's effectiveness in ensuring uniform environmental conditions. Conventional operations at lower extraction rates (2 m^3/min) failed to maintain uniform comfort, while high rates (20 m^3/min) resulted in elevated energy usage. The optimal fan operation notably achieves a balanced performance, demonstrating minimal gradients in temperature and CO_2 concentration with optimized energy efficiency.

Scientific Implications and Reliability

The demonstrated performance of the MOO framework signifies its effectiveness for practical HVAC control in dynamic environments. Its reliance on a validated CFD model (Sect. 5.3.2) underscores the importance of accurate modeling and cali-

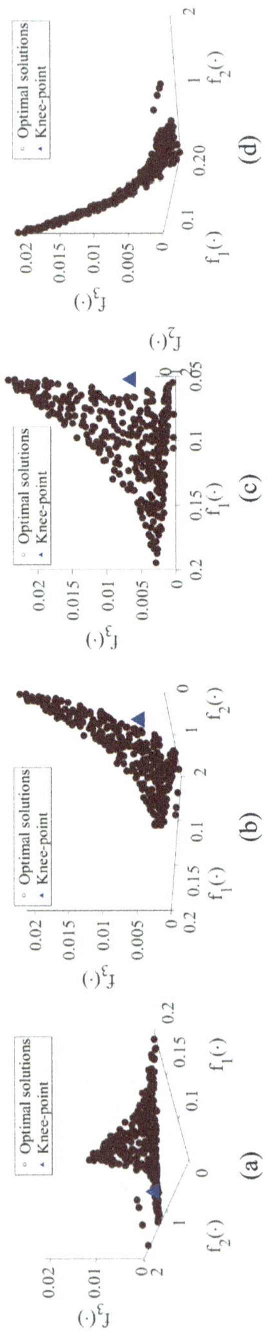

Fig. 5.11 3D Approximate Pareto Front for Scenario-2A, highlighting the selected optimal knee-point solution [63]

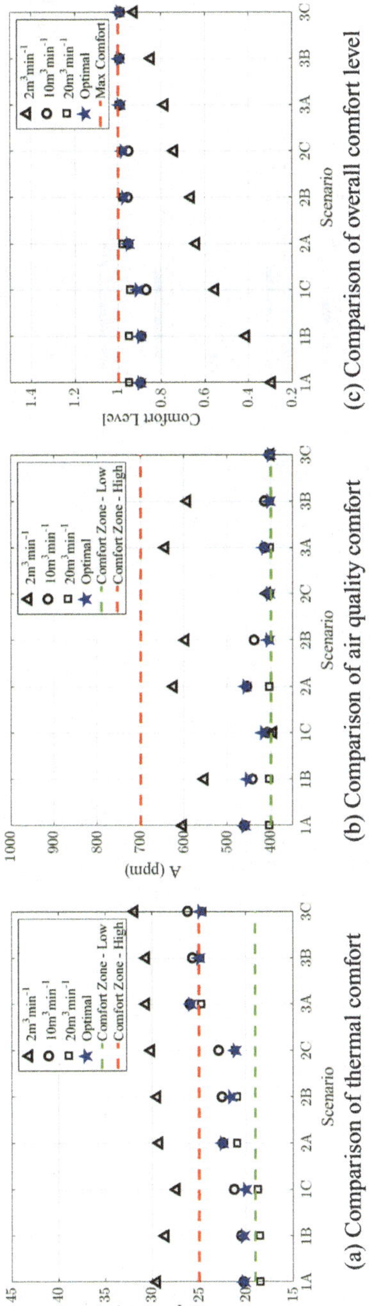

Fig. 5.12 Comparison of overall comfort [63]

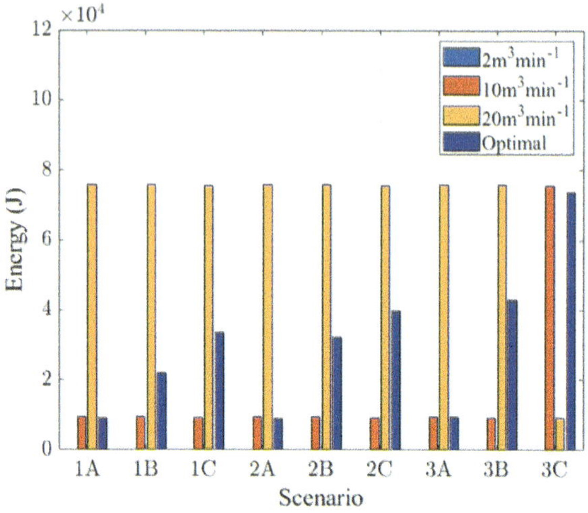

Fig. 5.13 Comparison of energy consumption for conventional versus optimal operations [63]

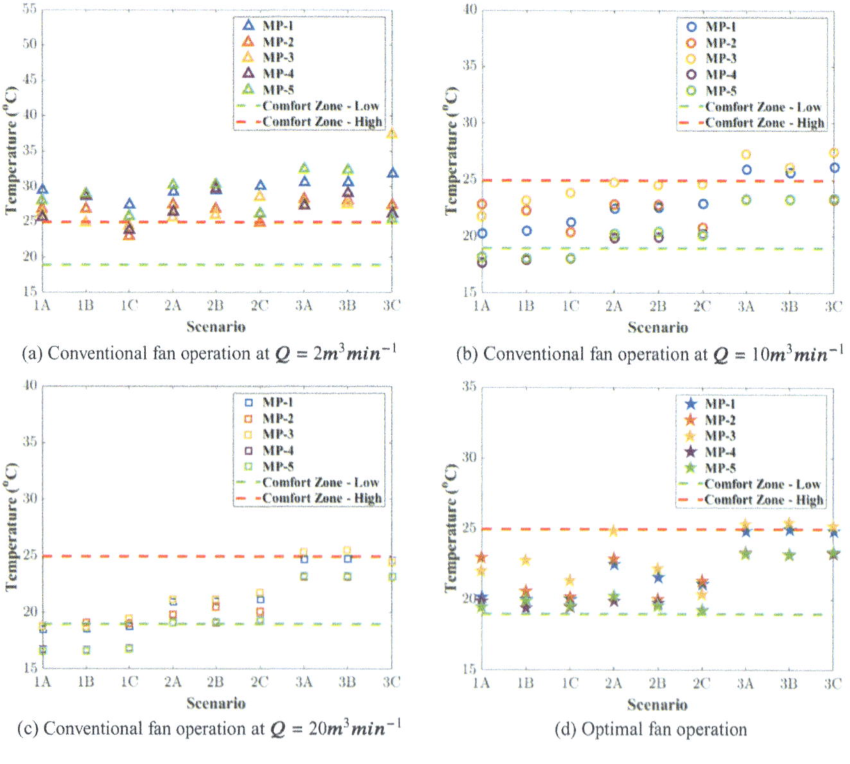

(a) Conventional fan operation at $Q = 2m^3min^{-1}$

(b) Conventional fan operation at $Q = 10m^3min^{-1}$

(c) Conventional fan operation at $Q = 20m^3min^{-1}$

(d) Optimal fan operation

Fig. 5.14 Temperature gradient comparison between conventional and optimal fan operations [63]

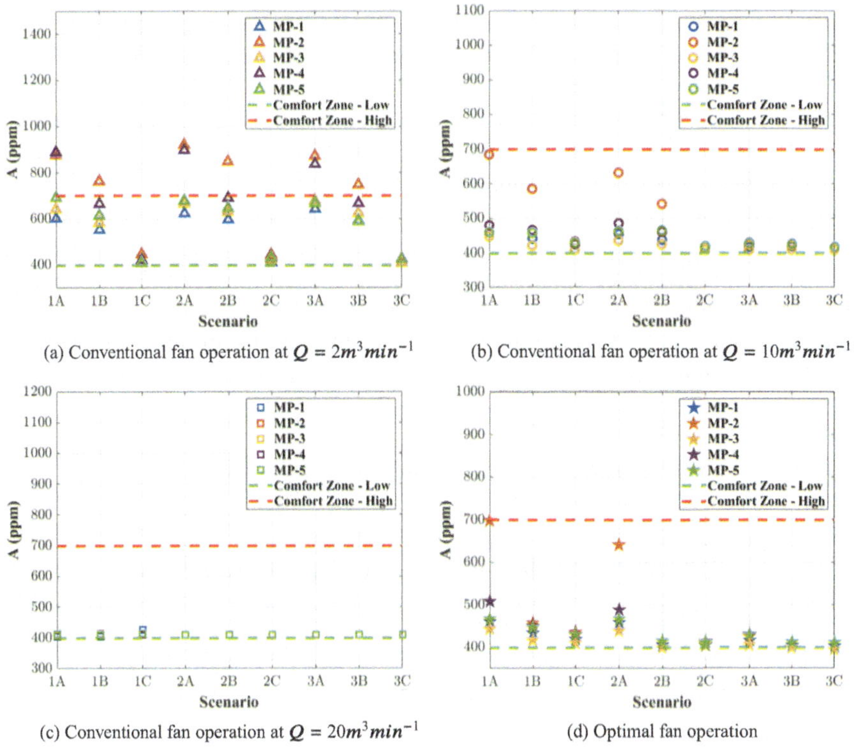

(a) Conventional fan operation at $Q = 2m^3min^{-1}$

(b) Conventional fan operation at $Q = 10m^3min^{-1}$

(c) Conventional fan operation at $Q = 20m^3min^{-1}$

(d) Optimal fan operation

Fig. 5.15 CO_2 concentration gradient comparison between conventional and optimal fan operations [63]

bration in real-world applications. The methodology employed provides robustness and adaptability across varying operational conditions, with broad implications for future research including the incorporation of occupant presence and integration with supplementary HVAC technologies such as air conditioning systems.

5.3.6 Discussions and Broader Implications

The second phase of this chapter presented a CFD-driven, multi-objective optimization (MOO) framework aimed at enhancing kitchen ventilation performance under dynamic thermal conditions. Through validated simulation and evolutionary optimization, it was demonstrated that comfort, air quality, and energy consumption can be effectively co-optimized using surrogate-assisted NSGA-II. Notably, the surrogate modeling substantially reduced computational time, enabling real-time or near-real-time decision-making capabilities for ventilation control.

This phase exemplified the potential of computational intelligence and physical modeling synergy in real-world settings. The proposed strategy not only enhanced uniformity in temperature and CO_2 distribution but also ensured measurable improvements in energy efficiency compared to conventional control logic. The broader implication lies in its generalizability across various built environments—highlighting how surrogate-based optimization, combined with CFD modeling, can serve as a robust foundation for next-generation building energy management systems (BEMS).

More importantly, the outcomes of Phase-II align with the overarching goal of this book—to unify advanced optimization, simulation, and decision-making strategies in smart buildings. This is particularly relevant in high-stakes environments where fluctuating internal loads and rapid decision-making needs are prevalent, such as commercial kitchens, medical labs, or industrial microclimates.

Scaling CFD Models with Adaptive Boundary Conditions and Real-Time Data

Despite its contributions, this phase does rely on certain assumptions, such as static occupancy and simplified boundary conditions. The CFD models, while high-fidelity, were validated under a narrow set of operational scenarios. To scale the approach, future developments should include adaptive boundary condition models and integration with real-time sensor data to close the loop between simulation and reality.

Looking Ahead

Building upon the surrogate-assisted MOO of Phase-II, Phase-III explores a novel direction wherein set-point optimization is dynamically adapted to real-time environmental conditions using multi-criteria decision-making (MCDM) techniques. The ensuing phase introduces a demand-responsive control strategy, moving from static optimization of specific environments to adaptive control of HVAC set-points in generalized settings using online Particle Swarm Optimization (PSO). This transition is motivated by the observed need to include Decision-Maker (DM) preferences explicitly and to develop a control logic that remains robust under uncertainty and temporal variability.

Phase-III is thus a natural continuation and evolution of the ideas seeded in Phase-II, aiming not only to optimize performance but also to democratize control decisions by embedding user comfort trade-offs, environmental variations, and real-time adaptivity.

Highlight

Optimizing HVAC control using MOO provides an efficient energy-management strategy, but a critical question remains: How do we incorporate human preferences and adaptive control into the system? Energy optimization must not only balance technical trade-offs but also align with occupant expectations, regulatory constraints, and dynamic environmental factors. The final phase of this research explores how Decision- Maker (DM) preferences can be integrated into intelligent BEMS through AI-driven control methodologies.

5.4 Phase-III: Incorporating Preferences of the DM in the MOO Frameworks

The preceding phases of this chapter explored the development and application of multi-objective optimization frameworks using data-driven surrogate models and high-fidelity simulations to enhance the performance of indoor environmental control systems. While these frameworks yielded optimal solutions across competing objectives such as comfort, energy use, and air quality, they operated under fixed or implicitly defined preferences, without direct input from a human Decision-Maker (DM).

To address this limitation and advance toward truly intelligent and human-centric control strategies, this phase introduces a preference-driven optimization framework that allows the DM to guide the search process based on evolving subjective priorities. This evolution is achieved through the incorporation of a first-principles dynamic model of an Air Handling Unit (AHU), which provides a physically meaningful and analytically tractable platform for assessing and comparing solutions in real-time. Unlike the CFD-based model of Phase II, which emphasized spatial fidelity, the use of a first-principles model here enables deeper exploration of temporal control and preference articulation mechanisms, bringing the optimization framework closer to real-world operational scenarios.

This final phase builds upon the foundations laid in the earlier parts of this chapter. Phase-I established a data-driven parameter estimation approach for a real commercial building using evolutionary algorithms. Phase-II extended these concepts to a highly dynamic kitchen environment, combining CFD validation with multi-objective optimization and surrogate modeling. Phase-III now deepens this progression by leveraging a first-principles modeling approach and introduces Decision-Maker (DM) preferences into the optimization loop. Together, these phases demonstrate a coherent trajectory—moving from model calibration, to control-oriented optimization, and finally to personalized decision support—reflecting the broader vision of this book in integrating computational intelligence with energy-efficient and human-centric building operation.

As building energy systems evolve toward smarter and more adaptive configurations, the demand for interpretable, physics-aligned, and computationally efficient models becomes critical. The earlier phases of this research presented in this chapter have explored data-driven modeling approaches (Phase-I) and CFD-integrated multi-objective frameworks (Phase-II) to optimize energy performance and indoor environmental quality. While these methods demonstrated high accuracy and robustness in dynamic environments, their real-time applicability in practical building control systems—especially in central HVAC infrastructure—requires a balance between model fidelity and control efficiency.

This need forms the foundation for Phase-III of this work, wherein we shift focus to first-principles based modeling of central Air Handling Units (AHUs), a key component in commercial HVAC systems. Unlike black-box or surrogate-based techniques, first-principles models are derived from thermodynamic and fluid flow equations that offer a transparent view of system behavior. Such models are particularly suited for control design, fault detection, and predictive diagnostics, thereby enhancing the intelligence and resilience of Building Energy Management Systems (BEMS).

The motivation for this phase stems from the increasing deployment of smart sensors and IoT infrastructure within commercial buildings, which enables the collection of real-time measurements needed to calibrate and operationalize physics-informed models. Moreover, by leveraging a reduced-order AHU model, this phase aims to bridge the gap between control-oriented modeling and multi-objective optimization, ensuring scalable and interpretable control under realistic building scenarios.

The core methodology presented in this phase is adapted from our published work [64], wherein a computationally lightweight but high-fidelity first-principles model was developed for a dual-zone AHU system. This model captures the dynamic interaction between indoor zones and the HVAC system across varying setpoints, thermal loads, and ambient fluctuations. Its integration into a multi-objective evolutionary framework provides actionable control strategies that co-optimize energy use, thermal comfort, and ventilation effectiveness.

5.4.1 Literature Review and Problem Context

The global energy landscape is witnessing increasing demand from residential and commercial buildings, which together constitute a major share of electricity consumption. Projections suggest that this demand will rise by over 30% by 2040 [65]. Although the adoption of renewable energy systems presents a promising pathway to reduce dependence on conventional sources, the transition is expected to span several decades. In the interim, there is a pressing need to enhance the efficiency of existing energy systems, particularly within the built environment.

To support this, several Demand Response (DR) based strategies have been proposed for Building Energy Management Systems (BEMS). These strategies enable intelligent scheduling and control of building loads under varying occupancy and operational conditions, contributing to improved energy efficiency [66–70].

Among building subsystems, Heating, Ventilation, and Air Conditioning (HVAC) systems account for roughly 35% of total energy use [65]. Consequently, several BEMS strategies emphasize HVAC optimization through adaptive control mechanisms. For example, Rocha et al. [71] contrasted static and adaptive set-point control methods, highlighting superior energy performance in the latter. Data-driven approaches such as machine learning [72], ANN-based prediction [73], and model-based cooling optimization [74] have furthered the field.

Nonetheless, most existing solutions optimize energy performance in isolation, neglecting thermal comfort, user-specific needs, or generalizability across different climate zones [75]. Addressing such limitations has prompted researchers to frame BEMS problems as multi-criteria optimization tasks [76–81].

Despite advances in balancing comfort and energy, the preferences of Decision Makers (DMs) are often excluded. For instance, Picard et al. [82] applied model predictive control to balance energy and comfort but did not account for DM priorities. Similarly, Jindal et al. [83] introduced a model-free heuristic for HVAC control, yet failed to incorporate non-linearities and user preferences explicitly.

In this phase, we propose a multi-criteria BEMS framework that incorporates DM preferences into the optimization loop. The framework adjusts the reference signals of an H_∞ controller and a boiler controller, which govern HVAC energy consumption and thermal comfort. Scalarized weight aggregation and ϵ-constraint methods are employed to achieve an optimal trade-off.

The contributions of the framework include:

- Development of a DM-aware multi-criteria optimization model integrating comfort and energy trade-offs.
- Proposal of a water mass flow-rate approximation method to support faster online optimization.
- Fitness landscape analysis highlighting the topological characteristics of the optimization surface and their sensitivity to ambient conditions.
- Validation using a non-linear AHU thermal model constructed from first-principles and tested with real-time Auckland weather data.

A wide spectrum of CI-based methods have demonstrated promising results in BEMS optimization. For instance, Manic et al. [76] advocated a CI-based architecture using EnergyPlus™ for optimal set-point control in HVAC systems. Research such as Lu [78] and Rezaei and Dagdougui [79] deployed MPC and direct load control methods to dynamically schedule HVAC loads in residential settings and microgrids.

Further advances include low-order thermal circuit models [84], dynamic set-point strategies [85], and seasonal set-point optimization using MPC [86]. Studies like

Wang et al. [87] integrated vision-based occupancy detection with comfort prediction models to optimize set-points in real-time.

While these methods introduce flexibility, they largely overlook the explicit incorporation of DM preferences. Solutions such as real-time thermal profiling [88] or activity recognition [89] are promising but not preference-driven.

This study builds on these advances by embedding DM preferences within a multi-criteria framework supported by rigorous control modeling.

Finally, from a methodological perspective, the decision support system is formulated using multi-criteria decision-making (MCDM) principles. MCDM approaches have seen widespread application in domains such as portfolio optimization [90, 91], sustainable development [92], and financial forecasting [93]. Techniques such as goal programming, compromise programming, and stochastic optimization have enabled tailored solutions to conflicting objective landscapes, often incorporating DM preferences explicitly [94–96].

Our framework contributes to this body of work by applying MCDM strategies to HVAC control, thereby bridging energy engineering with personalized decision analytics for intelligent building operations.

The rest of this phase of the chapter is organized as follows: Sect. 5.4.2 describes the thermal modeling of the AHU system based on first principles and outlines the control system architecture. The subsequent sections (Sects. 5.4.3 and 5.4.4) focus on the development of an online optimizer, its search strategy, objective estimation, and real-time preference integration. Finally, the last parts of the phase (Sects. 5.4.5 and 5.4.6) comprise of results and discussion—with Sect. 5.4.7 reflecting on the summary and conclusions.

5.4.2 AHU System Modeling Using First-Principles

To facilitate the integration of Decision-Maker (DM) preferences into building energy control strategies, it is imperative to adopt a modeling approach that balances fidelity, interpretability, and computational efficiency. In contrast to data-driven or black-box models, this phase employs a first-principles modeling strategy to represent the thermal dynamics of an Air Handling Unit (AHU), a core component in centralized HVAC systems.

The AHU thermal model used in this study is derived from established thermodynamic laws, capturing the energy exchange mechanisms between the water loop, heating coil, and the supplied air stream. The resulting model not only facilitates the design of robust control strategies but also enables real-time evaluation of comfort and energy trade-offs under varying ambient and operational conditions.

Figure 5.16 shows the schematic of the Heating Coil Unit (HCU), which is a subsystem of the AHU plant. The corresponding thermal dynamics of the system are governed by the following differential equations [97]:

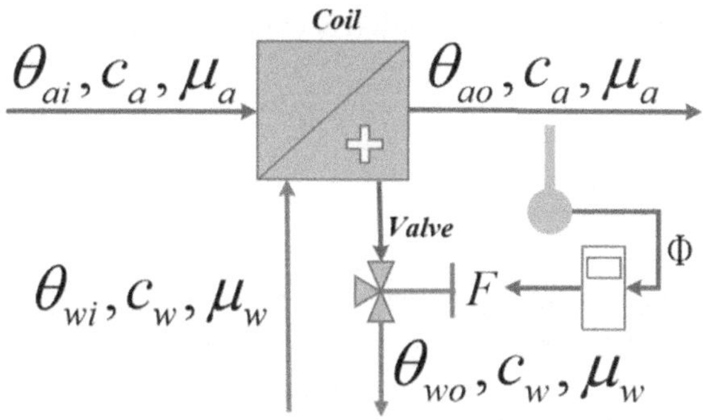

Fig. 5.16 Thermal subsystem layout of the AHU Heating Coil Unit (HCU), depicting key heat exchange paths between water, coil metal, and supply air, as considered in the first-principles thermal model [64] (adapted from [97])

$$\begin{aligned} C_w\dot{\theta}_{wo} &= \mu_w c_w(\theta_{wi} - \theta_{wo}) - Q_w \\ C_m\dot{\theta}_m &= Q_w - Q_a \\ \text{where} \quad Q_w &= U_w A_w(\theta_{wo} - \theta_m), \quad Q_a = U_a A_a(\theta_m - \theta_{ao}) \end{aligned} \tag{5.10}$$

Here, the key state variables represent temperatures of inflow and outflow water (θ_{wi}, θ_{wo}), coil metal (θ_m), and outflow air (θ_{ao}). The terms Q_w and Q_a denote heat transfer between water-coil and coil-air respectively, while C_w and C_m indicate the thermal capacities of water and metal components.

To express θ_{ao} more explicitly under quasi-steady state, an instantaneous heat transfer approximation is employed:

$$Q_a = \mu_a c_a(\theta_{ao} - \theta_{ai}) \tag{5.11}$$

Combining Eq. (5.11) with the definition of Q_a above yields an algebraic expression for the outflow air temperature:

$$\theta_{ao} = \frac{\mu_a c_a \theta_{ai} + U_a A_a \theta_m}{\mu_a c_a + U_a A_a} \tag{5.12}$$

Thermal conductances U_w and U_a depend on flow regimes and temperature, reflecting the dynamic nature of heat exchange. Specifically, for turbulent flow ($Re \geq 2300$), the water-side heat transfer coefficient is

$$U_w\big|_{Re\geq 2300} = (5.823 + 1.153 \times 10^{-1}\theta_{wo} - 1.48 \times 10^{-4}\theta_{wo}^2)\left(\frac{\mu_w}{n_{tb}d_i^{2.25}}\right)^{0.8} \tag{5.13}$$

For laminar flow ($Re < 2300$), a different empirical relation applies:

$$U_w\Big|_{Re<2300} = \frac{4.36}{d_i} \times (0.5702 + 1.79 \times 10^{-3}\bar{\theta}_m - 6.7714 \times 10^{-6}\bar{\theta}_{wm}^2) \quad (5.14)$$

$$\text{where} \quad \bar{\theta}_{wm} = \frac{\theta_{wo} + \theta_m}{2}$$

The water-side conductance is taken as the maximum of the two to smoothly transition across flow regimes:

$$U_w = \max\left(U_w|_{Re\geq 2300},\ U_w|_{Re<2300}\right) \quad (5.15)$$

The air-side conductance, U_a, varies with the coil metal temperature θ_m:

$$U_a = 60.37 + 140.3 \times \exp\left(\frac{28.62 - \theta_m}{9.796}\right) \quad (5.16)$$

Control of the water flow through the heating coil is achieved using a valve, whose dynamic behavior is captured via:

$$\tau_a \dot{\mu}_w = \Psi \mu_{wd} - \mu_w \quad (5.17)$$

where Ψ represents the valve installed characteristic as a function of its inherent profile and authority $\mathcal{V}$:

$$\Psi = \gamma \times [\gamma^2(1 - \mathcal{V}) + \mathcal{V}]^{-1/2}, \quad (5.18)$$

$$\gamma = f_o^{(1-\mathcal{F})} \quad (5.19)$$

Finally, the measured outflow air temperature is represented using a first-order lag sensor model:

$$\tau_s \dot{\Phi} = \theta_{ao} - \Phi \quad (5.20)$$

where Φ is the temperature sensor output. This compact yet expressive model enables both controller design and real-time optimization by capturing the essential physics of heat transfer within the AHU. In the subsequent sections, this model will be embedded into a preference-aware multi-criteria optimization framework to dynamically adjust control set-points as per prevailing environmental and DM conditions.

Control Loop Architecture for AHU Regulation

The thermal model of the AHU subsystem, as mentioned earlier, operates in conjunction with a dual-loop control architecture designed to regulate water flow and inflow temperature. As illustrated in Fig. 5.17, the system comprises two dedicated controllers: an H_∞ controller and a boiler controller.

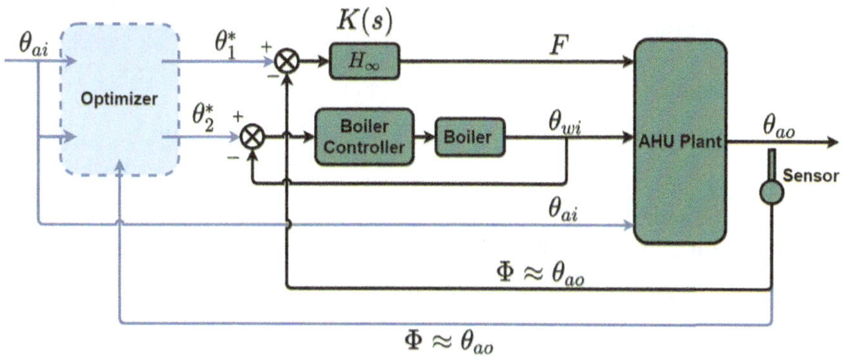

Fig. 5.17 Control system structure of the AHU plant, highlighting the dual-loop strategy involving the H_∞ controller for flow modulation and the boiler controller for inlet water temperature regulation [64]

The H_∞ controller, denoted $K(s)$, adjusts the fractional valve-stem position to modulate the mass-flow rate of water through the coil, μ_w. This acts as the fast inner loop, responding to fluctuations in thermal demand. The outer loop involves the boiler controller, which sets the inflow water temperature θ_{wi} based on a predefined set-point, typically assumed to be instantaneously available due to negligible actuation delay.

Together, these controllers determine the air outlet temperature, θ_{ao}, thereby influencing both occupant thermal comfort and energy consumption. Their coordinated action allows the system to dynamically adapt to environmental and internal heat load changes. In the subsequent sections, this control layer is integrated into a preference-aware optimization framework that tunes set-points (θ_1, θ_2) to align with the occupant or operator priorities.

5.4.3 *Problem Formulation with Preference Integration*

Building on the AHU system model and the thermal comfort estimation techniques developed in previous sections, this section formulates a multi-objective optimization problem that is both performance-driven and preference-aware. Specifically, we seek to identify optimal control set-points for the AHU that balance energy efficiency with thermal comfort, while allowing Decision-Maker (DM) preferences to guide the selection from among trade-off solutions.

Let $\Theta_p = \{\theta_{p,1}, \theta_{p,2}\}$ denote the decision variables, representing the target outflow air temperature and the inflow water temperature set-points, respectively. Each candidate Θ_p is evaluated based on two competing objectives:

1. Energy Cost Function

The energy required to raise the water temperature is modeled as

$$J_1(\Theta_p) \approx \mu_w c_w(\theta_{p,2} - \theta_{p,1}) \tag{5.21}$$

where μ_w is the approximated water mass-flow rate (see Eq. (5.30)), and $\theta_{p,1}$ approximates the return water temperature. This cost reflects the total heating energy supplied to the system.

2. Thermal Comfort Deviation

The discomfort level, as a function of deviation from the target air temperature, is defined as

$$J_2(\Theta_p) = \left| \frac{\theta_{ao} - \theta_{p,1}}{\theta_{p,1}} \right| \tag{5.22}$$

Here, θ_{ao} is estimated algebraically using Eq. (5.29), enabling rapid evaluation without simulating full plant dynamics.

Approximation Scheme for Fast Objective Evaluation

To enable real-time optimization without compromising model fidelity, we propose an approximation scheme that simplifies the evaluation of the key output variable θ_{ao} and the internal control variable μ_w. This scheme leverages the quasi-steady-state behavior of the AHU system and avoids repeated numerical integration of system dynamics, thus significantly reducing computational burden.

At any given optimization iteration, it is assumed that the system reaches a near-stationary state. Accordingly, the temperature dynamics from Eq. (5.10) are simplified under the assumption:

$$\dot{\theta}_{wo} \approx 0; \qquad \dot{\theta}_m \approx 0 \tag{5.23}$$

This leads to the following simplifications:

$$Q_w \approx \mu_w c_w(\theta_{wi} - \theta_{wo}) \tag{5.24}$$

$$Q_a \approx Q_w \approx \mu_a c_a(\theta_{ao} - \theta_{ai}) \tag{5.25}$$

From the above, the outlet air temperature θ_{ao} can be algebraically expressed as

$$\theta_{ao} \approx \frac{\mu_w c_w}{\mu_a c_a}(\theta_{wi} - \theta_{wo}) + \theta_{ai} \tag{5.26}$$

To simplify the evaluation of θ_{ao}, we adopt an ideal tracking assumption based on the control structure described earlier. It is assumed that both the boiler controller and the H_∞ controller operate with negligible delay, allowing us to approximate:

$$\theta_{ao} \approx \theta_{wo} \approx \theta_{p,1}, \quad \theta_{wi} \approx \theta_m \approx \theta_{p,2} \tag{5.27}$$

This assumption enables a direct substitution of system temperatures with the decision variables at each optimization step. Therefore, under ideal controller tracking assumptions (see Eq. (5.27)), we substitute

$$\theta_{wi} \approx \theta_{p,2}, \quad \theta_{wo} \approx \theta_{p,1} \tag{5.28}$$

This yields the final form used during optimization:

$$\theta_{ao} \approx \frac{\mu_w c_w}{\mu_a c_a}(\theta_{p,2} - \theta_{p,1}) + \theta_{ai} \tag{5.29}$$

However, μ_w is governed by the internal H_∞ controller and is not directly available. Simulating the control system at each iteration would incur high computational costs. To address this, we introduce a lightweight mass-flow rate approximation scheme:

Mass-Flow Rate Approximation

To reduce computational effort during optimization, we bypass full simulation of the H_∞ controller by estimating the water mass-flow rate μ_w using its steady-state approximation:

$$\mu_w \approx \Psi(\mathcal{F}) \cdot \mu_{wd} \tag{5.30}$$

where Ψ is the valve installed characteristic (see Eq. (5.18)), and $\mathcal{F}$ is inferred from the decision variable $\theta_{p,1}$. This direct algebraic evaluation allows rapid objective function computation while retaining sufficient model fidelity.

By using the approximated μ_w in the earlier expression for θ_{ao}, the thermal response of the AHU can be evaluated efficiently at each candidate point. This is especially beneficial in population-based meta-heuristics, where large numbers of evaluations are required.

The approximation framework is conceptually illustrated in Fig. 5.18.

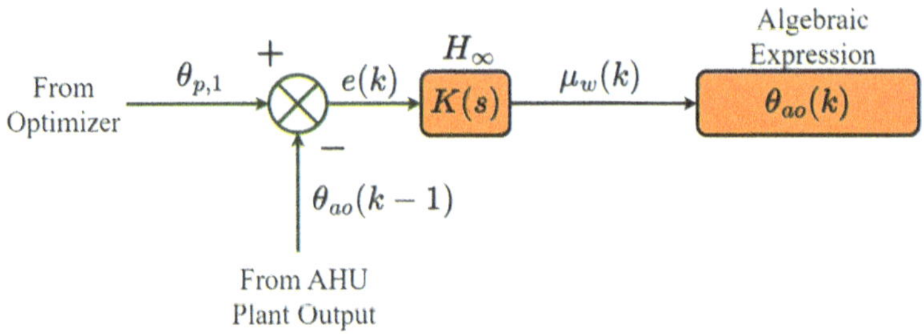

Fig. 5.18 Block diagram of the proposed approximation framework for fast evaluation of θ_{ao} and μ_w under quasi-steady-state assumptions. The structure bypasses detailed controller simulation and enables lightweight integration within real-time optimization loops

Preference-Aware Optimization

Rather than scalarizing or pre-weighting the objectives, we employ a *Pareto-based search*, which produces a diverse front of non-dominated solutions. This supports *interactive or contextual preference integration* in later stages. In this formulation, the multi-objective problem is

$$\min_{\Theta_p \in \mathcal{D}} \boldsymbol{J}(\Theta_p) = \{J_1(\Theta_p),\ J_2(\Theta_p)\} \tag{5.31}$$

subject to the admissible decision space:

$$\mathcal{D} = \left\{\Theta_p \mid \theta_{ao}^{\min} \le \theta_{p,1} \le \theta_{ao}^{\max},\ \theta_{wi}^{\min} \le \theta_{p,2} \le \theta_{wi}^{\max}\right\} \tag{5.32}$$

The choice of bounds in $\mathcal{D}$ reflects physical and operational constraints of the AHU plant and user-defined comfort limits.

DM Preference Embedding

A key novelty of this phase is the ability to *embed DM preferences* not during the optimization, but in the *solution interpretation* phase. This decouples the search from rigid weighting assumptions and allows real-time user input, context-based filtering (e.g., time of day, energy tariff), or learned preference profiles to guide final set-point selection from the Pareto front.

Summary

This problem formulation enables energy-efficient and comfort-oriented control of HVAC systems, while retaining flexibility in aligning with occupant or operational preferences. The next section describes the search algorithm employed, followed by the runtime preference integration logic used to select optimal solutions adaptively.

5.4.4 Real-Time Multi-objective Optimization with Preference Embedding

Having established the thermal model of the AHU system, the formulation of performance objectives, and an approximation strategy for computational efficiency, we now present the complete multi-objective optimization (MOO) framework. The objective of this framework is to support real-time selection of AHU control set-points that reflect both operational performance and contextual user preferences.

The core architecture of the framework is shown in Fig. 5.19. It consists of four key stages:

1. **Input Layer:** This module receives contextual inputs including outside weather conditions, current indoor thermal state, and occupant or Decision-Maker preferences. Additionally, a dynamic weight profile or priority structure (e.g., energy-saving vs. comfort) can be injected based on time-of-day, tariff windows, or adaptive learning.
2. **Objective Evaluation Engine:** For each candidate set-point pair $\Theta_p = \{\theta_{p,1}, \theta_{p,2}\}$, the discomfort index $J_2(\Theta_p)$ and energy cost $J_1(\Theta_p)$ are computed using the algebraic approximations introduced earlier (Eqs. (5.29), (5.21)). This eliminates the need for full dynamic simulation or controller emulation at each evaluation step.
3. **Pareto Search Layer:** A population-based meta-heuristic algorithm, Particle Swarm Optimization (PSO), is used to explore the decision space and generate a Pareto-optimal front. PSO is well-suited for continuous optimization problems such as HVAC control, and its fast convergence and simple implementation make it ideal for real-time applications. Each point on the Pareto front offers a trade-off between comfort and energy use, enabling informed decision-making.
4. **Preference Filter and Selector:** The final module implements a preference filter that either selects a single solution (e.g., via weighted aggregation, thresholding, or rule-based heuristics) or presents a reduced Pareto set to the Decision-Maker for manual or AI-assisted selection. This layer is extensible to include reinforcement learning or decision support tools.

The key novelty of this framework lies in the integration of *preference-awareness* at runtime. Unlike traditional MOO formulations where preferences are embedded as fixed weights during optimization, this architecture allows preferences to be separated from the search process. As a result, the optimizer remains flexible and can accommodate changing priorities or contextual policies without retraining or re-tuning.

To accommodate subjective priorities of the Decision Maker (DM), we incorporate two classic multi-criteria strategies within the optimization loop: scalarized weighted aggregation (also called Conventional Weighted Aggregation or CWA) and the ε-constraint method. These formulations allow the optimization process to be steered toward energy-efficiency or thermal comfort objectives depending on the prevailing operational context or user preferences.

- **In the scalarized (CWA) formulation**, the two objectives–energy cost J_1 and discomfort J_2–are combined into a single scalar objective:

$$J(\Theta_p) = \lambda_1 J_1(\Theta_p) + \lambda_2 J_2(\Theta_p) \tag{5.33}$$

  where $\lambda_1, \lambda_2 \geq 0$ and $\lambda_1 + \lambda_2 = 1$. The weights reflect the DM's *a priori* preferences.
- **In the ε-constraint formulation**, one objective is optimized (e.g., energy cost), while the other (e.g., discomfort) is restricted by a soft threshold:

$$J(\Theta_p) = J_1(\Theta_p) + \mathcal{P} \quad \text{where} \quad \mathcal{P} = \alpha \cdot \max(0, J_2(\Theta_p) - \varepsilon) \tag{5.34}$$

  Here, α is a penalty multiplier and ε denotes the maximum allowable discomfort, set by the DM.

These formulations support adaptive trade-offs in real-time operation. Depending on user priorities (e.g., time-of-day or tariff window), the system can seamlessly switch between these formulations during runtime.

Furthermore, the proposed use of analytical estimators for μ_w and θ_{ao} makes the optimizer suitable for near real-time deployment, a feature typically absent in simulation-intensive HVAC control approaches. The Decision-Maker (DM) remains in the loop, and preferences can be expressed explicitly (via weights or rules) or implicitly (via learned models).

This closes the loop between physical model fidelity, computational tractability, and human-centered decision-making—addressing the core goals of this book.

5.4.5 Implementation Architecture and Simulation Setup

The implementation architecture of the proposed optimization framework adheres to the layered structure shown earlier in Fig. 5.19, encapsulating the key building blocks required for real-time preference-aware control. It includes contextual input processing, fast objective evaluation using approximations, a PSO-based search engine, and a dynamic preference filter. For detailed information on the software workflow, data

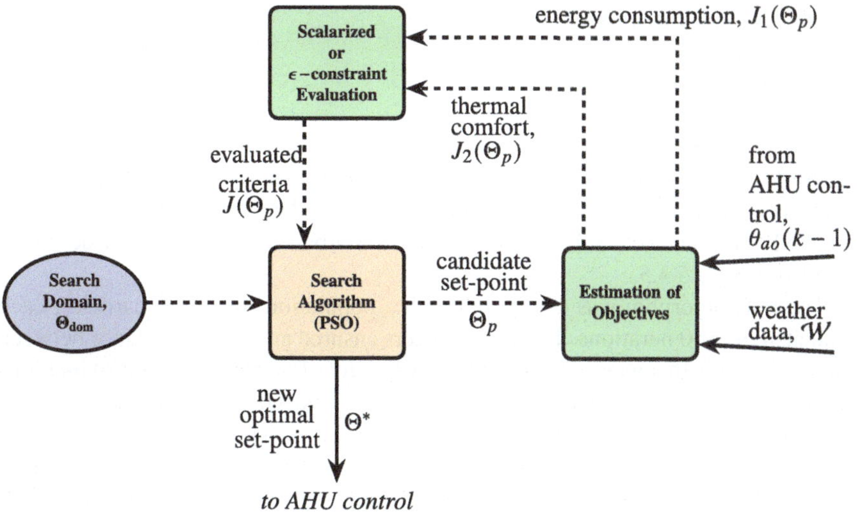

Fig. 5.19 Proposed search framework for multi-criteria selection of set-points (Θ) at a particular time-step k [64]

flow, and internal structure of the simulation loop, refer to Sects. 4 and 5 of our prior work [64]. The overall control and optimization logic were programmed in MATLAB®, while the thermal response and actuator dynamics were modeled using Simulink®. This allowed for seamless integration between algorithmic decision-making (e.g., PSO loops, objective evaluation) and the plant-side simulation of AHU behavior. To ensure realistic environmental inputs, actual meteorological data from *NIWA (National Institute of Water and Atmospheric Research), New Zealand* was used. The weather data covered an entire calendar year and included hourly dry-bulb temperature, humidity, and solar irradiation-enabling full seasonal testing of the control strategy under diverse conditions. Realistic environmental inputs were generated using this weather data to enable comprehensive testing. The optimization variables include the air outflow temperature set-point $\theta_{p,1}$ and the water inflow temperature set-point $\theta_{p,2}$, constrained within operational bounds of $\theta_{p,1} \in [21\,°C, 25\,°C]$ and $\theta_{p,2} \in [35\,°C, 55\,°C]$, with a resolution of 0.5 °C. The PSO optimizer is configured with a population size of 20 and 50 iterations per run, clamped within feasible domains as per the ε-constraint formulation. PSO configurations and model coupling strategy remain consistent with our previous work, ensuring replicability grounded in validated computational infrastructure.

5.4.6 *Results and Analysis*

Runtime Performance Evaluation

The computational efficiency and effectiveness of the proposed preference-aware optimization framework were evaluated comprehensively using simulations spanning an entire calendar year. The simulations employed hourly weather data sourced from NIWA, New Zealand, and utilized the PSO-based optimization structure as detailed in Sect. 5.4.5.

The PSO algorithm was configured with a population size of 20 particles, each evaluated over 50 iterations. These parameters ensured an effective balance between computational efficiency and optimization quality. The algebraic approximations described in Sect. 5.4.3 facilitated significantly reduced computational demand, enabling timely execution suitable for practical, near-real-time applications.

Table 5.6 summarizes the annual performance metrics across various Decision-Maker (DM) preference scenarios, adapted directly from our prior work [64]. It presents the total annual energy consumption of the AHU system (E_{AHU}) and occupant discomfort quantified as the discomfort ratio ($\mathcal{D}_{ratio}$), along with the corresponding percentage improvements relative to the baseline scenario.

As evident from Table 5.6, significant reductions in annual energy use and occupant discomfort were achieved across various DM scenarios. Specifically, the energy-focused ϵ-constraint scenario demonstrated the highest energy savings (49.20%) relative to the baseline, whereas the comfort-focused ϵ-constraint scenario yielded the greatest improvement in occupant comfort (89.00%).

Table 5.6 Annual runtime performance evaluation under different DM preference [64]

Scenario	Energy		Discomfort	
	E_{AHU}	Improvement Over Baseline† (%)	$\mathcal{D}_{ratio}$	Improvement Over Baseline†† (%)
CWA + *comfort focused*‡	13949.55	40.85	0.81	87.10
CWA + *balanced*	13062.07	44.62	0.76	75.30
CWA + *energy focused*	12374.59	47.53	0.67	55.04
ϵ−constraint +*comfort focused*‡‡	14280.22	39.45	0.82	89.00
ϵ−constraint +*balanced*	12791.95	45.76	0.78	79.88
ϵ−constraint +*energy focused*	11980.97	49.20	0.63	45.39

† The annual energy consumption with the baseline framework is $E_{AHU} = 23{,}584.76$

†† The annual discomfort ratio with the baseline framework is $\mathcal{D}_{ratio} = 0.43$

‡ *Comfort-focused*: $\{\lambda_1, \lambda_2\} = \{0.1, 0.9\}$, *Balanced*: $\{\lambda_1, \lambda_2\} = \{0.5, 0.5\}$, *Energy-focused*: $\{\lambda_1, \lambda_2\} = \{0.9, 0.1\}$

‡‡ *Comfort-focused*: $\epsilon = 0.1$, *Balanced*: $\epsilon = 0.5$, *Energy-focused*: $\epsilon = 0.9$

These quantified performance improvements validate the operational effectiveness and adaptability of the proposed optimization framework, confirming its suitability for practical real-time decision-making in Building Energy Management Systems (BEMS). Detailed discussions of scenario-specific behaviors and seasonal adaptability are presented in following section.

Preferential Realization and Case Results

The preceding sections presented the theoretical framework and computational efficiency of the proposed preference-aware optimization methodology. In this section, we explore how decision-maker (DM) preferences influence real-world outcomes and demonstrate these effects using documented results from our previous detailed evaluation [64].

The preference realization aspect of this research explicitly illustrates how different preference configurations yield distinct operational outcomes. Specifically, two distinct decision-making formulations were examined: scalarized weighted aggregation (CWA) and the ϵ-constraint method. Each of these methodologies provided flexibility in balancing the objectives of energy efficiency and occupant comfort, depending on the DM's real-time priorities.

Figure results (see Figs. 5.20, 5.21, 5.22, 5.23, 5.24, 5.25, 5.26, and 5.27) and annual summaries from our prior study [64] clearly illustrate how preferences were realized into tangible operational decisions:

- **Scalarized (CWA) Scenarios**: Under the comfort-focused preference ($\lambda = 0.1, 0.9$), significant improvements in occupant comfort (up to 87.10% improvement over baseline) were prioritized, leading to moderate energy savings (40.85%). Conversely, under the energy-focused scenario ($\lambda = 0.9, 0.1$), energy consumption was reduced by approximately 47.53%, with a comparatively moderate improvement in occupant comfort.

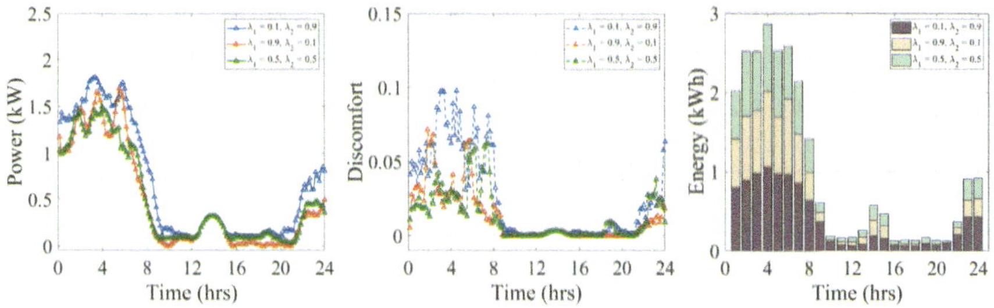

(a) AHU Power consumption (b) Discomfort level (c) Energy consumption

Fig. 5.20 The results of set-point optimization using the CWA method for a typical *summer day* in Auckland city. The results are shown for three different preference scenarios: *comfort-focused* with $\{\lambda_1, \lambda_2\} = \{0.1, 0.9\}$, *balanced* with $\{\lambda_1, \lambda_2\} = \{0.5, 0.5\}$, and *energy-focused* with $\{\lambda_1, \lambda_2\} = \{0.9, 0.1\}$

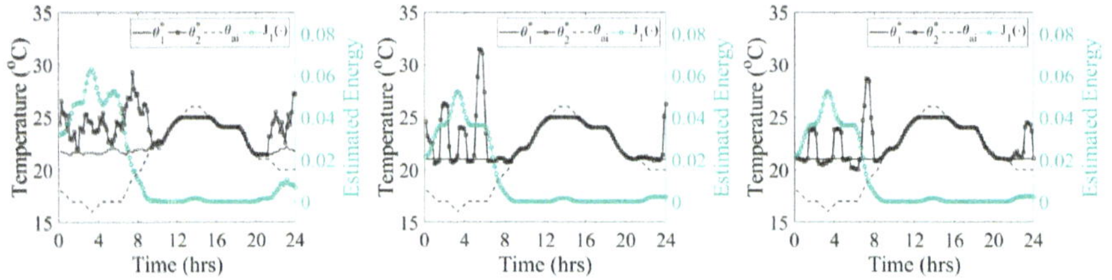

(a) *comfort-focused* scenario (b) *energy-focused* scenario (c) *balanced* scenario

Fig. 5.21 The results of set-point optimization using the CWA method for a typical *summer day*. **a–c** show correlation between the *estimated energy consumption* (J_1) and the optimal set-points (θ_1^* and θ_2^*) under different preferences

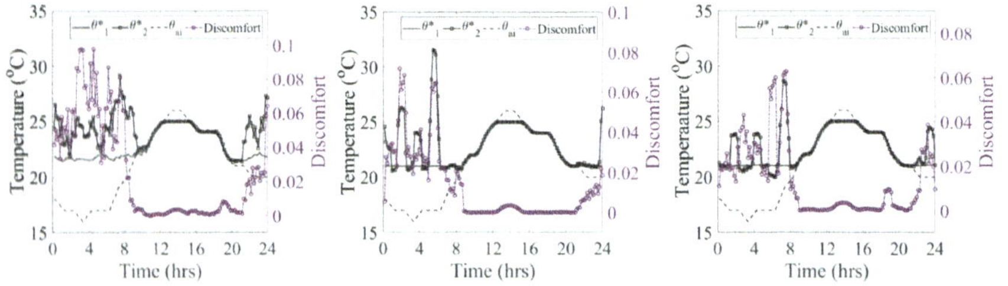

(a) *comfort-focused* scenario (b) *energy-focused* scenario (c) *balanced* scenario

Fig. 5.22 The results of set-point optimization using the CWA method for a typical *summer day*. **a–c** show correlation between the *discomfort* (J_2) and the optimal set-points (θ_1^* and θ_2^*) under different preferences

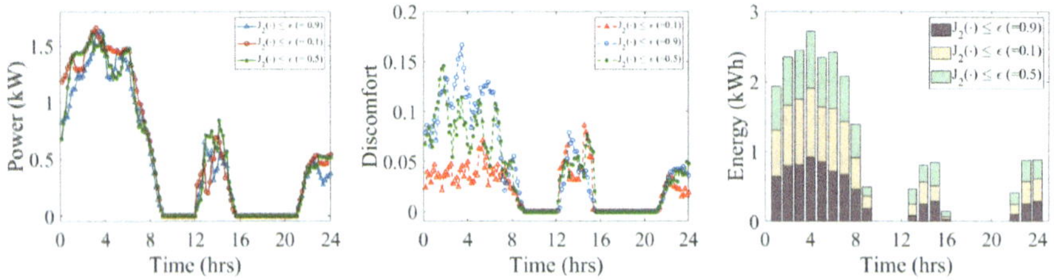

(a) AHU power consumption (b) Discomfort level (c) Energy consumption

Fig. 5.23 The results of set-point optimization using the $\epsilon-$constraint method for a typical *summer day* in Auckland city. The results are shown for three different preference scenarios: *comfort-focused* with $\epsilon = 0.1$, *balanced* with $\epsilon = 0.5$, and *energy-focused* with $\epsilon = 0.9$

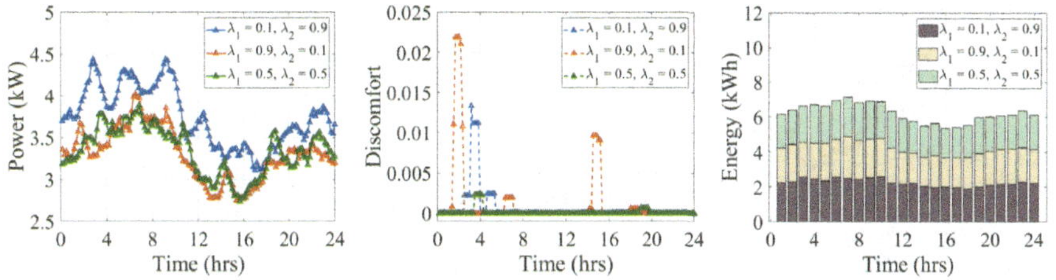

(a) AHU Power consumption (b) Discomfort level (c) Energy consumption

Fig. 5.24 The results of set-point optimization using the CWA method for a typical *winter day* in Auckland city. The results are shown for three different preference scenarios: *comfort-focused* with $\{\lambda_1, \lambda_2\} = \{0.1, 0.9\}$, *balanced* with $\{\lambda_1, \lambda_2\} = \{0.5, 0.5\}$, and *energy-focused* with $\{\lambda_1, \lambda_2\} = \{0.9, 0.1\}$

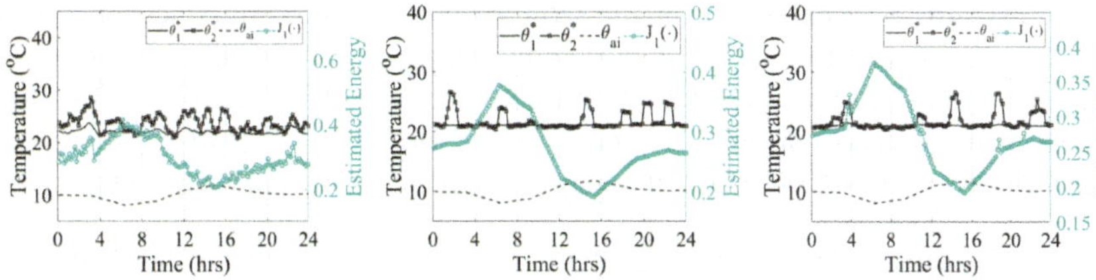

(a) *comfort-focused* scenario (b) *energy-focused* scenario (c) *balanced* scenario

Fig. 5.25 The results of set-point optimization using the CWA method for a typical *winter day*. **a–c** show correlation between the *estimated energy consumption* (J_1) and the optimal set-points (θ_1^* and θ_2^*) under different preferences

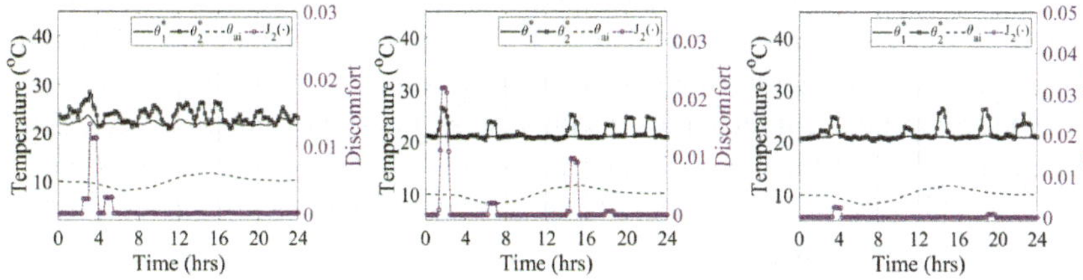

(a) *comfort-focused* scenario (b) *energy-focused* scenario (c) *balanced* scenario

Fig. 5.26 The results of set-point optimization using the CWA method for a typical *winter day*. **a–c** show correlation between the *discomfort* (J_2) and the optimal set-points (θ_1^* and θ_2^*) under different preferences

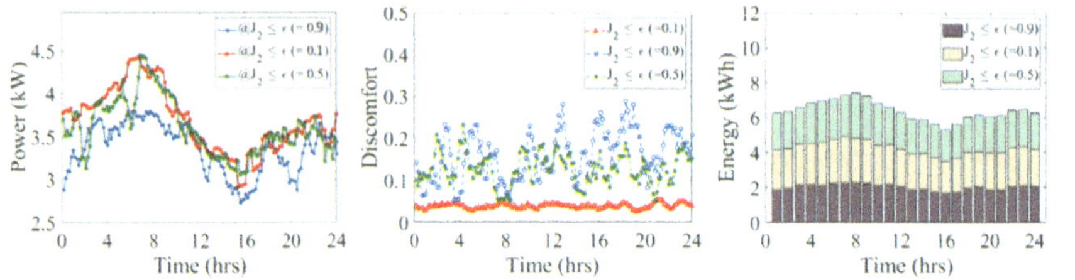

(a) AHU power consumption (b) Discomfort level (c) Corresponding energy consumption

Fig. 5.27 The results of set-point optimization using the $\epsilon-$constraint method for a typical *winter day* in Auckland city. The results are shown for three different preference scenarios: *comfort-focused* with $\epsilon = 0.1$, *balanced* with $\epsilon = 0.5$, and *energy-focused* with $\epsilon = 0.9$

- **ϵ-Constraint Scenarios**: Under the stringent comfort constraint scenario ($\epsilon = 0.1$), occupant comfort achieved substantial improvement (89.00%), demonstrating the framework's capability to prioritize occupant well-being effectively. Conversely, the scenario prioritizing energy reduction ($\epsilon = 0.9$) realized maximum energy savings (49.20%) with relatively moderate comfort improvements.

These documented case results highlight how the decision-making mechanisms translate preferences into explicit operational set-point decisions, impacting the performance outcomes significantly. The flexibility demonstrated through the scalarized and ϵ-constraint scenarios underscores the framework's adaptability to various operational priorities, such as comfort-focused or energy-focused approaches.

Moreover, the adaptability of set-points to seasonal variations and changing external conditions was rigorously demonstrated, showing consistent performance across different weather scenarios. This adaptability confirms the framework's suitability for dynamic real-time applications in practical building energy management.

In summary, the documented results underscore the significant potential of preference embedding within optimization frameworks for real-world decision-making, aligning seamlessly with the broader vision of developing robust, adaptive, and intelligent energy management systems.

5.4.7 Summary and Key Takeaways of Phase-III

This concluding section reflects on the broader context and scientific underpinnings of Phase-III in relation to the overall objectives of this book. The goal of this chapter has been to advance the intelligent control of building systems by systematically incorporating preference-aware optimization methods into HVAC operation.

This phase addressed a critical research question: how can Decision-Maker preferences be embedded in the control decision process in a computationally efficient yet operationally meaningful way? Toward this, a rigorous modeling and control framework was proposed for an Air Handling Unit (AHU) system, culminating in a preference-integrated multi-objective optimization scheme. Distinct formulations based on scalarization and ϵ-constraint methods were employed to reflect varying control priorities, and their real-world impact was validated using realistic building simulation data.

Notably, this study introduced an efficient thermal approximation strategy, allowing the framework to operate with practical runtime feasibility without compromising optimization quality. This computational tractability enabled integration with a Particle Swarm Optimization (PSO) engine and facilitated real-time adaptability to changing weather and user-defined conditions. Extensive simulations across comfort- and energy-focused scenarios demonstrated not only tangible performance improvements but also the capability of the proposed method to generalize across seasonal and operational conditions.

The outcomes of this phase must also be viewed as a logical progression of the prior two phases of this book chapter. In Phase-I, data-driven estimation of thermal parameters was conducted on a real commercial building, laying the groundwork for modeling accuracy. In Phase-II, the focus shifted to optimization of forced air extraction systems using NSGA-II, where multiple objectives were explicitly defined and solved in parallel. Phase-III built on both foundations by generalizing these capabilities: it introduced the concept of preference, and demonstrated how such preferences can drive system behavior in realistic settings.

Together, these three phases move toward a central thesis of this work: that energy-efficient building control must evolve from static rule-based systems to intelligent, preference-sensitive, and computationally tractable control frameworks. Phase-III closes this loop by validating that preference-aware decision-making can be made both responsive and robust—thereby meeting the dual demands of comfort and efficiency, as dictated by real-world Decision-Makers.

This culmination opens promising avenues for future work-such as extending the framework to coordinated multi-zone control, integrating renewable energy systems, and investigating long-horizon planning strategies—to further evolve intelligent Building Energy Management Systems (BEMS).

Highlight

By systematically integrating thermal parameter estimation, Multi-Objective Optimization (MOO), and preference-aware meta-heuristic control, the author's Ph.D. research—and by extension, this chapter—has established a robust and scalable framework for next-generation intelligent BEMS. Future research may further investigate real-time adaptability, cloud-based Digital Twin integration for predictive control, and large-scale deployment within interconnected smart building and urban energy ecosystems.

5.5 Conclusion

This chapter presented a structured exploration into the design and realization of intelligent, adaptive, and preference-aware optimization frameworks for Building Energy Management Systems (BEMS). Through a staged development approach, three distinct but interlinked phases were articulated—each advancing the capability and scope of control strategies for HVAC systems in increasingly complex and realistic settings.

Phase-I focused on data-driven thermal parameter estimation for a real commercial building, establishing a foundation for physically meaningful and empirically validated modeling. Phase-II extended this foundation by applying multi-objective optimization via NSGA-II to forced extraction systems, introducing rigorous performance trade-off handling under realistic operating constraints. Phase-III then generalized the control paradigm by embedding Decision-Maker preferences directly into the optimization loop, using scalarization and ϵ-constraint methods. The integration of a PSO engine, paired with thermal approximation techniques, enabled real-time feasible optimization without compromising accuracy or robustness.

Across all three phases, the research consistently demonstrated the value of combining principled control modeling with Computational Intelligence (CI) to enable systems that are not only efficient but also responsive to user-centric and context-aware demands. By validating these frameworks through extensive simulations, including annual-scale performance assessments and seasonal adaptability tests, this work provides a strong empirical and methodological foundation for advancing modern BEMS.

This chapter has further demonstrated how intelligent control strategies—particularly meta-heuristic optimization, multi-objective formulations, and *AI-driven decision-making*—can revolutionize the operation of building energy systems. Importantly, it reinforces the central premise of this book: that the future of building control lies in adaptive, preference-sensitive, and computationally efficient frameworks that harmonize occupant comfort and energy efficiency.

Looking forward, several directions offer promise for future research. These include real-time integration of Machine Learning (ML) models, cloud-based *Digital Twin ecosystems* for predictive analytics, and federated learning approaches that leverage distributed data while preserving privacy. These avenues hold significant potential to realize scalable, secure, and context-aware energy optimization strategies for intelligent buildings at large.

In conclusion, the ideas and techniques presented in this chapter set a foundational benchmark for how preference-aware optimization can be holistically designed, evaluated, and practically implemented within next-generation intelligent BEMS.

References

1. Wani MH (2022) Intelligent control strategies for efficient building energy management systems (BEMS). Department of Electrical and Electronic Engineering, The University of Auckland, PhD thesis
2. Energy Information Administration (EIA) (2019) Consumption & Efficiency. [Online]. https://www.eia.gov/consumption/, [Accessed: 03-09-2019]
3. Wang Z, Wang L, Dounis AI, Yang R (2012) Multi-agent control system with information fusion based comfort model for smart buildings. Appl Energy 99:247–254
4. De Felice M, Yao X (2011) Short-term load forecasting with neural network ensembles: a comparative study [Application notes]. IEEE Comput Intell Mag 6(3):47–56
5. Evans G (1991) Solving home automation problems using artificial intelligence techniques. IEEE Trans Consum Electron 37(3):395–400
6. Ramchurn S, Vytelingum P, Rogers A, Jennings NR (2012) Putting the "Smarts" into the smart grid: a grand challenge for artificial intelligence. Commun ACM 55(4):86–97
7. Loveday D, Virk G (1992) Artificial intelligence for buildings. Appl Energy 41(3):201–221
8. Alibabaei N, Fung AS, Raahemifar K, Moghimi A (2017) Effects of intelligent strategy planning models on residential HVAC system energy demand and cost during the heating and cooling seasons. Appl Energy 185:29–43
9. Liu Y, Yu N, Wang W, Guan X, Xu Z, Dong B, Liu T (2018) Coordinating the operations of smart buildings in smart grids. Appl Energy 228:2510–2525
10. Adhikari R, Pipattanasomporn M, Rahman S (2018) An algorithm for optimal management of aggregated HVAC power demand using smart thermostats. Appl Energy 217:166–177
11. Wang Z, Chen Y (2019) Data-driven modeling of building thermal dynamics: methodology and state of the art. Energy Build 109405
12. Png E, Srinivasan S, Bekiroglu K, Chaoyang J, Su R, Poolla K (2019) An internet of things upgrade for smart and scalable heating, ventilation and air-conditioning control in commercial buildings. Appl Energy 239:408–424
13. Drgoňa J, Picard D, Kvasnica M, Helsen L (2018) Approximate model predictive building control via machine learning. Appl Energy 218:199–216
14. Kalema T, Haapala T (1995) Effect of interior heat transfer coefficients on thermal dynamics and energy consumption. Energy Build 22(2):101–113
15. Shi Z, Zhang X (2011) Analyzing the effect of the longwave emissivity and solar reflectance of building envelopes on energy-saving in buildings in various climates. Sol Energy 85(1):28–37
16. Ozel M (2012) The influence of exterior surface solar absorptivity on thermal characteristics and optimum insulation thickness. Renew Energy 39(1):347–355
17. Liu J, Heidarinejad M, Gracik S, Srebric J (2015) The impact of exterior surface convective heat transfer coefficients on the building energy consumption in urban neighborhoods with different plan area densities. Energy Build 86:449–463
18. Li W, Tian Z, Lu Y, Fu F (2018) Stepwise calibration for residential building thermal performance model using hourly heat consumption data. Energy Build 181:10–25
19. Yang Z, Li X, Bowers CP, Schnier T, Tang K, Yao X (2011) An efficient evolutionary approach to parameter identification in a building thermal model. IEEE Trans Syst Man Cybern Part C (Appl Rev) 42(6):957–969
20. Rabl A (1988) Parameter estimation in buildings: methods for dynamic analysis of measured energy use. Sol Energy Eng 110:52–66
21. Dewson T, Day B, Irving A (1993) Least squares parameter estimation of a reduced order thermal model of an experimental building. Build Environ 28(2):127–137
22. Bouache T, Limam K, Bosschaerts W (2015) New thermal parameters identification approach applied to the thermal renovation of buildings. Energy Build 104:156–164
23. Wang S, Xu X (2006) Parameter estimation of internal thermal mass of building dynamic models using genetic algorithm. Energy Convers Manage 47(13–14):1927–1941
24. Radecki P, Hencey B (2012) Online building thermal parameter estimation via unscented Kalman filtering. In: American control conference (ACC), pp 3056–3062. IEEE

25. Castillo L, Enríquez R, Jiménez M, Heras M (2014) Dynamic integrated method based on regression and averages, applied to estimate the thermal parameters of a room in an occupied office building in Madrid. Energy Build 81:337–362
26. Chen T, Athienitis A (2003) Investigation of practical issues in building thermal parameter estimation. Build Environ 38(8):1027–1038
27. Asadi S, Geem ZW (2015) Sustainable building design: a review on recent metaheuristic methods. In: Recent advances in swarm intelligence and evolutionary computation. Springer, pp 203–223
28. Liang JJ, Qin AK, Suganthan PN, Baskar S (2006) Comprehensive learning particle swarm optimizer for global optimization of multimodal functions. IEEE Trans Evol Comput 10(3):281–295
29. DOE US (ver. 9.1.0, 2019) EnergyPLus: getting started. [Online]. https://energyplus.net/, [Accessed: 03-06-2019]
30. Wani M, Hafiz F, Swain A, Ukil A (2021) Estimating thermal parameters of a commercial building: a meta-heuristic approach. Energy Build 231:110537
31. Lai C (2005) Assessment of side exhaust systems for residential kitchens in Taiwan. Build Serv Eng Res Technol 26(2):157–166
32. Wei P, Zhou B, Tan M, Li F, Lu J, Dong Z, Xu M, Wang G, Xiao Y (2017) Study on thermal comfort under non-uniform thermal environment condition in domestic kitchen. Procedia Eng 205:2041–2048
33. Liu S, Cao Q, Zhao X, Lu Z, Deng Z, Dong J, Lin X, Qing K, Zhang W, Chen Q (2020) Improving indoor air quality and thermal comfort in residential kitchens with a new ventilation system. Build Environ 180:107016
34. Chen Z, Xin J, Liu P (2020) Air quality and thermal comfort analysis of kitchen environment with Cfd simulation and experimental calibration. Build Environ 172:106691
35. Cao C, Gao J, Wu L, Ding X, Zhang X (2017) Ventilation improvement for reducing individual exposure to cooking-generated particles in Chinese residential kitchen. Indoor Built Environ 26(2):226–237
36. ASHRAE (2017) Ashrae standard: thermal environmental conditions for human occupancy. ISSN 1041-2336
37. Wang H, Xiang Z, Wang L, Jing S, Lou S, Tao S, Liu J, Yu M, Li L, Lin L et al (2018) Emissions of volatile organic compounds (VOCs) from cooking and their speciation: a case study for Shanghai with implications for China. Sci Total Environ 621:1300–1309
38. Du B, Gao J, Chen J, Stevanovic S, Ristovski Z, Wang L, Wang L (2017) Particle exposure level and potential health risks of domestic Chinese cooking. Build Environ 123:564–574
39. Lizhen L (2007) Detection and evaluation of air pollution in kitchen. Gas & Heat 27(5):40–42
40. GB-T 18883–2002, (2002) Indoor air quality standard. Standardization Administration of China, Beijing
41. Zhao Y, Li A, Gao R, Tao P, Shen J (2014) Measurement of temperature, relative humidity and concentrations of CO, CO_2 and TVOC during cooking typical Chinese dishes. Energy Build 69:544–561
42. Jinhui Z, Yujiao Z (2015) Residential kitchen smoke pollution field tests and numerical simulation. Contam Control Air-Cond Technol
43. Kuchen E, Fisch MN (2009) Spot monitoring: thermal comfort evaluation in 25 office buildings in winter. Build Environ 44(4):839–847
44. Li Y, Hao L, Wang S, Hou L, Zhang J, Qi J (2009) An experimental study on removal efficiency of bio-particles in an airtight decontamination chamber. Build Environ 44(11):2270–2275
45. Chang L, Zhang X, Cai Y (2016) Experimental determination of air inleakage to pressurized main control room caused by personnel entering. Build Environ 99:142–148
46. Nielsen PV, Restivo A, Whitelaw J (1978) The velocity characteristics of ventilated rooms. J Fluid Eng 100(3):291–298
47. Li Y, Delsante A (1996) Derivation of capture efficiency of kitchen range hoods in a confined space. Build Environ 31(5):461–468

48. Livchak A, Schrock D, Sun Z (2005) The effect of supply air systems on kitchen thermal environment. ASHRAE Trans 111(1):748–754
49. Cheng Y, Niu J, Gao N (2012) Thermal comfort models: a review and numerical investigation. Build Environ 47:13–22
50. Rupp RF, Vásquez NG, Lamberts R (2015) A review of human thermal comfort in the built environment. Energy Build 105:178–205
51. Wang Z, de Dear R, Luo M, Lin B, He Y, Ghahramani A, Zhu Y (2018) Individual difference in thermal comfort: a literature review. Build Environ 138:181–193
52. Enescu D (2017) A review of thermal comfort models and indicators for indoor environments. Renew Sustain Energy Rev 79:1353–1379
53. Yang L, Ye M et al (2014) Cfd simulation research on residential indoor air quality. Sci Total Environ 472:1137–1144
54. Putra JCP (2017) A study of thermal comfort and occupant satisfaction in office room. Procedia Eng 170:240–247
55. Poshtiri AH, Mohabbati SM (2017) Performance analysis of wind catcher integrated with shower cooling system to meet thermal comfort conditions in buildings. J Clean Prod 148:452–466
56. Atthajariyakul S, Leephakpreeda T (2004) Real-time determination of optimal indoor-air condition for thermal comfort, air quality and efficient energy usage. Energy Build 36(7):720–733
57. Leitner AV, Kajtár L (2018) Thermal environment assessment of gas stove surroundings. Slovak J Civ Eng 26(4):26–31
58. Graphics M (2019) Wilsonville, Oregon, 19.47.3 Simcenter Flovent
59. International Electrotechnical Commission (IEC) (2020) Cooking Fume Extractors–Methods for Measuring Performance, Standard BS EN IEC 61591. The British Standard Institution, London, United Kingdom
60. Chiu WY, Yen GG, Juan TK (2016) Minimum Manhattan distance approach to multiple criteria decision making in multi-objective optimization problems. IEEE Trans Evol Comput 20(6):972–985
61. Hafiz F, Swain A, Mendes EMAM, Aguirre LA (2020) Multiobjective evolutionary approach to grey-box identification of buck converter. IEEE Trans Circuits Syst I Regul Pap 67(6):2016–2028. https://doi.org/10.1109/TCSI.2020.2970759
62. Hafiz F, Swain A, Mendes E (2020) Multi-objective evolutionary framework for non-linear system identification: a comprehensive investigation. Neurocomputing 386:257–280
63. Wani M, Swain A, Ukil A, Ploder M, Koole R (2022) Optimizing the overall performance of forced extraction systems: a multi-objective framework. Build Environ 221:109214. https://doi.org/10.1016/j.buildenv.2022.109214, https://www.sciencedirect.com/science/article/pii/S0360132322004504
64. Wani M, Hafiz F, Swain A et al (2025) Balancing energy consumption and thermal comfort in buildings: a multi-criteria framework. Ann Oper Res 346:1841–1867. https://doi.org/10.1007/s10479-023-05747-y
65. Chakraborty N, Mondal A, Mondal S (2019) Multi-objective optimal scheduling framework for HVAC devices in energy-efficient buildings. IEEE Syst J 13(4):4398–4409
66. Yu L, Jiang T, Zou Y (2017) Online energy management for a sustainable smart home with an HVAC load and random occupancy. IEEE Trans Smart Grid 10(2):1646–1659
67. Ma K, Hu G, Spanos CJ (2016) Energy management considering load operations and forecast errors with application to HVAC systems. IEEE Trans Smart Grid 9(2):605–614
68. Moazeni F, Khazaei J, Asrari A (2021) Step towards energy-water smart microgrids; buildings thermal energy and water demand management embedded in economic dispatch. IEEE Trans Smart Grid 12(5):3680–3691
69. Aparicio-Ruiz P, Barbadilla-Martín E, Guadix J, Cortés P (2021) KNN and adaptive comfort applied in decision making for HVAC systems. Ann Oper Res 303:217–231
70. Nyawa S, Gnekpe C, Tchuente D (2023) Transparent machine learning models for predicting decisions to undertake energy retrofits in residential buildings. Ann Oper Res 1–29

71. Rocha P, Siddiqui A, Stadler M (2015) Improving energy efficiency via smart building energy management systems: a comparison with policy measures. Energy Build 88:203–213
72. Manjarres D, Mera A, Perea E, Lejarazu A, Gil-Lopez S (2017) An energy-efficient predictive control for HVAC systems applied to tertiary buildings based on regression techniques. Energy Build 152:409–417
73. Reynolds J, Ahmad MW, Rezgui Y, Hippolyte JL (2019) Operational supply and demand optimisation of a multi-vector district energy system using artificial neural networks and a genetic algorithm. Appl Energy 235:699–713
74. Kang WH, Yoon Y, Lee JH, Song KW, Chae YT, Lee KH (2021) In-situ application of an ANN algorithm for optimized chilled and condenser water temperatures set-point during cooling operation. Energy Build 233:110666
75. Mariano-Hernández D, Hernández-Callejo L, Zorita-Lamadrid A, Duque-Pérez O, García FS (2021) A review of strategies for building energy management system: model predictive control, demand side management, optimization, and fault detect & diagnosis. J Build Eng 33:101692
76. Manic M, Wijayasekara D, Amarasinghe K, Rodriguez-Andina JJ (2016) Building energy management systems: the age of intelligent and adaptive buildings. IEEE Ind Electron Mag 10(1):25–39
77. Tsui KM, Chan SC (2012) Demand response optimization for smart home scheduling under real-time pricing. IEEE Trans Smart Grid 3(4):1812–1821
78. Lu N (2012) An evaluation of the HVAC load potential for providing load balancing service. IEEE Trans Smart Grid 3(3):1263–1270
79. Rezaei E, Dagdougui H (2020) Optimal real-time energy management in apartment building integrating microgrid with multi-zone HVAC control. IEEE Trans Industr Inf 16(11):6848–6856
80. Shakouri H, Kazemi A (2017) Multi-objective cost-load optimization for demand side management of a residential area in smart grids. Sustain Cities Soc 32:171–180
81. Li D, Chiu WY, Sun H, Poor HV (2017) Multiobjective optimization for demand side management program in smart grid. IEEE Trans Industr Inf 14(4):1482–1490
82. Picard D, Kvasnica M, Helsen L (2018) Approximate model predictive building control via machine learning. Appl Energy
83. Jindal A, Kumar N, Rodrigues JJ (2018) A heuristic-based smart HVAC energy management scheme for university buildings. IEEE Trans Industr Inf 14(11):5074–5086
84. Nikovski D, Xu J, Nonaka M (2013) A method for computing optimal set-point schedules for HVAC systems. In: Proceedings of the 11th Rehva world congress Clima. Citeseer
85. Asad HS, Yuen RKK, Huang G (2016) Degree of freedom based set-point reset scheme for HVAC real-time optimization. Energy Build 128:349–359
86. Afram A, Janabi-Sharifi F, Fung AS, Raahemifar K (2017) Artificial neural network based model predictive control and optimization of HVAC systems: a state of the art review and case study of a residential HVAC system. Energy Build 141:96–113
87. Wang Z, Calautit J, Wei S, Tien PW, Xia L (2022) Real-time building heat gains prediction and optimization of HVAC setpoint: an integrated framework. J Build Eng 49:104103
88. Ghahramani A, Jazizadeh F, Becerik-Gerber B (2014) A knowledge based approach for selecting energy-aware and comfort-driven HVAC temperature set points. Energy Build 85:536–548
89. Tien PW, Wei S, Calautit JK, Darkwa J, Wood C (2020) A vision-based deep learning approach for the detection and prediction of occupancy heat emissions for demand-driven control solutions. Energy Build 226:110386
90. Abdelaziz FB, Aouni B, El Fayedh R (2007) Multi-objective stochastic programming for portfolio selection. Eur J Oper Res 177(3):1811–1823
91. Aouni B (2009) Multi-attribute portfolio selection: new perspectives. INFOR: Inf Syst Oper Res 47(1):1–4
92. Jayaraman R, Colapinto C, La Torre D, Malik T (2015) Multi-criteria model for sustainable development using goal programming applied to the United Arab Emirates. Energy Policy 87:447–454

93. Hafiz F, Broekaert J, La Torre D, Swain A (2023) Co-evolution of neural architectures and features for stock market forecasting: a multi-objective decision perspective. Decis Support Syst 174:114015
94. Abdelaziz FB, Mallek RS (2018) Multi-criteria optimal stopping methods applied to the portfolio optimisation problem. Ann Oper Res 267:29–46
95. Razmak J, Aouni B (2015) Decision support system and multi-criteria decision aid: a state of the art and perspectives. J Multi-Criteria Decis Anal 22(1–2):101–117
96. Colapinto C, Jayaraman R, Ben Abdelaziz F, La Torre D (2020) Environmental sustainability and multifaceted development: multi-criteria decision models with applications. Ann Oper Res 293(2):405–432
97. Underwood C (2000) Robust control of HVAC Plant I: modelling. Build Serv Eng Res Technol 21(1):53–61

Chapter 6
EnergyPlus™ Primer

Abstract This chapter serves as a structured primer on EnergyPlus™, tailored for students, researchers, and practitioners aiming to bridge the gap between traditional building energy modeling and modern co-simulation frameworks. The chapter begins with a concise comparison of popular simulation tools—including EnergyPlus™ and IES-VE—framing the rationale for selecting EnergyPlus™ as the computational engine for this work/book. Key modeling workflows are introduced using OpenStudio® and SketchUp, including space assignment, template-driven zoning, and interactive shading visualization. Building upon this foundation, the chapter delves into EnergyPlus™ Input Data Files (IDFs), parametric control mechanisms, and Python-based automation using the console. Advanced users are introduced to the `PythonPlugin` object suite, enabling real-time simulation control via custom Python™ scripts. To contextualize these capabilities, a functional research framework—**`PyE+`**—is introduced as an open-source Python-EnergyPlus co-simulation platform. `PyE+` integrates baseline simulation, Multi-Objective Optimization (via `NSGA-II`), and surrogate modeling using `Machine Learning (ML)` to accelerate analysis and design iteration. Code listings, IDF excerpts, and GUI walkthroughs are provided throughout, enabling readers to apply concepts using real files and replicate control loops, parameter sweeps, and result interpretation. By the end of the chapter, the reader will have gained both theoretical and practical insight into developing intelligent, modular, and scalable building simulation workflows using EnergyPlus™ and Python™.

Keywords Energy modeling · EnergyPlus™ · IES-VE · IDF · OpenStudio® · Plug-in · SketchUp

6.1 Introduction to Energy Modeling

In the evolving landscape of smart buildings and sustainable infrastructure, the role of simulation has expanded far beyond simple energy estimation. Today, simulation engines serve as the digital laboratory where complex interactions between occupants, systems, climate, and controls are explored, tested, and optimized. Among

M. H. Wani et al., *Digital Twins for Building Optimization: Enhancing Energy and Comfort*, Studies in Infrastructure and Control,
https://doi.org/10.1007/978-3-032-23274-8_6

these engines, EnergyPlus™ stands out as a rigorously validated, open-source platform capable of modeling building physics at high spatial and temporal resolution.

This chapter introduces EnergyPlus™ within the context of whole-building energy modeling and intelligent control. It forms a bridge between static modeling approaches and dynamic, data-integrated workflows—where buildings are not merely simulated but actively optimized and monitored through co-simulation and feedback loops. With the advent of advanced controls, optimization algorithms, and real-time data acquisition, EnergyPlus™ has become a foundational tool for Digital Twin development, control-centric modeling, and energy-efficiency research.

Through this chapter, readers will be guided from fundamental modeling workflows in OpenStudio® and IDF configuration to advanced scripting, plugin-based control, and the integration of surrogate modeling via the `PyE+` framework. Practical code listings, examples, and annotated files accompany the discussion, enabling hands-on engagement with EnergyPlus™ as both a simulation engine and a real-time co-simulation environment.

To be precise, this chapter sets the stage for turning buildings into dynamic cyber-physical systems—responsive to their environment, shaped by data, and governed by intelligence.

6.2 Prominent Energy Modeling Software in the Market

Building Energy Modeling Tools

Building Energy Modeling (BEM) tools form the computational backbone for evaluating and simulating energy performance, HVAC operation, internal gains, and thermal comfort within a building. Several tools exist today—ranging from compliance-focused GUI environments to research-grade simulation engines. Among the most prominent are EnergyPlus™ and Integrated Environmental Solutions—Virtual Environment (IES-VE).

EnergyPlus™ is an open-source simulation engine developed by the U.S. Department of Energy. It operates on a text-based input model (IDF) and provides deep access to internal variables, enabling customized simulation and control strategies. IES-VE, by contrast, is a commercial, GUI-rich tool that is widely used by practitioners for visualizing thermal performance and satisfying building regulations through certified workflows.

Why EnergyPlus™? A Comparison with IES-VE

Among the various simulation tools available, **EnergyPlus™** and **IES-VE** stand out as widely adopted platforms for building performance evaluation, Table 6.1 provides a high-level comparison. IES-VE offers a highly visual and integrated GUI for

Table 6.1 Comparison of EnergyPlus™ and IES-VE modeling platforms

Aspect	EnergyPlus™	IES-VE
License Type	Open-source (BSD-style)	Proprietary
Cost	Free	Paid (license required)
Customizability	High (supports custom plugins, IDF edits)	Moderate (some scripting via MacroFlo, APIs limited)
Scripting Support	Excellent (Python, EMS, BCL, JSON)	Limited (visual tools, less code exposure)
Plugin/API Access	Yes (PythonPlugin, ExternalInterface, FMU)	Restricted (custom API access limited to vendors)
GUI and Visualization	Moderate (via OpenStudio, third-party GUIs)	Excellent (integrated 3D GUI, real-time shading)
Compliance/Certification Tools	Limited (manual integration required)	Built-in (LEED, ASHRAE, etc.)
Learning Curve	Steep	Gentler for GUI users
Community & Documentation	Extensive (DOE, Big Ladder, GitHub)	Good (commercial support, user forums)
Use in Research & Academia	Widely used (academic papers, reproducibility)	Less common due to closed architecture

thermal comfort analysis, daylighting, and compliance checks, with drag-and-drop workflow features appealing to architectural designers and consultants. However, its reliance on proprietary components, limited openness for customization, and constrained control over low-level simulation loops can be restrictive for researchers seeking fine-grained or programmatic control.

In contrast, EnergyPlus™ is open-source, script-friendly, and extremely extensible—particularly when paired with Python-based co-simulation workflows or tools like OpenStudio©. Although its learning curve is steeper due to the text-based IDF structure and sparse GUI support, EnergyPlus™ offers unparalleled flexibility and transparency. Crucially, it allows real-time control of zone parameters, actuator states, and internal gains using the `PythonPlugin` interface, enabling deep integration with optimization and Machine Learning (ML) frameworks.

Given the scope of this book and the author's PhD research (detailed in Chap. 5)—which explores intelligent control, co-simulation, and surrogate modeling—EnergyPlus™ was selected as the simulation engine. Its architecture aligns with research-oriented applications, where customization and reproducibility outweigh convenience. The use of EnergyPlus™ also facilitates integration with version-controlled scripts, parametric testing pipelines, and advanced scheduling logic—capabilities foundational to the `PyE+` framework [1] introduced later in this chapter.

EnergyPlus™ Versus IES-VE: A Feature Comparison

6.3 Whole-Building Simulations and EnergyPlus™

Whole-Building Energy Simulation (WBES) refers to the process of modeling and numerically simulating a building's thermal and electrical performance over time, accounting for all relevant systems and interactions: building envelope, occupancy, schedules, internal loads, HVAC systems, lighting, controls, and local weather. This process enables engineers and researchers to evaluate energy consumption, comfort compliance, and design alternatives before physical implementation.

EnergyPlus™, developed by the U.S. Department of Energy, stands as the leading open-source engine for whole building simulation. Unlike simpler tools that perform steady-state or zone-level approximations, EnergyPlus™ employs a detailed, timestep-based calculation engine that captures transient dynamics across thermal zones, HVAC loops, and control systems. It supports custom weather data (e.g., TMY3, EPW), multiple simulation periods, high-resolution schedules, and modular plant loop configurations.

A core strength of EnergyPlus™ lies in its ability to simulate the interaction between different subsystems—lighting schedules affecting cooling loads, or infiltration driven by occupancy and wind pressure, for example. The platform's Input-Data-File (IDF) structure allows for granular control of materials, schedules, systems, and geometry. Furthermore, EnergyPlus™ supports a suite of advanced capabilities including demand-controlled ventilation, daylight harvesting, thermal storage, radiant slabs, and mixed-mode ventilation, making it suitable for both code compliance and advanced research modeling.

From a pedagogical perspective, EnergyPlus™ challenges students and researchers to engage deeply with the assumptions, parameters, and structures underpinning whole building energy models. Unlike GUI-based tools where models are abstracted behind visual interfaces, EnergyPlus™ requires explicit definition of object relationships, schedule logic, and control loops. This transparency makes it an ideal engine for experimentation, validation, and control research—particularly when combined with scripting environments such as Python™ or platforms like OpenStudio®.

Within the scope of this book, EnergyPlus™ provides the simulation foundation for intelligent building control strategies explored in other chapters of this book. Its timestep engine is leveraged not only for evaluating baseline building performance, but also for dynamic reconfiguration through plugin scripting, co-simulation, and optimization. The open and reproducible nature of EnergyPlus™ aligns well with the broader research goals of transparency, scalability, and extensibility.

In subsequent sections, we demonstrate how EnergyPlus™ can be extended via Python™ scripts, parametric workflows, and the `PythonPlugin` interface. The reader will be guided through practical examples—including the `PyE+` framework—which show how EnergyPlus™ can evolve from a passive simulation tool into an active, intelligent component of a cyber-physical control loop.

6.3.1 *Workflow of EnergyPlus™ Modeling*

The EnergyPlus™ workflow includes several software tools, as summarized in Table 6.2 [2, 3].

The installation procedure for each application, summarized in Table 6.2, is straightforward and can be followed from the respective developers' websites or sources. It is important to note that the installation sequence should be as follows: SketchUp Pro > OpenStudio® Application > EnergyPlus™ > OpenStudio® SketchUp Plug-in. Modelers must ensure that they check the compatibility matrix of these software before downloading and installing them [4].

Figure 6.1 illustrates a simple block diagram that demonstrates the workflow. In particular, the process of modeling a building envelope and an HVAC system in the EnergyPlus™ platform involves the following key steps:

- **Drawing Geometry**: The initial step in EnergyPlus™ involves defining the building geometry, encompassing walls, floors, roofs, and windows. This can be achieved either by creating a 3D representation of the building from scratch using SketchUp (further detailed in Sect. 6.3.2) or by importing the AutoCAD DXF of

Table 6.2 Software tools which form a collective part of EnergyPlus™ workflow [2, 3]

Software	Developer	Description
SketchUp Pro	Trimble® Inc.	A 3D modeling software which allows modelers to create and manipulate 3D models of buildings, landscapes, furniture, and other objects
OpenStudio® Application	NREL, ANL, LBNL, ORNL, and PNNL†	A fully featured Graphical User Interface (GUI) to EnergyPlus™ models including envelope, loads, schedules, and HVAC system
EnergyPlus™	NREL, and various DOE† National Laboratories, academic institutions and private firms	A comprehensive building energy simulation program utilized by engineers, architects, and researchers to model energy consumption—for heating, cooling, ventilation, lighting, and plug and process loads—as well as water use in buildings
OpenStudio® SketchUp Plug-in	OpenStudio® Coalition	The SketchUp Plug-in is an extension to the SketchUp 3D modeling tool, which enables users to efficiently generate the geometry required for EnergyPlus™

† Please refer to the Acronyms section for these abbreviations

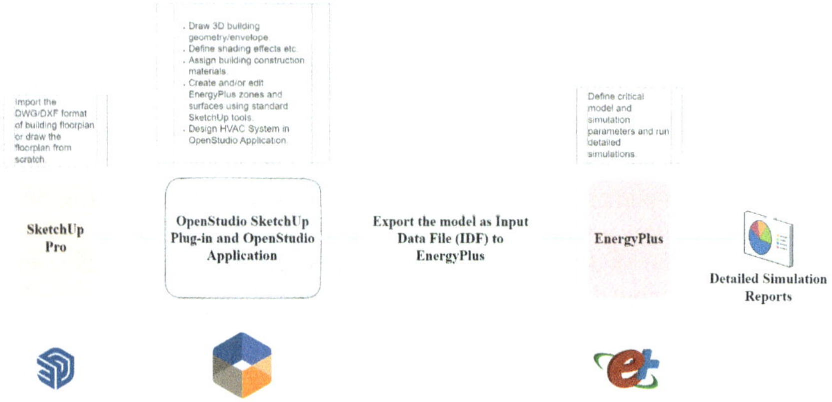

Fig. 6.1 EnergyPlus™ Workflow

the building floor plan into SketchUp and constructing the building envelope from there.

- **Surface Matching**: The next step involves matching the surfaces (walls, roofs, etc.) to their corresponding thermal zones. Each thermal zone represents a distinct area within the building. This step is crucial for setting the boundary conditions for each space within the building model and will be explained in more detail in Sect. 6.4.
- **Assigning Spaces**: Upon completion of surface matching, spaces (such as rooms and corridors) can be associated with their respective thermal zones. Subsequently, internal loads—including occupancy, lighting, and equipment—must be specified for each space. This process can be executed either manually or through the utilization of templates available in the OpenStudio® plug-in, as elaborated in Sect. 6.4.3.
- **Assigning Constructions**: Following the assignment of spaces, the building envelope must be defined by assigning constructions (insulation, glazing, etc.) to surfaces. These constructions impact heat transfer, as each construction material comes with pre-assigned thermal parameters, which can be modified as needed.
- **Assigning Thermal Zones**: Grouping spaces into thermal zones based on their heating and cooling requirements is essential. Zones may share similar conditions, such as offices located on the same floor.
- **HVAC System Modeling in OpenStudio®**: HVAC system modeling and design are typically performed using the OpenStudio® application, as elaborated in Sect. 6.4. This tool allows modelers to specify equipment, such as chillers, boilers, and air handlers, and connect them to thermal zones.
- **Exporting the IDF to EnergyPlus™**: Upon completion of the preceding workflow steps, the final task for the modeler is to export the input data, including geometry, constructions, and HVAC system(s), to an EnergyPlus™ Input Data File (IDF). EnergyPlus™ will then simulate the building's energy performance based on this

file. Additionally, minor details and parameters can be refined within EnergyPlus™ using the IDF Editor.

Note that, the above workflow simplifies the process and there may be additional details and considerations.

6.3.2 *Modeling Building Geometry and Envelope Using SketchUp Pro*

While the OpenStudio® plug-in is active, modeling is conducted within SketchUp. However, the information is subsequently converted into text files (or IDFs) by OpenStudio®, which is implemented in Ruby, an interpreted high-level general-purpose programming language. This process is abstracted from the modeler, who does not need to concern themselves with the underlying software development unless they choose to do so [5].

6.3.2.1 Copying Levels and Surfaces During Modeling

One of the most common techniques which modelers follow while modeling a building envelope, especially those comprising of multiple levels, is copying the levels. This has tendency to result in errors and/or warnings once the "Surface Matching" script is run in OpenStudio®. In order to avoid this, there are several steps one can follow (which is based on author's personal experience with OpenStudio®-SketchUp plug-in):

i Just select the level(s) you want to copy (there shall be no windows or any sub-surfaces in such levels).
ii Select the "Move" command from palette and press "Ctrl" key from keyboard to duplicate/replicate/copy levels.
iii After you are done with copying the selected levels, you are safe to add sub-surfaces to the building envelope.
iv Here, it is worth adding that, always run the "Surface Matching" script before adding the sub-surfaces (windows, etc.)

Furthermore, regularly save your.osm model at each step and consistently check the warning/error log (indicated by the orange triangle with an exclamation mark) to ensure everything is proceeding correctly. This practice allows you to revert to a previously saved .osm file if the current one encounters errors.

6.3.2.2 Adjusting the Dimensions of Levels and Spaces

Another instance where modelers usually run into issues is while adjusting the dimensions of the building levels and/or spaces. In author's experience, this can be achieved using one of the following three commands within SketchUp:

I **Move Command (Press "M"):** This command facilitates the translation of edges, faces, or vertices, thereby altering the spatial dimensions.
II **Scale Command (Press "S"):** This command is employed to resize the entire space or specific components thereof. For uniform scaling, one should click and drag the corner handle of the selected space. To scale in a single direction, click and drag the center handle along the desired axis.
III **Push/Pull Command (Press "P"):** This command allows for the extension or contraction of faces within the space. By clicking on a face and dragging, one can adjust it to the required dimension.

For any of the above methods, the modeler has to ensure that the following two steps are fulfilled first:

i **Selecting the Space:** Use the select tool within SketchUp to click on the space modeler wants to modify. This highlights the space in the model.
ii **Entering the Edit mode:** Double-click the selected space. This allows modeler to manipulate the geometry of the space.

For precise adjustments, modelers can enter specific dimensions. For example, when using the *Move* command, initiate the movement of the element, input the exact distance, and press "Enter". Similarly, with the *Push/Pull* command, start the action, specify the desired dimension change, and press "Enter". Based on empirical observations, the *Push/Pull* command is generally more reliable, as the other two commands are prone to errors.

Modelers should note that it is preferable to perform space dimension adjustments before adding sub-surfaces such as *windows* and *doors*. Once the space dimensions are set, modelers can safely proceed with adding or modeling sub-surfaces and/or fenestrations.

6.3.3 OpenStudio®: A Graphical User Interface (GUI) for EnergyPlus™

OpenStudio® serves as an intuitive graphical interface that significantly enhances the accessibility and usability of EnergyPlus™—particularly beneficial for those new to energy modeling or those who prefer visual interaction over text-based configuration. While EnergyPlus™ excels in detailed and granular modeling through its robust IDF (Input Data File) framework, OpenStudio® bridges the complexity of direct code manipulation, offering modelers a more user-friendly pathway to manage and visualize complex building simulation parameters.

The real advantage of using OpenStudio® lies in its ability to replicate most EnergyPlus™ functionalities through an accessible, interactive environment. Modelers can conveniently manage geometry, materials, schedules, internal loads, HVAC systems, and weather data within an organized visual framework. However, it is crucial to recognize that while OpenStudio® streamlines workflows, the underlying IDF files remain richer and more detailed, affording deeper control to advanced users willing to engage directly with the text-based code. Thus, OpenStudio® acts not only as an introductory gateway but also as a complementary tool that seamlessly integrates visual modeling with the precision and extensibility of EnergyPlus™ IDF structures.

For students, researchers, and practitioners, OpenStudio® dramatically reduces the initial barrier to effective building modeling. It allows rapid prototyping and scenario exploration, facilitating quick visual assessments and iterative refinements. Yet, the interface's simplicity does not detract from the potential complexity of the resulting models. By maintaining direct linkage to EnergyPlus™ simulations, OpenStudio® ensures that models developed within its GUI retain their compatibility with more sophisticated scripting and co-simulation approaches detailed later in this chapter, thereby fully supporting the book's overarching goal of intelligent, adaptive, and computationally efficient building energy management.

6.4 OpenStudio® Application (Plugin)

The OpenStudio® SketchUp Plugin serves as a graphical modeling interface layered on top of SketchUp, allowing users to create and edit building geometry specifically for EnergyPlus™ simulations. It integrates seamlessly with the OpenStudio® Application and extends SketchUp's modeling environment with building energy modeling capabilities.

Installation Guidelines:

To install and correctly configure the plugin:

1. First, download and install the `OpenStudio® Application` (not the SDK) from the OpenStudio Coalition GitHub page.
2. Next, install `SketchUp Pro` (ensure compatibility with your OpenStudio® version).
3. From within SketchUp, install the `OpenStudio® SketchUp Plugin` via the Extension Warehouse or by using the provided `.rbz` file.
4. Upon first launch, SketchUp will prompt for the `OpenStudio® Root Directory`—this must be set to the path where the OpenStudio® Application (e.g., `C:\openstudioapplication-1.6.0`) is installed.

5. Ensure that Ruby bindings and internal files such as `openstudio_modeleditor.rb` and `openstudio.so` are correctly recognized; these are bundled with the Application (not the SDK).

Correct installation ensures the plugin can fully communicate with EnergyPlus™ through the OpenStudio® Application backend, enabling successful model editing, surface matching, zone assignment, and HVAC system integration. For scripting and AI algorithm interaction, integration with the EnergyPlus™ Python™ API is recommended.

Note

Although a detailed exposition of the capabilities of the OpenStudio® Application exceeds the scope of this book, the primary intent is to inform energy modelers, students and EnergyPlus™ practitioners of the critical insights that proved beneficial during author's research.

6.4.1 Modeling Multiple Buildings in One .osm File

OpenStudio® allows users to model multiple spatially distinct structures—such as separate buildings, annexes, or pavilions—within a single `.osm file`. This functionality is particularly useful for comparative analysis, microclimate interaction studies, or shared system modeling within a campus or compound setting. However, it is essential to understand that OpenStudio® and EnergyPlus™ define the entire model as a single Building object, even if multiple physical structures exist in the file. As such, all geometry and HVAC equipment are technically associated with one building entity at the simulation level.

To model multiple buildings in one `.osm file`, users must treat each physical structure as a group of uniquely named spaces and thermal zones. These can be distinguished using naming conventions such as Building-1: ZoneA and Building-2: ZoneB. This nomenclature not only helps in organizing the model but also supports downstream processes such as sub-metering, performance comparison, or reporting.

Although EnergyPlus™ does not natively support multiple Building objects within a single IDF, users can simulate distinct buildings as thermal pavilions (or wings) within the same IDF. The OpenStudio® interface supports this by allowing the construction of separate, unconnected geometry sets in SketchUp, each representing an individual building. These can be individually zoned and connected—or deliberately left unconnected—to model buildings that operate independently or share HVAC resources like chillers or boilers. To obtain additional assistance and insights on the subject of modeling multiple buildings in an `.osm file`, readers are encouraged to consult the discussion available on author's `Unmet Hours` exchange.[1]

[1] Unmet Hours: How can I define new buildings in a SketchUp/OpenStudio Model?.

When more granular analysis is required, such as monitoring the performance of each building separately, users can introduce sub-metering strategies. While OpenStudio® does not offer a GUI-based meter-assignment per space or zone by default, it can be achieved through scripting (using the OpenStudio® SDK or Measures) or post-processing via the EnergyPlus™ IDF file. Sub-metering each structure's energy consumption facilitates detailed reporting and comparison under different operating scenarios.

Furthermore, from a microclimatic standpoint, keeping multiple buildings in the same file enables shading and view factor interactions—allowing the simulation to capture solar gains, shadowing, and radiative exchanges between buildings. This can be essential in studies where inter-building effects influence thermal performance.

For true urban-scale simulation or when separate Building objects are required, platforms like `URBANopt™` [6] or co-simulation frameworks using multiple IDFs and external orchestration (e.g., Python™) are recommended. These allow each building to be modeled with its own simulation context while enabling interaction at the level of shared systems or environmental feedbacks.

In summary, while OpenStudio® does not allow the formal creation of multiple Building objects in one model, with thoughtful structuring and naming, users can effectively simulate and study the behavior of multiple buildings within a unified .osm framework. This approach offers practical benefits for early-stage analysis, prototyping, and comparative performance assessments.

6.4.2 Spatial Templates, Shadow Casting, and Shading Configuration in OpenStudio®

OpenStudio®, when used as a front-end for EnergyPlus™, provides a user-friendly interface for defining building geometry, assigning space properties, and visualizing solar exposure. In particular, proper naming and tagging of spaces is essential during zone layout and envelope definition, as these identifiers carry over to downstream workflows, including IDF-based scripting and PythonPlugin-driven control logic.

Figure 6.2 shows multiple buildings modeled within one OpenStudio® model—inclusive of the shadows casted by the buildings. While OpenStudio® provides multiple ways to assign names to spaces, it is essential to understand the distinction between SketchUp's native controls and those introduced by the OpenStudio® plugin. Specifically, space names can be assigned via either:

- The **Entity Info** panel—a native SketchUp property window, or
- The **OS Inspector window**—which reflects and modifies OpenStudio-specific attributes.

Best practice recommends using the *OS Inspector* window, which updates the energy model consistently and ensures that the space naming conforms to the required format for EnergyPlus™ IDF export. It also synchronizes naming with the SketchUp-side visual representation (Entity Info panel), making the editing workflow seamless.

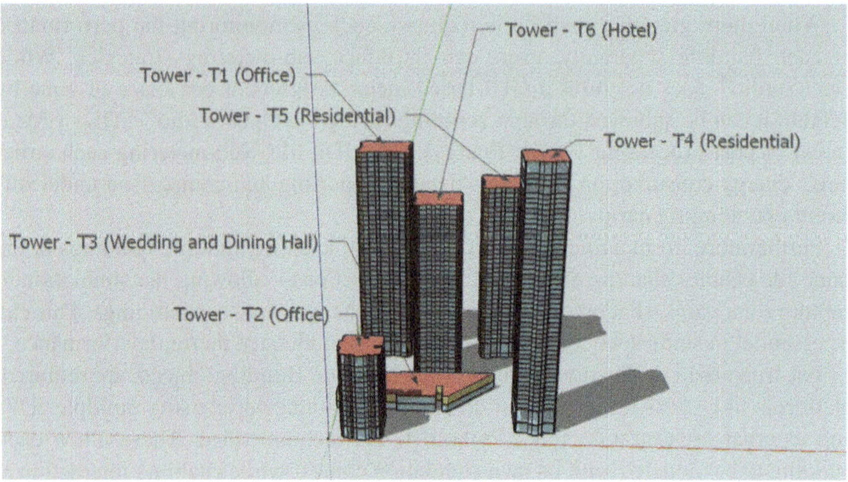

Fig. 6.2 Multiple Buildings Modeled within one OpenStudio® Model

Beyond space assignment, OpenStudio® facilitates visualization of solar gains and shading using its native GUI. Shadow casting and diurnal shading patterns can be toggled via the "`Shadows`" option and enhanced further through the "`Show Shadow Settings Dialog`" tool. This allows users to visualize solar geometry impacts in real time, similar to daylight simulation tools like IES-VE.

Note
It is imperative for modelers to ensure that the shadows of building and environmental objects are accurately represented in simulations. This requires specifying the "`Solar Distribution`" field of the "`Building`" object, along with defining the "`ShadowCalculation`" object within the IDF in EnergyPlus™.

Importantly, OpenStudio® supports assignment of *space types and thermal templates*, further detailed in Sect. 6.4.3, which allow predefinition of Lighting Power Density (LPD), Plug Load Density (EPD), occupancy schedules, and internal gains. This not only improves workflow efficiency but also enforces standardization across similar zones. These templates, once assigned, propagate relevant parameters through the EnergyPlus object tree (e.g., `Lights`, `People`, `ElectricEquipment`, and `Schedule:Compact` objects), ensuring consistent simulation behavior.

Finally, it's worth noting that some elements in the model are maintained entirely within SketchUp, while others are OpenStudio® plugin-specific. Hence, users are advised to always save both the active `.skp` (SketchUp) file and the linked `.osm` (OpenStudio®) file in tandem (.skp first) before exporting to EnergyPlus™. Likewise to open the model correctly, the modeler is advised to always load the active `.skp`

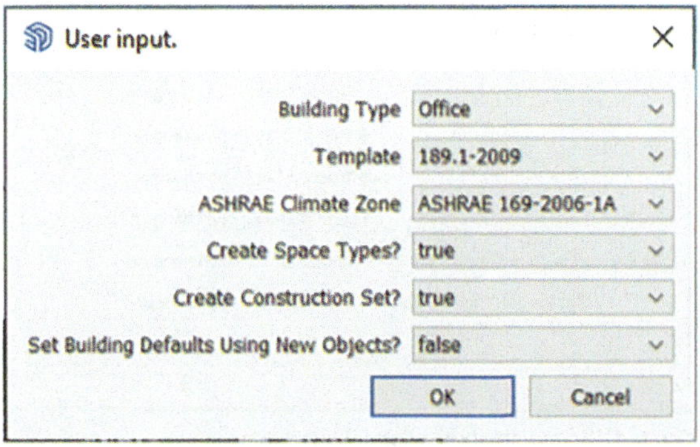

Fig. 6.3 Space type and construction wizard

file first, click on "`Zoom Extents`" option (or Shift+Z), and then open the corresponding `.osm` file. This practice prevents misalignment between geometry and energy model data. For further insights and guidance on modeling-related queries, feel free to visit and follow author's page on `Unmet Hours`[2]—The author has engaged with experts there to keep track and record valuable information.

The ability to preview and interactively explore shadow movement throughout the day provides meaningful insight into envelope design and passive strategies—and, when combined with scripting and plugin control (as discussed in Sect. 6.6), enables a complete simulation-to-automation pipeline.

6.4.3 Assigning Spaces and Thermal Zones to Building(s)

Defining space properties and thermal zones accurately are crucial components of EnergyPlus™ modeling. If these elements are not set correctly, the model will not function correctly. In fact, EnergyPlus™ operates by segmenting the building model into individual thermal zones. EnergyPlus™ performs computations for each zone and then aggregates the results to determine the overall performance of the model. Therefore, in general, the workflow (see Sect. 6.3.1) begins with the allocation of space characteristics, which means specifying the type and properties of the space, and is followed by the segregation of these spaces into *thermal zones*.

The process of assigning spaces to a building model involves the following steps:

- **Step I—Adding/Generating Space Types from Templates:** The initial step involves adding all the required space types to the model. This is accomplished through the "Space Type and Construction Set Wizard" found under the "Extensions" tab in SketchUp. The exact pathway in SketchUp is: *Extensions > OpenStudio User Scripts > On Demand Template Generators > Space Type and Construction Set Wizard*. Upon accessing this option, a window will open (see Fig. 6.3),

[2] https://unmethours.com/users/78306/mhw/.

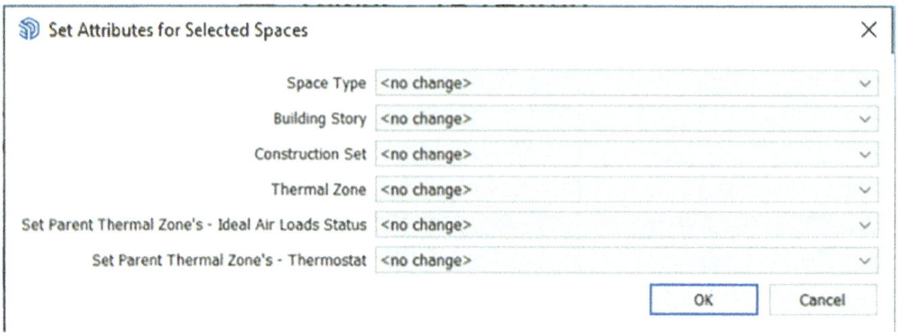

Fig. 6.4 Space attributes for selected spaces

allowing the modeler to select various parameters such as building type (e.g., Midrise Apartment, Large Hotel, Office, Small Hotel, etc.) and climate zone. Once a building type is added to the model from the templates, SketchUp will notify the user that new "space types" and "construction sets" have been incorporated into the model.

- **Step II—Applying Attributes to the Spaces:**
  Following the completion of Step I, the next stage involves applying space attributes to the model. This is done via the "Set Attributes for Selected Spaces" option accessible on the OpenStudio® SketchUp Plug-in toolbar. Upon selection, a dialog box will appear (refer to Fig. 6.4), allowing the modeler to specify various space attributes. These attributes encompass

  – *Space Type*: Previously added from templates,
  – *Building Story*: Referring to the floor or level number, and
  – *Construction Set*: The construction template corresponding to the selected space types.

  It is important to note that in order to explicitly assign `thermal zones` within OpenStudio®, the user must first select the relevant spaces directly in the SketchUp workspace. Upon highlighting the target spaces, the user can use the `Set Attributes for Selected Spaces` dialog box to assign thermal zones. This can be either an existing/defined zone or a new one by selecting the "`New Thermal Zone`" option from the `Thermal Zone` dropdown in the dialog box. Completing the process by clicking "OK" firmly establishes the linkage between spatial geometry and thermal zones, a crucial connection for organizing HVAC zoning and ensuring the integrity of energy simulation workflows.

It should be noted that "Step I" can be repeated multiple times to incorporate space types from different templates into the model. A modeler also has the option to create their own template from scratch. It is advisable to thoroughly review and adjust the aspects of the space types added from templates, such as schedules and loads, as these can vary from one model to another.

6.4.4 *OpenStudio® Rendering Toolbars*

The OpenStudio® plugin for SketchUp not only enables geometry creation and zone tagging, but also offers a suite of rendering and diagnostic tools that assist with model validation, troubleshooting, and presentation. These tools—available via the OpenStudio® Toolbars—allow users to visualize various thermal and spatial attributes of the model without needing to launch a full simulation. Users can render the building byspace type, thermal zone, surface boundary condition, surface type (wall, floor, roof, window), and construction assignment. These rendering modes apply color overlays to the SketchUp model, making it easy to detect misassigned or missing data. For example, visualizing by surface type can help confirm whether all fenestrations have been properly modeled, or whether interior partitions are correctly distinguished from external walls.

These tools are especially valuable during early model development as they provide a fast and visual method to audit data inputs. When combined with the OS Inspector and Entity Info panel (discussed earlier), they form a powerful triad for ensuring model fidelity.

Figure 6.5 demonstrates examples of OpenStudio rendering by `surface type`, `boundary condition`, `construction` and so on—highlighting how such overlays can reveal issues such as "unzoned" spaces, incorrectly assigned/tagged surfaces, or duplicate constructions.

Used iteratively, the rendering toolbar becomes more than a visual aid: It becomes a diagnostic canvas to catch issues before they propagate into simulation errors or skewed results.

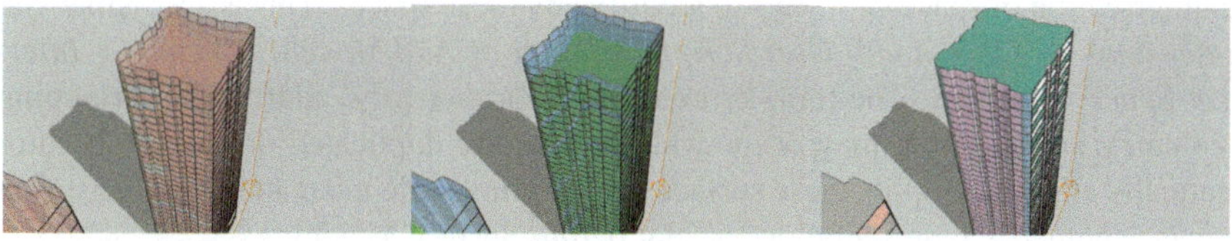

(a) Rendering by surface type (b) Rendering by boundary condition (c) Rendering by construction

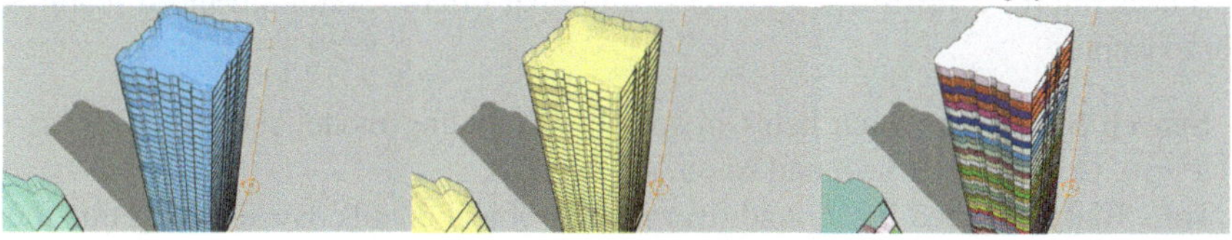

(d) Rendering by space type (e) Rendering by thermal zone (f) Rendering by building story

Fig. 6.5 OpenStudio®: Visualization based on different spatial and construction attributes

Note

In the domain of HVAC system design utilizing OpenStudio® and subsequently EnergyPlus™, it is imperative to initially assess attributes such as rendering by surface, space and boundaries. These components are essential in ensuring the efficiency and efficacy of the design process.

6.4.5 OpenStudio® Tools

The OpenStudio® application acts as a graphical user interface that connects an intuitive modeling environment with the powerful simulation features of EnergyPlus™. It excels with its comprehensive suite of diagnostic and preparatory tools that transform simple geometric models into a ready-to-simulate workflow. These built-in tools guarantee precision and consistency, allowing modelers to create detailed and accurate representations fit for advanced co-simulation environments and Digital Twin applications. A key tool in this suite is `Surface Matching`, see Fig. 6.7, crucial for model accuracy. It ensures that connected spaces have aligned adjacent surfaces, enabling EnergyPlus™ to accurately calculate heat transfer across shared boundaries. Neglecting Surface Matching can result in incorrectly designated internal partitions, affecting simulation results. It is recommended to perform Surface Matching after finalizing building geometry but before adding sub-surfaces like windows. This tool is found in the '`OpenStudio Tools`' toolbar, and users are advised to first perform an intersection of space geometry to identify any loose spaces. This is accessible via: *Extensions → OpenStudio User Scripts → Alter or Add Model Elements → Intersect Space Geometry*. The tool, `Intersect Geometry`, addresses overlapping geometric surfaces that may occur when spaces are duplicated or altered. It automatically aligns and splits these surfaces, preventing node mismatches and ensuring precise geometric representation, crucial before Surface Matching.

Furthermore, `OpenStudio® tools` toolbar offers several advanced features, for example:

- **Search Surfaces:** which helps in searching building model surfaces by `Name`, `Type`, `Construction` and so on, see Fig. 6.8.
- **Info Tool:** designed for a rapid overview of object details. By selecting this tool and hovering over the model, users can swiftly access information such as surface type, space name, coordinates, and more.
- **Show Errors and Warnings:** Modelers are recommended to utilize this OpenStudio® tool regularly as a 'sanity check' on the building model, prior to commencing HVAC modeling.

These tools support broader aims in intelligent modeling and co-simulation as described in this book. By preparing accurate, detailed, and thermodynamically consistent models, OpenStudio® sets a foundational framework for implementing control strategies using PythonPlugin, Multi-Objective Optimization (MOO), and Digital Twin methodologies, as illustrated by the `PyE+` framework introduced later. Ultimately, mastering OpenStudio® tools not only ensures successful simulations but also facilitates meaningful analysis, bridging the gap between standalone building models and integrated, intelligence-driven simulation workflows.

6.4.6 Modeling the HVAC System

One of the critical capabilities of the `OpenStudio® Application` is its robust and modular environment for modeling HVAC systems. Beyond just assigning loads and schedules, OpenStudio® enables users to define, configure, and visualize complete HVAC systems with a level of detail that integrates seamlessly with the thermal zone architecture established in previous steps. This includes the ability to model central plant components—such as chillers, boilers, cooling towers, and pumps—as well as terminal units like fan coil units, VAV boxes, and packaged rooftop units.

The `OpenStudio® Application` can be accessed via *Extensions → OpenStudio → Launch OpenStudio*, or alternatively, users may directly open it from `OpenStudio® tools` option present within the toolbar. Once the application opens, the HVAC modeling in OpenStudio® is performed through the "`HVAC Systems`" tab, where users can access a drag-and-drop component library. Figure 6.6 illustrates a sample layout that users may observe during the process of HVAC system modeling in `OpenStudio® Application`. The intricacy and specifics of the layout are expected to differ across various models, reflecting the unique requirements and inputs of individual users. This interface facilitates the intuitive assembly of HVAC loops: both plant and air-side. The plant loop editor allows for the creation of chilled water, hot water, or condenser loops, where primary equipment like chillers and boilers can be placed in the supply side and connected to thermal loads on the demand side. Similarly, the air loop editor enables users to construct Air Handling Units (AHUs), add supply and return fans, coils, and outdoor air systems, and link these systems directly to thermal zones via zone equipment.

Connections between HVAC components and thermal zones are essential. Once thermal zones are created and defined (as discussed in Sect. 6.4.3), HVAC systems can be assigned to them either manually or automatically using templates. The system-level modeling logic ensures that thermal zones are effectively served by the correct loop, enabling detailed simulation of heating, cooling, and ventilation energy demands.

OpenStudio® also supports the assignment of performance curves, control strategies, setpoints, and schedules associated with HVAC operation, allowing for detailed energy modeling and operational tuning. These capabilities are particularly useful when performing comparative analysis, load balancing, or energy optimization simulations.

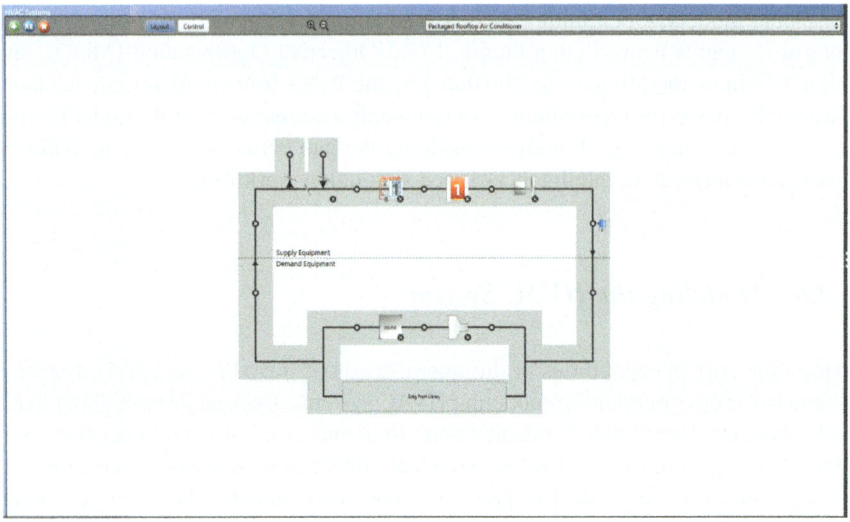

Fig. 6.6 Typical layout showing a *Packaged Rooftop Unit* as viewed from the OpenStudio® Application within `HVAC Systems` tab. The image, viewed in a clockwise direction, includes an HVAC Air Loop, nodes, a DX cooling coil, a heating coil, a CAV fan, and an air terminal duct, which can be connected to specific thermal zones

Ultimately, the HVAC modeling capabilities of OpenStudio® provide a bridge between geometry-based thermal zoning and the complex thermal dynamics simulated by EnergyPlus™, ensuring that energy consumption, equipment performance, and indoor comfort metrics can all be realistically evaluated in a unified environment.

6.4.7 Exporting the Input Data File (IDF) to EnergyPlus™

Once a building model has been constructed and configured using the OpenStudio® interface—complete with geometry, thermal zones, internal loads, and HVAC systems—it can be exported as an EnergyPlus™ Input Data File (IDF) for advanced editing or direct simulation.

EnergyPlus™ IDF is an ASCII file containing the data describing the building and HVAC system to be simulated. It serves as the primary input format for EnergyPlus™, written in a structured text-based format that lists objects and their parameters in a hierarchical fashion. This format makes the file both human-readable and machine-parseable, allowing advanced users to perform manual modifications or automate changes through scripting.

To export the IDF from the OpenStudio® Application, navigate to *Extensions → OpenStudio → Export → Export EnergyPlus IDF*. This generates a complete IDF representation of the model, capturing all OpenStudio-configured data in

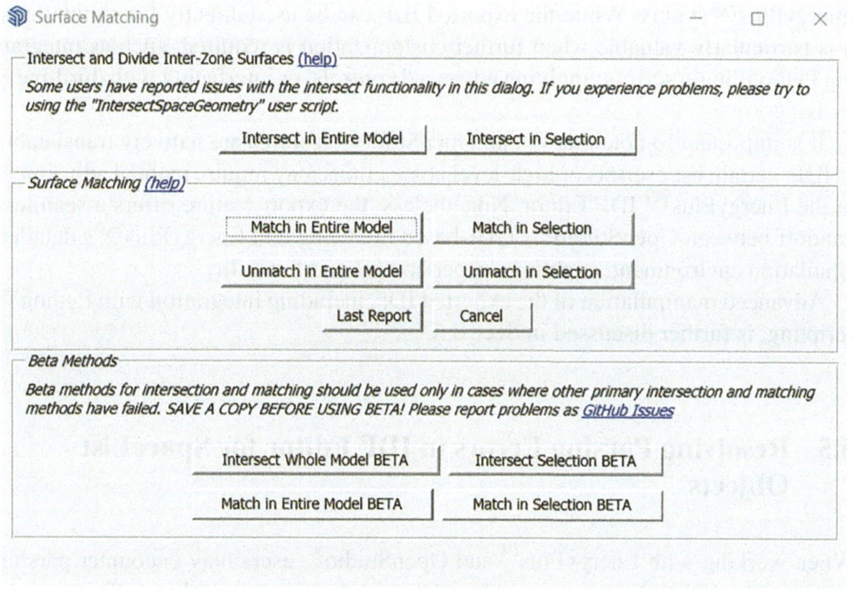

Fig. 6.7 Surface matching tool interface in OpenStudio®

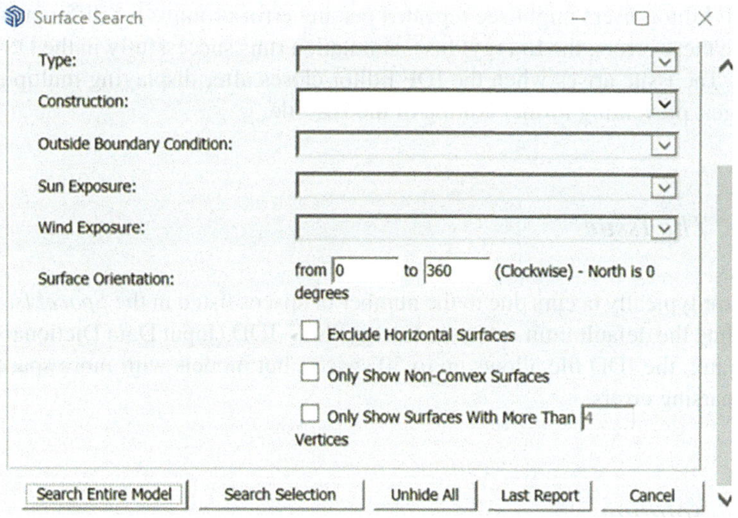

Fig. 6.8 Surface search tool in OpenStudio®

EnergyPlus™ syntax. While the exported IDF can be used directly for simulations, it is particularly valuable when further customization is required, such as integrating PythonPlugin scripts, applying advanced controls, or interfacing with third-party tools.

It is important to note that not all OpenStudio® features are natively translatable to IDF; certain user scripts or high-level abstractions may require manual adjustment in the EnergyPlus™ IDF Editor. Nonetheless, the export feature offers a seamless handoff between OpenStudio®'s GUI-based modeling and EnergyPlus™'s detailed simulation environment, enabling deeper control and flexibility.

Advanced manipulation of the exported IDF, including integration with Python™ scripting, is further discussed in Sect. 6.6.

6.5 Resolving Parsing Errors in IDF Editor for SpaceList Objects

When working with EnergyPlus™ and OpenStudio®, users may encounter parsing errors in the IDF Editor, particularly related to the *SpaceList* object. This section outlines a common issue and provides a step-by-step solution.

After exporting an OpenStudio® (OS) model as an IDF file and opening it in the IDF Editor, users might see repeated parsing error prompts for different spaces. Despite these errors, the EnergyPlus™ simulation runs successfully in the OS application. The issue arises when the IDF Editor closes after displaying multiple error messages, preventing further editing of the IDF file.

6.5.1 *The Issue*

The error typically occurs due to the number of spaces listed in the *SpaceList* object exceeding the default limit set in the EnergyPlus™ IDD (Input Data Dictionary) file. By default, the IDD file allows up to 50 spaces, but models with more spaces will cause parsing errors.

6.5.2 *Solution*

To resolve this issue, follow these steps:

- **Open the IDD File:** Locate the EnergyPlus™ IDD file and open it in a text editor like Notepad.
- **Find the SpaceList Object:** Search for the *SpaceList* object definition. It will look something like this:

```
A1 , \field Name
     \note Name of the SpaceList.
     \note This name must be unique across Zone, Space, ZoneList, and SpaceList names.
     \required-field
     \type alpha
     \reference SpaceListNames
     \reference SpaceAndSpaceListNames
     \reference ZoneAndZoneListAndSpaceAndSpaceListNames
A2 , \field Space 1 Name
     \begin-extensible
     \required-field
     \type object-list
     \object-list SpaceNames
...
A50; \field Space 49 Name
     \type object-list
     \object-list SpaceNames
```

- **Extend the SpaceList Object:** Remove the semicolon (;) after *A50* and replace it with a comma (,). Add additional fields to accommodate the number of spaces in your model. For example, if your model has 87 spaces, extend the list as follows:

```
A50, \field Space 49 Name
     \type object-list
     \object-list SpaceNames
A51, \field Space 50 Name
     \type object-list
     \object-list SpaceNames
...
A88; \field Space 87 Name
     \type object-list
     \object-list SpaceNames
```

- **Save the IDD File:** Save the changes to the IDD file and close the text editor.
- **Open the IDF File in the IDF Editor:** The parsing errors should no longer occur, allowing you to edit the file as needed.

By extending the *SpaceList* object in the EnergyPlus™ IDD file, users can resolve parsing errors in the IDF Editor. This solution ensures that models with a large number of spaces can be edited without issues, facilitating smoother workflows in building energy modeling.

6.6 Co-simulating EnergyPlus™ and Python™

6.6.1 Example-1: Calling EnergyPlus™ IDF from a Python™ Script

EnergyPlus™ is a console-based simulation engine that reads input from text-based files such as IDF and EPW, and writes output in a similarly structured manner. This design makes it especially well-suited for automation via scripting languages, particularly Python™, which excels at executing command-line processes and managing text I/O.

Listing 6.1 demonstrates how a complete EnergyPlus™ simulation can be programmatically triggered from Python™. The script first checks the availability of the EnergyPlus™ executable, prepares simulation directories, and constructs a command-line string to invoke EnergyPlus™ with appropriate arguments (e.g., IDF path, weather file, and output directory). After simulation execution, the script proceeds to parse the resulting CSV output using `pandas` and visualize air temperature data. This integration of subprocess control, output parsing, and visualization illustrates the foundational workflow of Python-EnergyPlus interaction.

Listing 6.2 adds an auxiliary utility function that is invoked from the main script. This modularization reflects good software design practice and encourages reusability and script chaining - a pattern commonly adopted in more complex co-simulation pipelines, such as in `PyE+`, discussed in Sect. 6.6.2.

Listing 6.3 displays the actual console output captured during execution. The listing shows EnergyPlus™ being initialized, input files being processed, and the simulation completing successfully. Notably, this log also captures the Python™ subprocess call itself, confirming that Python™ can be used as the master controller for launching, monitoring, and post-processing EnergyPlus™ simulations.

Together, these three listings demonstrate a minimal, fully functional example of triggering EnergyPlus™ from Python™. This scripting pattern forms the basis for more advanced frameworks such as PyE+, where Python™ is used not only to launch EnergyPlus™ runs, but also to loop over parametric variations, train machine learning models, and evaluate performance metrics—all in an automated and repeatable workflow.

```python

import subprocess
import os
import pandas as pd
import matplotlib.pyplot as plt
import utility_script

# Part 1: Verify EnergyPlus executable
energyplus_executable_path = r"C:/EnergyPlusV23-2-0/energyplus.exe"
command = [energyplus_executable_path, "--help"]
print(command)
result = subprocess.run(command, capture_output=True, text=True)
print('STDOUT:', result.stdout)
print('STDERR:', result.stderr)

# Part 2: Ensure the sim directory exists
```

```python
sim_directory = r'C:/Users/wanimh/PyCharmProjects/pythonProject/.venv/Scripts
    /sim'
if not os.path.exists(sim_directory):
    os.makedirs(sim_directory)

# Part 3: Define paths for IDF and EPW files (Now using absolute paths)
idf_filepath = os.path.join(sim_directory, '1ZoneUncontrolled.idf')
epw_filepath = os.path.join(sim_directory, 'USA_CA_San.Francisco.Intl.AP
    .724940_TMY3.epw')

# Correct the EnergyPlus installation directory
energyplus_install_dir = r"C:/EnergyPlusV23-2-0"

# Construct the command line string for EnergyPlus simulation
cl_st = (
    f'"{energyplus_install_dir}\\EnergyPlus" '
    + '--readvars '
    + f'--output-directory "{sim_directory}" '
    + f'--weather "{epw_filepath}" '
    + f'"{idf_filepath}"'
)
print(cl_st)

# Change the current working directory to sim_directory before running the
    EnergyPlus simulation
original_directory = os.getcwd()  # Save the original working directory
os.chdir(sim_directory)

# Part 4: Execute the EnergyPlus simulation
try:
    result = subprocess.run(cl_st, capture_output=True, text=True, shell=True
    )
    print('---ARGS---\n', result.args)
    print('---RETURNCODE---\n', result.returncode, '(SUCCESS)' if result.
    returncode == 0 else '(FAIL)')
    print('---STDOUT---\n', result.stdout)  # Removed .decode() because of
    text=True
    print('---STDERR---\n', result.stderr)  # Removed .decode() because of
    text=True

finally:
    # Change back to the original directory after running the simulation
    os.chdir(original_directory)

# Part 5: Analyze the Results

def parse_energyplus_datetime_string(st, year=2021):
    st = st.strip()
    month = int(st[0:2])
    day = int(st[3:5])
    hour = int(st[7:9]) - 1
    minute = int(st[10:12])
    second = int(st[13:15])  # Convert string to integer
    if not hour == 24:
        dt = pd.Timestamp(year, month, day, hour, minute, second)
    else:
        hour = 0
        dt = pd.Timestamp(year, month, day, hour, minute, second)
        dt += pd.Timedelta('1 day')
    return dt

# Read the CSV without explicitly parsing dates to use the custom function
df = pd.read_csv(os.path.join(sim_directory, 'eplusout.csv'), dtype={'Date/
    Time': 'object'})

# Convert the 'Date/Time' column
```

```
df['Date/Time'] = df['Date/Time'].apply(lambda x:
    parse_energyplus_datetime_string(x, year=2021))

# Set the 'Date/Time' column as the index
df.set_index('Date/Time', inplace=True)

# Plotting
fig, ax = plt.subplots(figsize=(16, 6))
ax.plot(df['ZONE ONE:Zone Mean Air Temperature [C](Hourly)'][48:], color='
    blue', linewidth=0.95)
ax.set_xlabel('Date')
ax.set_ylabel('Air temperature ($(^o)C$)')
plt.show()

# Part 6: Script within Script (Just for Demo)

utility_script.utility_function()
```

Listing 6.1 Calling EnergyPlus™ IDF using a Python™ script

```
    def utility_function():
        print("******* All tasks accomplished successfully *******")
```

Listing 6.2 Calling another Python™ Script from that provided in Listing 6.1

```

C:\Python311\python.exe -X pycache_prefix=C:\Users\wanimh\AppData\Local\
    JetBrains\PyCharmCE2023.3\cpython-cache "C:/Users/wanimh/AppData/Local/
    Programs/PyCharm Community/plugins/python-ce/helpers/pydev/pydevd.py" --
    multiprocess --qt-support=auto --client 127.0.0.1 --port 54777 --file C
    :\Users\wanimh\PycharmProjects\pythonProject\.venv\Scripts\RunThisNow.py
Connected to pydev debugger (build 233.13135.95)
Backend TkAgg is interactive backend. Turning interactive mode on.
['C:/EnergyPlusV23-2-0/energyplus.exe', '--help']
STDOUT: EnergyPlus, Version 23.2.0-7636e6b3e9, YMD=2024.06.04 22:09
PythonLinkage: Linked to Python Version: "3.8.10 (tags/v3.8.10:3d8993a, May
    3 2021, 11:48:03) [MSC v.1928 64 bit (AMD64)]"
Built on Platform: Windows_x86_64

Usage: C:/EnergyPlusV23-2-0/energyplus.exe [OPTIONS] [input_file]

Positionals:
  input_file TEXT:FILE          Input file (default: in.idf in current
    directory)

Options:
  -h,--help                     Print this help message and exit
  -v,--version                  Display program version information and exit
  -a,--annual Excludes: --design-day
                                Force annual simulation
  -D,--design-day Excludes: --annual
                                Force design-day-only simulation
  -d,--output-directory DIR     Output directory path (default: current
    directory)
  -i,--idd IDD                  Input data dictionary path (default: Energy+.
    idd in executable directory)
  -m,--epmacro                  Run EPMacro prior to simulation
  -p,--output-prefix PRE        Prefix for output file names (default: eplus)
  -r,--readvars                 Run ReadVarsESO after simulation
  -c,--convert                  Output IDF->epJSON or epJSON->IDF, dependent on
     input file type
  --convert-only                Only convert IDF->epJSON or epJSON->IDF,
    dependent on input file type. No simulation
  -s,--output-suffix SUFFIX     Suffix style for output file names (default: L)
                                   L: Legacy (e.g., eplustbl.csv)
                                   C: Capital (e.g., eplusTable.csv)
                                   D: Dash (e.g., eplus-table.csv)
```

```
  -j,--jobs N                  Multi-thread with N threads; 1 thread with no
     arg. (Currently only for G-Function generation)
  -w,--weather EPW             Weather file path (default: in.epw in current
     directory)
  -x,--expandobjects           Run ExpandObjects prior to simulation


Example: energyplus -w weather.epw -r input.idf

STDERR:
"C:/EnergyPlusV23-2-0\EnergyPlus" --readvars --output-directory "C:/Users/
     wanimh/PyCharmProjects/pythonProject/.venv/Scripts/sim" --weather "C:/
     Users/wanimh/PyCharmProjects/pythonProject/.venv/Scripts/sim\USA_CA_San.
     Francisco.Intl.AP.724940_TMY3.epw" "C:/Users/wanimh/PyCharmProjects/
     pythonProject/.venv/Scripts/sim\1ZoneUncontrolled.idf"
---ARGS---
 "C:/EnergyPlusV23-2-0\EnergyPlus" --readvars --output-directory "C:/Users/
     wanimh/PyCharmProjects/pythonProject/.venv/Scripts/sim" --weather "C:/
     Users/wanimh/PyCharmProjects/pythonProject/.venv/Scripts/sim\USA_CA_San.
     Francisco.Intl.AP.724940_TMY3.epw" "C:/Users/wanimh/PyCharmProjects/
     pythonProject/.venv/Scripts/sim\1ZoneUncontrolled.idf"
---RETURNCODE---
 0 (SUCCESS)
---STDOUT---
 EnergyPlus Starting
EnergyPlus, Version 23.2.0-7636e6b3e9, YMD=2024.06.04 22:10
Adjusting Air System Sizing
Adjusting Standard 62.1 Ventilation Sizing
Initializing Simulation
Reporting Surfaces
Beginning Primary Simulation
Initializing New Environment Parameters
Warming up {1}
Warming up {2}
Warming up {3}
Warming up {4}
Warming up {5}
Warming up {6}
Warming up {7}
Warming up {8}
Warming up {9}
Warming up {10}
Warming up {11}
Warming up {12}
Warming up {13}
Warming up {14}
Warming up {15}
Warming up {16}
Warming up {17}
Warming up {18}
Warming up {19}
Warming up {20}
Warming up {21}
Warming up {22}
Starting Simulation at 12/21 for DENVER CENTENNIAL  GOLDEN   N ANN HTG 99%
     CONDNS DB
Initializing New Environment Parameters
Warming up {1}
Warming up {2}
Warming up {3}
Warming up {4}
Warming up {5}
Warming up {6}
Warming up {7}
Warming up {8}
Warming up {9}
Warming up {10}
```

```
Warming up {11}
Warming up {12}
Warming up {13}
Warming up {14}
Warming up {15}
Warming up {16}
Starting Simulation at 07/21 for DENVER CENTENNIAL  GOLDEN   N ANN CLG 1% CONDNS DB=>MWB
Initializing New Environment Parameters
Warming up {1}
Warming up {2}
Warming up {3}
Warming up {4}
Warming up {5}
Warming up {6}
Warming up {7}
Warming up {8}
Warming up {9}
Warming up {10}
Warming up {11}
Warming up {12}
Warming up {13}
Warming up {14}
Warming up {15}
Warming up {16}
Warming up {17}
Warming up {18}
Starting Simulation at 01/01/2013 for RUN PERIOD 1
Updating Shadowing Calculations, Start Date=01/21/2013
Continuing Simulation at 01/21/2013 for RUN PERIOD 1
Updating Shadowing Calculations, Start Date=02/10/2013
Continuing Simulation at 02/10/2013 for RUN PERIOD 1
Updating Shadowing Calculations, Start Date=03/02/2013
Continuing Simulation at 03/02/2013 for RUN PERIOD 1
Updating Shadowing Calculations, Start Date=03/22/2013
Continuing Simulation at 03/22/2013 for RUN PERIOD 1
Updating Shadowing Calculations, Start Date=04/11/2013
Continuing Simulation at 04/11/2013 for RUN PERIOD 1
Updating Shadowing Calculations, Start Date=05/01/2013
Continuing Simulation at 05/01/2013 for RUN PERIOD 1
Updating Shadowing Calculations, Start Date=05/21/2013
Continuing Simulation at 05/21/2013 for RUN PERIOD 1
Updating Shadowing Calculations, Start Date=06/10/2013
Continuing Simulation at 06/10/2013 for RUN PERIOD 1
Updating Shadowing Calculations, Start Date=06/30/2013
Continuing Simulation at 06/30/2013 for RUN PERIOD 1
Updating Shadowing Calculations, Start Date=07/20/2013
Continuing Simulation at 07/20/2013 for RUN PERIOD 1
Updating Shadowing Calculations, Start Date=08/09/2013
Continuing Simulation at 08/09/2013 for RUN PERIOD 1
Updating Shadowing Calculations, Start Date=08/29/2013
Continuing Simulation at 08/29/2013 for RUN PERIOD 1
Updating Shadowing Calculations, Start Date=09/18/2013
Continuing Simulation at 09/18/2013 for RUN PERIOD 1
Updating Shadowing Calculations, Start Date=10/08/2013
Continuing Simulation at 10/08/2013 for RUN PERIOD 1
Updating Shadowing Calculations, Start Date=10/28/2013
Continuing Simulation at 10/28/2013 for RUN PERIOD 1
Updating Shadowing Calculations, Start Date=11/17/2013
Continuing Simulation at 11/17/2013 for RUN PERIOD 1
Updating Shadowing Calculations, Start Date=12/07/2013
Continuing Simulation at 12/07/2013 for RUN PERIOD 1
Updating Shadowing Calculations, Start Date=12/27/2013
Continuing Simulation at 12/27/2013 for RUN PERIOD 1
Writing tabular output file results using comma format.
Writing tabular output file results using tab format.
Writing tabular output file results using text format.
```

```
Writing tabular output file results using HTML format.
Writing tabular output file results using XML format.
 ReadVarsESO program starting.
 ReadVars Run Time=00hr 00min  0.42sec
 ReadVarsESO program completed successfully.
 ReadVarsESO program starting.
 ReadVars Run Time=00hr 00min  0.16sec
 ReadVarsESO program completed successfully.
EnergyPlus Run Time=00hr 00min  2.51sec

---STDERR---
 EnergyPlus Completed Successfully.

******* All tasks accomplished successfully *******

Process finished with exit code 0
```

Listing 6.3 Python™ Console Output for Scripts given in Listings 6.1 and 6.2

6.6.2 Example-2: Co-simulating EnergyPlus™ with Python™ in Real-time

6.6.2.1 Establishing a Co-simulation Interface Using the Python™Application Peripheral Interface (API)

PyE+: An Educational and Research-Grade Co-simulation Framework

To enhance the reader's grasp of Python's integration capabilities with EnergyPlus™, **PyE+**[3] emerges as an accessible framework for optimization and co-simulation. Developed by the lead author,[4] PyE+ allows seamless interaction between Python™ and EnergyPlus™, supporting baseline simulations, Multi-Objective Optimization (MOO), and Machine Learning (ML)-driven surrogate modeling [1]. PyE+ builds a dynamic bridge between `EnergyPlus™` and `Python™`, enabling users to manipulate parameters in real-time while optimizing and evaluating performance in terms of energy and comfort metrics.

The framework is designed for students, researchers, and practitioners who wish to advance from passive EnergyPlus™ simulation to active control and intelligent building design. PyE+ is openly available under the MIT license on GitHub:

> **Repository Access:**
> The PyE+ framework and all supporting materials are publicly available at: https://github.com/MubashirWani/PyE-Plus

---

[4] Website: https://drmubashirwani.com/.

Included Components

- A graphical user interface (`Graphics.py`) to launch baseline simulations and optimization workflows.
- Python™ backend scripts for simulation interfacing (`Pilot_Interface.py`), multi-objective optimization (`NSGA_II_Algorithm.py`), and surrogate modeling (`ml_surrogates.py`).
- Sample IDF files that demonstrate Python™ Plugin usage, including dynamic control of lighting, plug loads, setpoints, and infiltration.
- Pre-configured batch files to assist with execution on Windows systems.

Educational Motivation

A common barrier for newcomers to EnergyPlus™ is bridging the conceptual gap between simulation and scripting. PyE+ was included in this book to

- Provide a tangible example of Python™– EnergyPlus™ co-simulation.
- Demonstrate the use of `PythonPlugin` and `ExternalInterface` objects within an IDF.
- Introduce parameter sweeping, surrogate-assisted optimization, and NSGA-II implementation in a building simulation context.

Suggested Workflow

Readers are encouraged to

1. Clone or download the `PyE+` repository from GitHub.
2. Follow the accompanying `README.md`, `SETUP_GUIDE.md`, and `USAGE_GUIDE.md` files.
3. Inspect the provided IDFs—particularly `RealBuilding_Test_MoreParameters_Wellington_Python.idf`—to examine the Python™ Plugin configuration.
4. Use the GUI to define bounds, run baseline simulations, and execute optimization tasks.

Learning Outcomes

Upon using PyE+, readers will gain

- Hands-on familiarity with the `pyenergyplus` API.
- Insight into timestep-wise co-simulation workflows.
- Understanding of the integration of surrogate models for computational efficiency.

Overall, PyE+ serves as both a learning scaffold and a research-enabling tool, providing a reproducible environment for exploring intelligent control strategies in building energy systems.

6.6.2.2 Integrating Python™ with EnergyPlus™: Key Components and Usage

When using the Python™ Plugin in EnergyPlus™, you typically do not need to use the "*ExternalInterface*" object in your IDF file. The Python™ Plugin allows you to write custom control logic directly in Python™, which interacts with EnergyPlus™ through the *Energy Management System (EMS)* framework.

Some key points to note are as follows:

- **Python™ Plugin**
  The Python™ Plugin uses the *PythonPlugin:Instance* and *PythonPlugin:Variables* objects specified in the IDF (Input Data File) to define the scripts and variables used during simulations. These components allow Python™ code to be executed seamlessly within EnergyPlus™.
- **ExternalInterface**
  The *ExternalInterface* object is generally used for connecting EnergyPlus™ to external programs or systems via sockets, such as co-simulation with other software. This differs from the Python™ Plugin, which runs Python™ scripts internally within EnergyPlus™. The *ExternalInterface* object is especially useful when co-simulating EnergyPlus™ with MATLAB™.
- **Python™ EMS**
  When using the Python™ Energy Management System (EMS), developers can create control logic using Python™ scripts. The "*EnergyManagementSystem:ProgramCallingManager*" object is a feature that allows these Python™ scripts to be executed at specific times during a simulation. This provides developers with powerful control and customization options. Furthermore, you don't need to manually define the *EnergyManagementSystem:ProgramCallingManager* in the IDF (Input Data File) because the Python™ Plugin handles this automatically. This simplifies the process and ensures that your Python™ scripts are seamlessly integrated into the simulation.

6.6.2.3 Python™ Plugin Objects in IDF

The energy modeler will need to make sure the following Python™ Plugin objects are properly configured, in addition to keeping the above points in mind (see Sect. 6.6.2.2). To enable Python-driven control within EnergyPlus™, several dedicated objects are required to define the runtime linkage, data flow, and variable exposure between EnergyPlus™ and external Python™ modules. The present framework, `PyE+` [1], provides a fully operational example of this linkage using the following core object types, each of which must be declared in the IDF file:

- **PythonPlugin:SearchPaths**
  This object specifies the directories where EnergyPlus™ will search for user-defined Python modules. In `PyE+` application [1], this path includes the directory containing `SixSigma.py`, which defines the custom plugin logic. Without correctly declaring the search path, EnergyPlus™ will not locate or execute the Python™ script.
- **PythonPlugin:Instance**
  This object is used to instantiate a specific Python™ class for use as a plugin controller. The object must provide both the module name (e.g., `SixSigma`) and the class name (e.g., `AllEnergyMeasures`) as defined in the Python™ file. During simulation, EnergyPlus™ loads this class and repeatedly calls its callback methods at predefined simulation events.
- **PythonPlugin:Variables**
  This object exposes user-defined Python™ global variables to the EnergyPlus™ simulation environment. These variables can be accessed or modified from within Python™ using the EnergyPlus™ API. In `PyE+`, such variables are employed to store and propagate lighting power density, plug load density, or shading deployment status between simulation timesteps.
- **PythonPlugin:OutputVariable**
  This object defines the output variables that EnergyPlus™ should report, based on the global variables computed or updated in the Python™ code. These outputs can be visualized like any standard EnergyPlus™ output using tools such as ReadVarsESO or postprocessed in CSV format. In `PyE+`, these outputs help track how lighting, plug loads, or shading are modulated by the Python™ plugin.

Each of these objects have been implemented in the IDF[5] file supplied with the initial release[6] of `PyE+` and connected to the core logic in `SixSigma.py`. Readers are encouraged to explore both the IDF and Python™ files together to understand how these objects enable co-simulation and runtime parameter control. These objects form the backbone of the "PythonPlugin" interface and are indispensable for any simulation workflow that integrates Python-driven logic within EnergyPlus™. In `PyE+`, they facilitate a flexible and modular control structure that handles lighting, infiltration, plug loads, and shading using real-time occupancy and environmental feedback.

Note

When using Python™ Plugins in EnergyPlus™, one common issue which users may encounter is related to missing or incorrectly defined *global variables*. For instance, if the variables controlling key aspects like heating or cooling set-points are not declared properly in the "*PythonPlugin:Variables*" section of the IDF file, the simulation will fail

[5] `RealBuilding_Test_MoreParameters_Wellington_Python.idf`.
[6] v1.0.0.

to retrieve these handles, leading to fatal errors. Always ensure that the global variables are properly initialized and referenced in both the EnergyPlus™ input data and the Python™ script. This helps avoid handle retrieval errors and ensures smooth plugin execution, especially when interacting with schedules or actuators. Regular debugging and cross-referencing schedule names between the IDF and script can save significant troubleshooting time.

6.7 EnergyPlus™ and Building Digital Twins

This chapter has provided a structured and practice-oriented primer on EnergyPlus™, guiding the reader from foundational concepts in building energy modeling through to advanced applications in simulation control and optimization. By juxtaposing EnergyPlus™ with other commercial platforms such as IES-VE, we highlighted the rationale for selecting an open-source, scriptable, and extensible simulation engine as the analytical backbone of this book.

Key workflows—from geometry creation in OpenStudio® and space tagging to weather data assignment, template mapping, and shadow visualization—were explored to demonstrate how a whole-building simulation is operationalized. We then transitioned into automation and co-simulation using Python™, demonstrating how the EnergyPlus™ console can be called programmatically and how simulation outputs can be parsed, visualized, and analyzed dynamically.

A major contribution of this chapter lies in the introduction of **`PyE+`**, a Python-EnergyPlus co-simulation and optimization framework. `PyE+` empowers users to configure parametric sweeps, invoke multi-objective optimizations, and build surrogate models that dramatically reduce simulation time while preserving analytical depth. The included PythonPlugin structure and IDF templates form a live reference architecture for building simulation co-design.

More broadly, this chapter sets the stage for the themes developed in the remainder of this book: the convergence of simulation, intelligent control, and real-time decision-making in building energy systems. These elements form the foundational layers of a Digital Twin—where the virtual building evolves in sync with real-world operations, enabling predictive analytics, adaptive optimization, and control-driven insight. Readers are encouraged to engage with the accompanying files and explore `PyE+` via the public repository, adapting it to their own research or design problems.

In doing so, we move beyond static simulations and toward a future where intelligent buildings respond dynamically—guided by models, enriched by data, and empowered by open, interoperable digital ecosystems.

References

1. Wani MH (2025) PyE+: A Python–EnergyPlus co-simulation and optimization framework with expedited computation using ML-based surrogate modeling. https://github.com/MubashirWani/PyE-Plus, MIT Licensed open-source software, version 1.0.0
2. OpenStudio® (2024) Openstudio. Retrieved from https://openstudio.net/
3. EnergyPlus™ (2024) Energyplus. Retrieved from https://energyplus.net/
4. OpenStudio Coalition (2024) Openstudio coalition user docs: introducing the openstudio coalition. Retrieved from https://openstudiocoalition.org/about/openstudio_coalition/
5. OpenStudio® (2024) Openstudio coalition user docs: current features. Retrieved from https://openstudiocoalition.org/about/features/
6. URBANopt™ (2024) URBANopt SDK documentation. Retrieved from https://docs.urbanopt.net/

Part III
Industrial Deployment of Intelligent Building Digital Twins

Chapter 7
Augmenting Building Digital Twins with Artificial Intelligence (AI)

Abstract This chapter examines how Artificial Intelligence (AI), and more specifically Computational Intelligence (CI) techniques, can augment Building Digital Twins (BDTs) to enable fast, interpretable, and decision-capable Building Energy Management (BEM). While physics-based simulation engines provide high-fidelity representations of building behavior, their computational latency often limits their applicability in time-sensitive operational contexts. To address this gap, the chapter advances a hybrid Digital Twin (DT) paradigm in which learning-enabled computational acceleration complements, rather than replaces, physically grounded models. A surrogate-assisted Digital Twin (DT) architecture is introduced and instantiated through the open-source `PyE+` framework, which integrates EnergyPlus™ with Machine Learning (ML) surrogates and evolutionary Multi-Objective Optimization (MOO). The framework supports low-latency evaluation of energy savings, Global Warming Potential (GWP) reduction, and occupant comfort through a conditional evaluation mechanism which dynamically switches between surrogate inference and high-fidelity EnergyPlus™ simulation. Detailed formulations of objectives, penalty handling, learning strategies, and Surrogate-Assisted (SA) NSGA-II optimization are presented to demonstrate how CI methods can be operationalized within realistic Digital Twin (DT) workflows. The performance of the proposed approach is evaluated through a comparative study against conventional simulation-based optimization. Results show that surrogate assistance reduces the number of true EnergyPlus™ simulations from 165 to 10, decreasing total computation time from approximately 197.5 minutes to 1.08 minutes while preserving *Pareto-front* quality and *knee-point* solutions. This represents a reduction in computational effort exceeding 99%, enabling optimization and scenario evaluation within time scales compatible with supervisory building control and Demand Response (DR) decision-making. By linking computational acceleration to practical deployment within Building Energy Management Systems (BEMSs), the chapter positions intelligent Digital Twins (DTs) as active decision-support entities within cyber-physical building ecosystems. The discussion concludes by outlining scalability considerations and motivating subsequent chapters, which examine iBEMS architectures as the physical substrate for Digital Twins (DTs) and explore their integration with distributed and consensus-based control strategies.

M. H. Wani et al., *Digital Twins for Building Optimization: Enhancing Energy and Comfort*, Studies in Infrastructure and Control,
https://doi.org/10.1007/978-3-032-23274-8_7

Keywords Building digital Twin · Cognitive cities · Machine learning · PyE+ · SA-NSGA-II · Surrogate modeling

7.1 Role of Intelligence in Building Digital Twins (BDTs)

Building Digital Twins (BDTs) have emerged as a powerful paradigm for representing, analyzing, and managing the behavior of complex building systems throughout their life cycle. At their core, BDTs combine physics-based simulation models, real-time or near-real-time data streams, and system-level representations of assets, subsystems, and control logic. While such models provide a high degree of physical fidelity, their practical deployment is often constrained by computational latency, scalability limitations, and the growing need for adaptive decision-making under uncertainty.

Traditional simulation-driven digital twins, particularly those relying on detailed Building Energy Simulation (BES) engines, are typically designed for offline analysis. They excel in evaluating design alternatives, testing retrofit scenarios, and assessing long-term energy performance; however, their computational cost makes them ill-suited for time-sensitive tasks such as operational optimization, Demand Response (DR) participation, or interactive decision support. As buildings evolve into active participants within smart grids and urban energy systems, this gap between model fidelity and actionable intelligence becomes increasingly pronounced.

Artificial Intelligence (AI),[1] and more specifically Computational Intelligence (CI) techniques, play a critical role in addressing these limitations [1, 2]. Rather than replacing physics-based Digital Twins (DTs), intelligent algorithms augment them by enabling rapid inference, learning from historical and simulated data, and supporting optimization under competing objectives. In this context, intelligence acts as a computational accelerator and decision-enabling layer, transforming static or slow-running digital replicas into responsive, decision-capable twins.

The application scope of intelligent Building Digital Twins (BDTs) spans the entire building life cycle. During early design and planning stages, digital twins can support virtual prototyping, parametric exploration, and performance-based evaluation of envelope, HVAC, and lighting systems. When integrated with Building Information Modeling (BIM) platforms [3], they can enhance visualization and coordination while enabling simulation-driven design decisions. In construction and commissioning phases, Digital Twins (DTs) could facilitate progress tracking, coordination between trades, and consistency checks between design intent and as-built conditions.

Once operational, BDTs equipped with intelligent analytics can enable continuous monitoring, `predictive maintenance` [4], and automated fault detection across HVAC, lighting, security, and ELV systems. From an energy perspective, they

[1] In this chapter, "AI" is used in the broad sense, the technical emphasis is on Computational Intelligence (CI) methods introduced in Chap. 4.

can support consumption forecasting, DR strategies, and coordination with on-site renewables and smart grids. Beyond energy, intelligent twins can contribute to space utilization analysis, occupant comfort management, safety and emergency response planning, and long-term asset management. At larger scales, building-level twins can increasingly act as "nodes" within district- or city-scale digital twin ecosystems, contributing to urban resilience, infrastructure planning, and sustainability assessment.

It is important to emphasize that these diverse applications do not require a monolithic or fully autonomous Artificial Intelligence (AI). Instead, they could rely on targeted intelligence embedded within specific decision loops—whether for optimization, prediction, or anomaly detection—*while remaining grounded in physically meaningful models*. This perspective aligns with the view that intelligence in Building Digital Twins (BDTs) is not an end in itself, but a means to bridge the gap between detailed simulation and actionable, real-world decision-making.

7.2 From Computational Intelligence (CI) to Decision-Capable Digital Twins (DTs)

Chapter 4 introduced a range of Computational Intelligence (CI) techniques—Evolutionary Algorithms (EAs), swarm-based methods, and learning systems—which have proven effective for modeling, optimization, and control in complex, nonlinear systems. While these methods are well established in theory, their true impact within the built environment emerges only when they are systematically embedded into Digital Twin (DT) architectures.

Evolutionary and population-based optimization methods, for example, are particularly well suited to building energy problems characterized by large, mixed-variable design spaces and conflicting objectives such as energy efficiency, emissions reduction, and occupant comfort. Within a Digital Twin, these algorithms enable structured exploration of design alternatives, retrofit options, and control strategies which would be infeasible to evaluate exhaustively using brute-force simulation alone. However, their effectiveness is often limited by the high computational cost of repeatedly evaluating detailed physics-based models.

Learning-based techniques, including Machine Learning (ML) surrogates, address this limitation by approximating the input–output behavior of high-fidelity simulation models [5, 6]. When trained on representative simulation or operational data, surrogate models can provide near-instantaneous predictions of energy use, comfort metrics, or emissions, reducing evaluation times from minutes to milliseconds. This capability is essential for enabling interactive optimization, real-time decision support, and adaptive control within a Digital Twin (DT) framework.

From a systems perspective, the integration of CI techniques transforms a Digital Twin (DT) from a passive mirror of the physical building into an active decision-support entity. Optimization algorithms define how alternative actions are explored;

surrogate models determine how quickly outcomes can be evaluated; and learning mechanisms govern how the twin adapts as new data become available. Together, these components establish feedback loops which are central to the operational concept of a Digital Twin (DT), linking prediction, evaluation, and adaptation in a coherent manner.

Crucially, the notion of "real-time" in the context of Building Digital Twins (BDTs) should be interpreted pragmatically [7]. Unlike fast dynamical systems such as electrical drives or aerospace platforms, buildings exhibit slow thermal and energy dynamics. As a result, sub-minute or near-real-time inference is often sufficient to support meaningful operational decisions. Computational Intelligence (CI) techniques which reduce simulation latency into this time scale therefore unlock new possibilities for responsive building operation without sacrificing model-based rigor.

This chapter adopts the view that decision-capable Building Digital Twins (BDTs) are best realized through hybrid architectures, where physics-based simulation models provide fidelity and interpretability, while Computational Intelligence (CI) supplies speed, adaptability, and optimization capability. In the following sections, this philosophy is instantiated through a concrete Surrogate-Assisted (SA) Digital Twin (DT) framework, illustrating how CI methods introduced earlier in the book can be operationalized within realistic Building Energy Management (BEM) workflows.

7.3 Surrogate-Assisted Digital Twin Architecture for Building Energy Systems

This section instantiates the philosophy developed in Sects. 7.1–7.2 by presenting a concrete, Surrogate-Assisted (SA) architecture which enables low-latency decision support within Building Digital Twins (BDTs). The intent is not to replace physics-based models, but to augment them with learning-enabled computational acceleration such that optimization and scenario evaluation become feasible within sub-minute time scales. The framework presented here—`PyE+`[2] [8, 9]—is designed around this principle and serves as a representative Digital Twin "`computational engine`" for simulation-driven building decision workflows.

7.3.1 System Overview and Architectural Design

`PyE+` is a modular Python™-based framework which integrates EnergyPlus™ with Evolutionary Algorithms (EAs) and Machine Learning (ML) surrogates to accelerate Multi-Objective Optimization (MOO) for building energy and comfort decision-making. The framework is positioned to support Digital Twin operation by enabling

[2] © 2025–2026 Mubashir Hussain Wani. Licensed under the MIT License. Accessible as an open-source project at: https://github.com/MubashirWani/PyE-Plus.

Fig. 7.1 High-level modular architecture of the `PyE+` application, illustrating the user interface, sampling strategy, surrogate modeling, optimization loop, and real-time visualization and logging capabilities

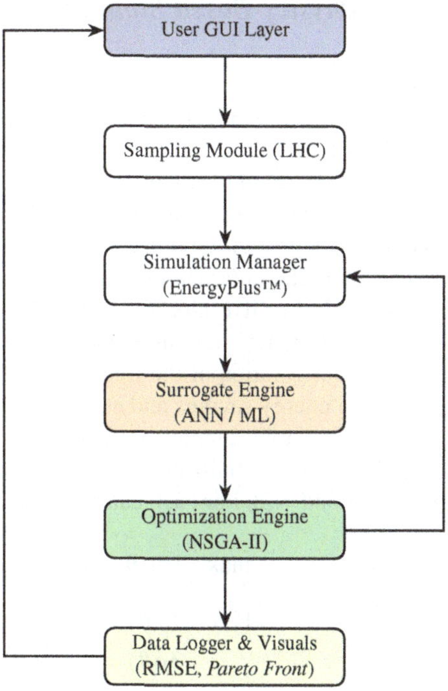

rapid inference and iterative optimization without abandoning physics-based rigor. This is aligned with the broader Digital Twin vision discussed earlier [10, 11] and with the need for low-latency decision workflows in grid-interactive and automation-driven buildings.

The framework begins at the *User GUI Layer*, see Fig. 7.1, providing an interface to define input parameters, configure simulation settings, and execute simulation and optimization runs. Once input parameters and simulation conditions are defined, the Sampling Module employs Latin Hyper Cube (LHC) sampling to generate an initial training dataset which covers the design space. These sampled scenarios are evaluated via the Simulation Manager, which orchestrates EnergyPlus™ subprocess execution. The resulting high-fidelity simulation data becomes the training set for the surrogate model.

The Surrogate Engine then trains a machine learning-based surrogate model (e.g., MLP) to approximate EnergyPlus™ outputs at low computational cost. This enables rapid evaluation of candidate solutions within an optimization loop. Central to the architecture is the Optimization Engine (NSGA-II), which uses the surrogate for most evaluations while selectively invoking EnergyPlus™ simulations for periodic correction and retraining. Finally, the Data Logger and visualization module tracks optimizer progress and surrogate credibility (e.g., RMSE) to support transparent, human-interpretable decision-making.

7.3.2 *Physics-Based Simulation and Data Generation*

A key feature of decision-capable Digital Twins (DTs) is the ability to remain grounded in physically meaningful models while providing operationally useful response times. In `PyE+`, EnergyPlus™ serves as the high-fidelity physics engine for generating ground-truth objective evaluations. The training dataset is constructed by sampling the design space via LHC, evaluating each sample with EnergyPlus™, and logging the resulting objective vectors for surrogate learning.

To initiate the optimization, the user specifies the bounds of each decision variable through the GUI interface. These 14 variables[3] comprise both `passive` and `active` parameters relevant to the building's envelope and operational profile. The baseline configuration of the EnergyPlus™ model serves as the reference point for calculating energy savings and carbon (GWP) reductions across all runs.

Design variable samples are generated either through Latin HyperCube (LHC) sampling during initialization or iteratively via the NSGA-II optimizer (see Algorithms 7.1 and 7.2). For each candidate, $\boldsymbol{\theta}$, a conditional evaluation is performed:

$$\begin{aligned} \max_{\boldsymbol{\theta}_i \in \Theta} \quad & [f_1(\boldsymbol{\theta}), f_2(\boldsymbol{\theta}), f_3(\boldsymbol{\theta})] \\ \text{where} \quad & f_1(\boldsymbol{\theta}) = \text{Energy Savings}, \\ & f_2(\boldsymbol{\theta}) = \text{GWP Reduction}, \\ & f_3(\boldsymbol{\theta}) = \text{Occupant Comfort Score/Index}. \end{aligned} \tag{7.1}$$

Here, $\boldsymbol{\theta} \in \mathbb{R}^d$ denotes a candidate[4] design vector (with $d = 14$ decision variables in this case), and Θ is the bounded feasible search space defined by the lower and upper limits $(\boldsymbol{\theta}_{\min}, \boldsymbol{\theta}_{\max})$. The conditional evaluation refers to computing the objective vector $\boldsymbol{f}(\boldsymbol{\theta}) = [f_1(\boldsymbol{\theta}), f_2(\boldsymbol{\theta}), f_3(\boldsymbol{\theta})]$ either via high-fidelity EnergyPlus™ simulation or via surrogate inference, depending on surrogate availability and reliability.

If the surrogate model is available and within uncertainty thresholds, predicted objective values, $\hat{\boldsymbol{f}}(\theta)$, are returned via ML inference. Otherwise, a full EnergyPlus™ simulation is executed, receiving the EPW weather file and the updated parameter vector via a Python™-managed subprocess.

Simulation outputs are parsed to extract key performance indicators, including total energy usage, GWP due to electricity and gas, and average thermal conditions in representative zones (e.g., guestroom and corridor). Occupant comfort is then computed as a normalized score based on the *Euclidean deviation* between the actual

---

[3] While traditionally considered fixed, passive building parameters—such as material absorptance, thermal conductivity, and glazing `U-values`—often change over the operational lifespan of a building due to environmental impacts, weathering, aging, and material degradation. Consequently, these parameters may require periodic re-assessment or re-optimization to maintain optimal building performance—particularly from *Digital Twin* perspective.

[4] "i" denotes the index of a candidate solution within the population or sample set, i.e., $\boldsymbol{\theta}_i$ is the i-th design vector considered during optimization, and $i \in \{1, 2, \ldots, N\}$.

average temperatures and an ideal setpoint specified by the user. This procedure is explained in more detail in Sect. 7.3.8.

Evaluation results are continuously logged and used to retrain the surrogate model at predefined intervals. The updated evaluations feed back into the NSGA-II search loop to iteratively evolve the *Pareto Front* toward high-performing and diverse design solutions.

7.3.3 Passive and Active Parameters in Twin-Assisted Optimization

The PyE+ framework integrates both passive and active building parameters into its optimization procedure (Tables 7.1, 7.2, and Fig. 7.2). Passive parameters influence the inherent thermal behavior and long-term performance of the building, while active parameters provide operational flexibility for responding to occupancy patterns, grid signals, and environmental variability. Optimizing both categories supports a holistic Digital Twin perspective, where design-time robustness and operational responsiveness are treated as complementary objectives.

7.3.4 Surrogate Modeling and Learning Components

Surrogate modeling enables fast inference of building performance metrics by learning a mapping from decision variables to objective outcomes. In PyE+, the surrogate model is initially trained using LHC samples and EnergyPlus™ simulation data, as outlined in Algorithm 7.1. The trained surrogate is then deployed within the Multi-Objective Optimization (MOO) loop, enabling rapid candidate evaluation while preserving accuracy through periodic retraining.

Table 7.1 Passive design variables: material properties

Variable	Parameter	Unit
θ_1	Thermal absorptance	–
θ_2	Solar absorptance	–
θ_3	Material thickness	m
θ_4	Thermal conductivity	$\mathrm{Wm^{-1}K^{-1}}$

Table 7.2 Active design variables: energy efficiency measures

Measure	Variable	Parameter	Unit
Dynamic thermostat setpoints	θ_5	Heating setpoint (guestroom)	°C
	θ_6	Cooling setpoint (guestroom)	°C
	θ_7	Heating setpoint (corridor)	°C
	θ_8	Cooling setpoint (corridor)	°C
Lighting optimization	θ_9	Lighting power density (guestroom)	Wm^{-2}
	θ_{10}	Lighting setpoint (corridor)	Wm^{-2}
Plug load scheduling	θ_{11}	Plug load density (guestroom)	Wm^{-2}
	θ_{12}	Plug load density (corridor)	Wm^{-2}
Infiltration mitigation	θ_{13}	Infiltration rate (guestroom)	m^3s^{-1}
	θ_{14}	Infiltration rate (corridor)	m^3s^{-1}

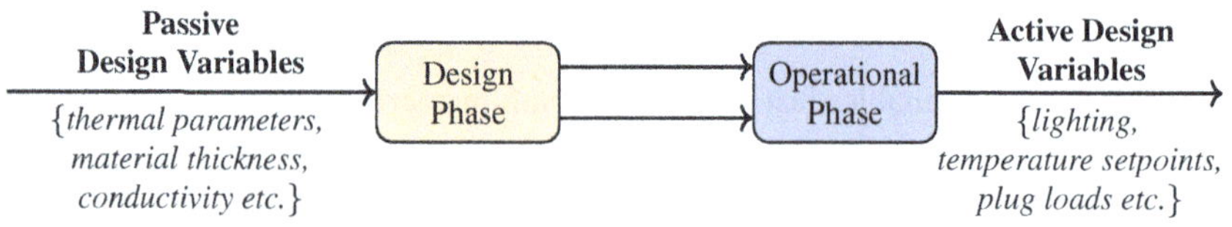

Fig. 7.2 Design parameters and typical phases during a building life cycle

7.3.5 Surrogate-Assisted Multi-objective Optimization Procedure

The core decision loop in `PyE+` is implemented through a Surrogate-Assisted (SA) variant of NSGA-II, where the optimizer evolves a population of candidate design vectors while reducing the number of expensive EnergyPlus™ calls. Conceptually, the procedure begins with an initial Design of Experiments (DOE) generated via Latin Hypercube (LHC) sampling to establish a physically grounded training dataset. A surrogate model, $\hat{f}(\cdot)$, is then trained to approximate the mapping from design variables to the objective vector defined in Eq. (7.1). During the evolutionary search, most offspring evaluations are performed using surrogate inference to maintain low latency, while selective high-fidelity simulations are triggered at a retraining interval I_r (or whenever surrogate reliability is deemed insufficient). The newly simulated samples are appended to the training set and used to periodically recalibrate $\hat{f}(\cdot)$, ensuring that computational speed is achieved without drifting away from physics-based accuracy. Algorithm 7.2 summarizes this switching mechanism and its integration into the standard NSGA-II selection and diversity-preservation steps.

Algorithm 7.1 Surrogate Model Training using Latin Hyper Cube (LHC) sampling

Require: Number of input variables d, number of samples N_{LHC}, bounds $\theta_{\min}$, $\theta_{\max}$
Ensure: Trained surrogate model $\rightarrow \hat{f}(\cdot)$
1: Generate N_{LHC} sample points $\{\theta_i\}_{i=1}^{N_{LHC}}$ using Latin Hyper Cube (LHC) sampling over d dimensions
2: **for all** sample θ_i **do**
3: Run EnergyPlus simulation to obtain true objective vector $f(\theta_i)$
4: Add $(\theta_i, f(\theta_i))$ to training dataset $\mathcal{D}$
5: **end for**
6: Normalize input features and objective values in $\mathcal{D}$
7: Select surrogate model type (e.g., ANN, Random Forest)
8: Train surrogate model, $\hat{f}(\cdot)$, on dataset $\mathcal{D}$ using chosen regression algorithm
9: Evaluate surrogate accuracy using cross-validation or hold-out set (e.g., RMSE)
10: **return** trained surrogate model $\hat{f}(\cdot)$

This operationalizes Eq. (7.1) into an executable decision loop which yields a *Pareto set* of trade-off solutions discussed in upcoming sections.

7.3.6 Multi-objective Optimization and Decision Loops

The Surrogate-Assisted (SA) optimization in `PyE+` identifies optimal building design vectors, $\boldsymbol{\theta}$, which simultaneously optimize energy savings, GWP reduction, and occupant comfort. The multi-objective formulation is stated in Eq. (7.1), and the optimization proceeds via NSGA-II and its Surrogate-Assisted variant (Algorithm 7.2). The 14-dimensional feasible *search space* or *design space* comprising all feasible combinations of `passive` and `active` parameters subject to physical and operational limits—lower (θ_{min}) and upper (θ_{max}) bounds of each parameter—are provided in Table 7.3. This formulation aligns directly with the Digital Twin requirement of handling trade-offs among competing operational goals while maintaining interpretability and feasibility.

Algorithm 7.2 Surrogate-Assisted NSGA-II for Building Energy Optimization

Require: Initial population size N, number of generations G, surrogate retraining interval I_r
Ensure: Pareto-optimal set of building design solutions
Generate N Latin Hyper Cube (LHC) samples and evaluate them using EnergyPlus
Train initial surrogate model, $\hat{f}(\cdot)$, on LHC data ▷ See Algorithm 7.1
for generation $g = 1$ **to** G **do**
  Generate offspring population using crossover and mutation
  for all individual θ_i in offspring **do**
    if g mod $I_r = 0$ **or** $\hat{f}(\cdot)$ not reliable **then**
      Evaluate $f(\theta_i)$ using EnergyPlus simulation
      Add $(\theta_i, f(\theta_i))$ to training set
    else
      Evaluate $\hat{f}(\theta_i)$ using surrogate model
    end if
  end for
  end
  Update the $\hat{f}(\cdot)$ if retraining criteria met ▷ See Algorithm 7.1
  Combine parent and offspring populations
  Perform non-dominated sorting and crowding distance assignment
  Select N best individuals for next generation
end for
end
return final Pareto Front

Table 7.3 Optimization search space, Θ, and corresponding baseline values for 14 design parameters used in the setup

	θ_1	θ_2	θ_3	θ_4	θ_5	θ_6	θ_7	θ_8	θ_9	θ_{10}	θ_{11}	θ_{12}	θ_{13}	θ_{14}
θ_{min}	0.75	0.55	0.1	1.0	15.0	20.5	15.0	20.5	0.1	0.01	0.01	0.01	0.01	0.01
θ_{max}	0.95	0.75	0.3	2.5	20.0	25.0	20.0	25.0	15.0	7.0	7.0	5.0	1.0	1.0
$\theta_{baseline}$	0.85	0.65	0.2	1.5	20.0	24.0	20.0	24.0	10.0	5.0	5.0	2.0	0.5	0.3

7.3.7 *Penalty Handling for Energy Savings and GWP Reduction*

For the objectives $f_1(\theta)$ (energy savings) and $f_2(\theta)$ (GWP reduction), scores are normalized between "0" and "1" with respect to their respective baseline values, E_{baseline} and $\text{GWP}_{\text{baseline}}$. However, when a candidate solution exhibits higher energy use or emissions than the baseline (i.e., $f_1(\theta) < 0$ or $f_2(\theta) < 0$), a soft penalty is applied to preserve optimizer sensitivity and discourage regression.

The penalty is computed using a smooth exponential decay function:

$$f_i(\theta) = -\left(1 - \exp\left[\frac{f_i(\theta)}{k}\right]\right), \quad i \in \{1, 2\} \tag{7.2}$$

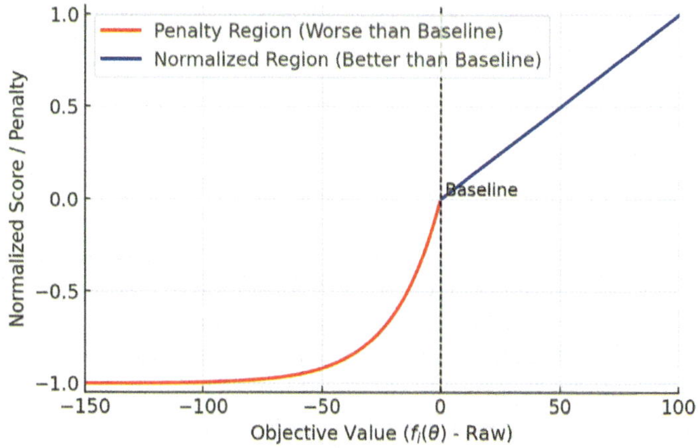

Fig. 7.3 Penalty transformation applied to energy savings, $f_1(\theta)$, and GWP reduction, $f_2(\theta)$, objectives. Solutions which underperform relative to the *baseline* receive exponential penalties (red), while improvements are linearly normalized (blue) for fair comparison and optimizer stability

where $k = \frac{|f_i^{\max}|}{5}$ is a scaling constant based on the baseline maximum value. This ensures that larger deviations receive proportionally stronger penalties while keeping the score bounded below by -1. For all other cases, the objective is normalized linearly:

$$f_i(\theta) = \frac{f_i(\theta) - f_i^{\min}}{f_i^{\max} - f_i^{\min}} \tag{7.3}$$

This piecewise approach ensures numerical stability and fairness across both objective scores during optimization. Figure 7.3 illustrates this graphically.

7.3.8 *Penalty Handling for Occupant Comfort*

The occupant comfort objective $f_3(\boldsymbol{\theta})$ is defined via the *Euclidean Distance*, $d(\boldsymbol{\theta})$, between actual and ideal average zone temperatures:

$$d(\boldsymbol{\theta}) = \|\boldsymbol{T}_{\text{actual}}(\boldsymbol{\theta}) - \boldsymbol{T}_{\text{ideal}}\|_2 \tag{7.4}$$

$$f_3(\boldsymbol{\theta}) = \begin{cases} 1 - \dfrac{d(\boldsymbol{\theta})}{d_{\max}}, \text{if } 0 \leq d(\boldsymbol{\theta}) \leq d_{\max} \\ \\ 0.01 \cdot \exp\left[-\lambda \cdot (d(\boldsymbol{\theta}) - d_{\max})\right], \text{otherwise} \end{cases} \tag{7.5}$$

Here, "$\boldsymbol{T}_{\text{actual}}$" is the vector of average measured temperatures (e.g., in guestroom and corridor), "$\boldsymbol{T}_{\text{ideal}}$" is the target temperature vector (e.g., 23 °C), and "d_{max}" is the maximum allowable comfort deviation computed during baseline setup. The decay rate "λ" controls how steeply "f_3" penalizes deviations beyond "d_{max}". This piecewise formulation ensures full reward for thermal neutrality and graceful degradation for comfort violations, thus preserving optimization sensitivity while maintaining thermal realism.

To reduce *computational overhead*, `PyE+` employs a Surrogate-Assisted NSGA-II (`SA-NSGA-II`) algorithm (Algorithm 7.2). A neural surrogate model, $\hat{f}(\cdot)$, is first trained on Latin Hyper Cube (LHC) samples (Algorithm 7.1) and used to estimate objective vectors, $\hat{\boldsymbol{f}}(\boldsymbol{\theta})$, for the majority of candidate evaluations. True EnergyPlus™ simulations are selectively invoked only when a retraining interval, I_r, is met or when prediction uncertainty exceeds a defined threshold.

This surrogate switching mechanism allows the optimizer to converge to high-quality Pareto Fronts efficiently, while retaining accuracy through periodic corrections. All objective functions are normalized relative to baseline simulation values to ensure fair trade-off handling and stable evolutionary behavior across generations.

Numerical Interpretation (Occupant Comfort Score/Index)

To enhance optimizer interpretability and performance, the occupant comfort score/index, $f_3(\theta)$, in `PyE+` is computed via a piecewise function normalized over the *Euclidean temperature deviation across two thermal zones* (guestroom and corridor)[5]. The following numerical example illustrates this formulation and, to aid in this understanding, the corresponding graphs are shown in Fig. 7.4.

Assumptions:

- Ideal zone temperature (setpoint): $\boldsymbol{T}_{\text{ideal}} = [23,\ 23]\,^{\circ}\text{C}$
- Comfort bounds per zone: $T_{\text{min}} = 18\,^{\circ}\text{C}$, $T_{\text{max}} = 28\,^{\circ}\text{C}$
- Worst-case deviation: $\boldsymbol{T}_{\text{worst}} = [28,\ 28]\,^{\circ}\text{C}$

$$\begin{aligned} d_{\text{max}} &= \|\boldsymbol{T}_{\text{worst}} - \boldsymbol{T}_{\text{ideal}}\|_2 \\ &= \sqrt{(5)^2 + (5)^2} = \sqrt{50} \approx 7.07 \end{aligned} \tag{7.6}$$

[5] The current application evaluates occupant comfort across two representative zones: a private space (e.g., guestroom) and a shared/common area (e.g., corridor). This abstraction reflects typical thermal diversity found in most building typologies. While the present implementation is fixed to two zones, the framework's underlying formulation generalizes to multi-zone evaluation with minimal extension.

Fig. 7.4 Occupant comfort score/index $f_3(\theta)$ as a function of Euclidean temperature deviation $d(\theta)$ across two thermal zones

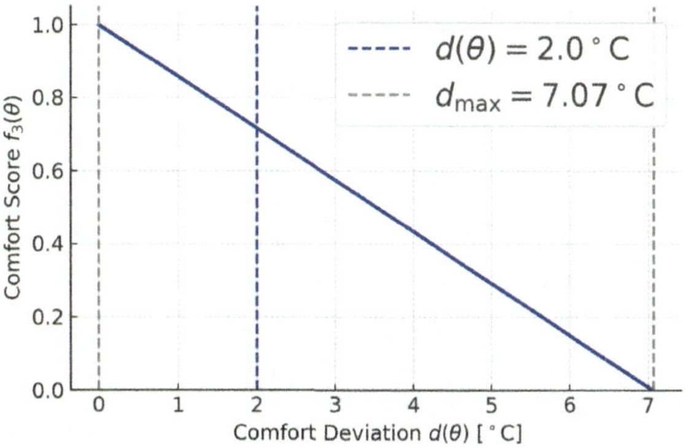

(a) Case 1: $0 \leq d(\theta) \leq d_{max}$

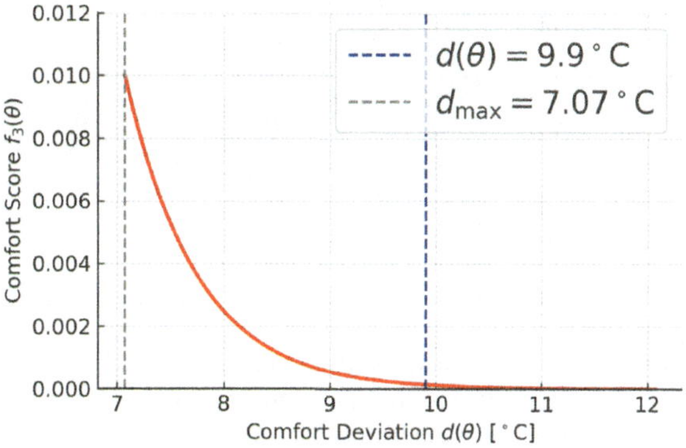

(b) Case 2: $d(\theta) > d_{max}$

Case 1—Deviation Within Acceptable Bounds

$$\begin{aligned} \boldsymbol{T}_{\text{actual}} &= [25.0,\ 23.0]^\circ\text{C} \\ d(\theta) &= \|[25, 23] - [23, 23]\|_2 = \sqrt{4} = 2.0 \\ f_3(\theta) &= 1 - \frac{2.0}{7.07} \approx 0.717 \end{aligned} \tag{7.7}$$

Case 2—Deviation Beyond $d_{\max}$

$$\begin{aligned} \boldsymbol{T}_{\text{actual}} &= [30.0,\ 30.0]^\circ\text{C} \\ d(\theta) &= \sqrt{(7)^2 + (7)^2} = \sqrt{98} \approx 9.90 \\ f_3(\theta) &= 0.01 \cdot \exp[-1.5 \cdot (9.90 - 7.07)] \\ &\approx 1.44 \times 10^{-4} \end{aligned} \tag{7.8}$$

The function is designed to create a smooth transition from perfect comfort, where "$f_3(\theta)$" is approximately "1", to states that are penalized. This ensures that there is

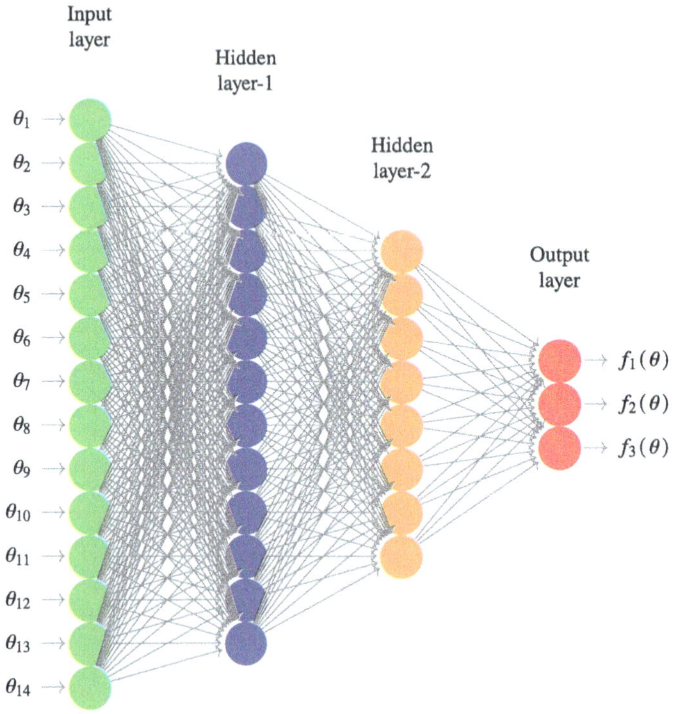

Fig. 7.5 The neural network architecture used in `PyE+` for surrogate modeling is a Multi-Layer Perceptron (MLP) with a default setup which includes 14 input features corresponding to design variables. It generates predictions for 3 target outcomes: energy savings, Global Warming Potential (GWP) reduction, and occupant comfort. The standard configuration comprises two hidden layers, representing the typical MLP structure applied in this tool

a gradual change rather than an abrupt shift, which is important for maintaining differentiable behavior. Such behavior is crucial for supporting optimization continuity. Additionally, the function effectively suppresses extreme discomfort, achieving this without the use of harsh thresholds (penalties) which could lead to undesirable behavior in the optimization process.

7.3.9 Learning Methods

In order to tailor the optimization workflow to specific problem characteristics, the `PyE+` application framework provides users with selectable surrogate modeling options: **(i)** *Multi-Layer Perceptron (MLP)*—a feedforward Neural Network (NN) architecture (Fig. 7.5) used as the default surrogate for its ability to capture complex nonlinear input–output relationships inherent in building energy dynamics; **(ii)**

Random Forest (RF)—an ensemble learning method which improves generalization and performs reliably in the presence of noise or irregular objective spaces; and **(iii)** *Gaussian Process Regression (GPR)*—a Bayesian surrogate which provides both prediction and confidence intervals, making it particularly useful for uncertainty-aware sampling and data-sparse scenarios. These options enable users to align surrogate fidelity, interpretability, and evaluation speed with the design or operational phase of their building energy optimization task.

7.3.10 Simulation Setup

Figure 7.6 shows the simulation setup of PyE+. To initiate the optimization procedure, the user specifies the bounds of each Decision Variable (DV) within the Graphical User Interface (GUI). The 14 DVs comprise both `passive` and `active` parameters associated with the building envelope and its operational schedule (see Table 7.3). The baseline EnergyPlus™ configuration, also reported in Table 7.3, serves as the reference scenario for calculating energy savings and reductions in Global Warming Potential (GWP) across all simulations.

As previously discussed in Sect. 7.3.2, design-variable configurations are obtained either through Latin Hypercube (LHC) sampling during the initialization stage or iteratively by the NSGA-II optimization algorithm (see Algorithms 7.1 and 7.2). Moreover, for each candidate solution, $\boldsymbol{\theta}$, a conditional evaluation procedure is employed, in which the *objective vector*, $\boldsymbol{f}(\boldsymbol{\theta})$, is determined either by running an EnergyPlus™ simulation or by performing surrogate inference.

> If a surrogate model is available and its predictive uncertainty remains within predefined admissible bounds, the objective values $\hat{\boldsymbol{f}}(\theta)$ are obtained directly via Machine Learning (ML) inference (Fig. 7.5). Otherwise, a full EnergyPlus™ simulation is executed, using the specified EPW weather file and the updated parameter vector, which is passed through a Python™-controlled subprocess.

The raw simulation outputs are post-processed to derive key performance indicators, including total energy consumption, GWP associated with electricity and natural gas use, and representative thermal comfort conditions in selected zones (e.g., guestroom and corridor). Occupant comfort is quantified via a normalized metric derived from the Euclidean distance between the realized average indoor air temperatures and a user-defined ideal setpoint. In this study, the ideal comfort temperature is fixed at $T_{\text{ideal}} = 23°\text{C}$, and thermal deviations are evaluated relative to a maximum allowable distance d_{max}, as defined in Sect. 7.3.8.

All evaluation results are continuously archived and employed to periodically recalibrate the surrogate model. The updated surrogate evaluations are then

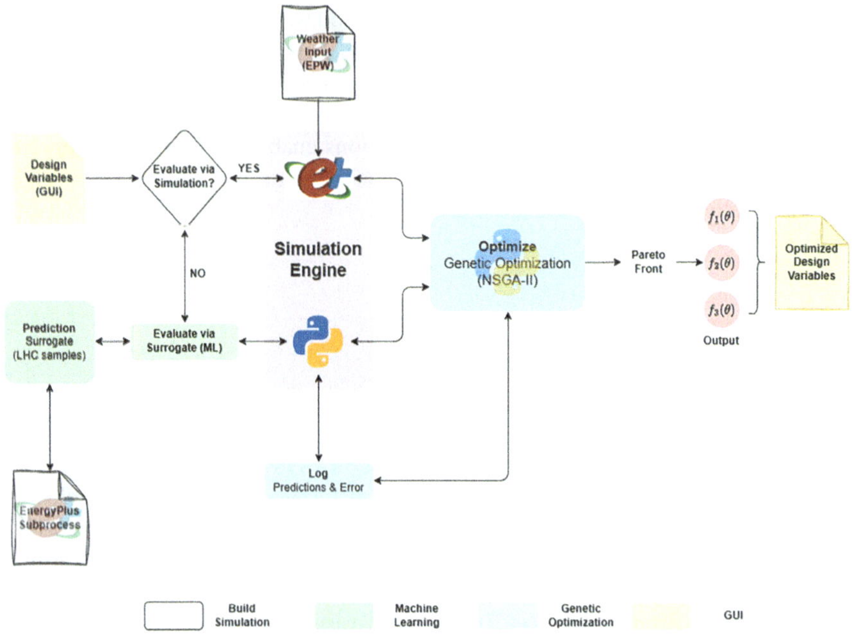

Fig. 7.6 Simulation setup for PyE+ showing the co-simulation loop between EnergyPlus™ and surrogate models. The system evaluates each design vector either through direct EnergyPlus™ simulation or a neural surrogate, with objective values used to evolve Pareto-optimal solutions

reintegrated into the NSGA-II search process, progressively guiding the evolution of the *Pareto Front* toward solutions that are simultaneously high-performing and diverse.

Upon convergence, the framework yields a non-dominated set of optimized parameter vectors $\boldsymbol{\theta}^{\star}$ and their corresponding objective values $\boldsymbol{f}(\theta^{\star})$. These results are visualized in the GUI and can be exported for subsequent analysis and post-processing.

7.3.11 Results and Discussion

This section evaluates the computational performance and optimization effectiveness of the Surrogate-Assisted NSGA-II (SA-NSGA-II) implemented within the PyE+ framework, compared with the conventional NSGA-II algorithm. Both algorithms were configured identically to ensure a fair comparison, with algorithm parameters set as shown in Table 7.4. Each optimization algorithm was executed for 10 generations with a population size of 15.

Table 7.4 Parameter settings for conventional NSGA-II and SA-NSGA-II used for comparing computational efficiency (SA-NSGA-II uses surrogate models to reduce computational cost)

Parameter	NSGA-II	SA-NSGA-II
Number of generations (G)	10	10
Population size (ps)	15	15
Max. estimated runs	165	–
Completed (true) runs	165	10
Initial DOE (DOE: Design of Experiments, which determines the number of initial candidate solutions generated through LHC sampling) sample	–	15
Infill fraction (Controls the percentage of candidate solutions in each generation selected for *true simulations*. Thus, for each generation: $10\% \times 15 = 1.5 \approx 1$ true run per generation) (%)	–	10
Retraining interval (I_r)	–	5
Surrogate model used	–	MLP regressor

A significant difference in computational efficiency was observed between the two optimization approaches. The conventional NSGA-II completed a total of 165 true EnergyPlus™ simulation runs, taking approximately 3 h 17 min 30 s ($\approx$ 197.5 min) to reach a convergent Pareto Front. In stark contrast, with an *infill fraction* of 10%, the simulation engine (SA-NSGA-II), leveraging surrogate model predictions, performed only 10 true EnergyPlus™ simulations alongside $\approx$ 140 surrogate evaluations. This resulted in a dramatically reduced computation time of approximately 64.8 s (1.08 min).

Figure 7.7 distinctly illustrates a visual comparison of the resulting Pareto Fronts from each method, with emphasis on the identified *knee-points*. These *knee-points* are determined using the widely recognized Minimum Manhattan Distance (MMD) technique, ensuring an unbiased identification of optimal compromise solutions. While this method is not a focus of our study, it has already been explained and referenced in prior research [12]. From Fig. 7.7, its clear that both methods yield a diverse Pareto Front, though `SA-NSGA-II` shows more consistent clustering around the "knee" region.

Table 7.5 summarizes the knee-point solutions, offering precise insights into the optimal parameter settings and corresponding objective values (energy savings, GWP reduction, and occupant comfort) for both algorithms used in `PyE+`.

To highlight the effectiveness of the SA-NSGA-II implemented in *PyE+*, we executed the SA-NSGA-II with an increased number of generations ($G = 100$) and a larger population size ($ps = 30$). The computation was completed in 3 h 38 min 10 s. Figure 7.8c illustrates the RMSE[6] progression for the surrogate model within

[6] **Root Mean Squared Error (RMSE)** quantifies the prediction error of the surrogate model relative to true simulation outputs. For a surrogate model $\hat{f}(\theta)$ and the actual objective function $f(\theta)$, RMSE is defined as

Fig. 7.7 Pareto Fronts and the corresponding *knee-points* evaluated by `PyE+` using NSGA-II and SA-NSGA-II, for parameter settings specified in Table 7.4

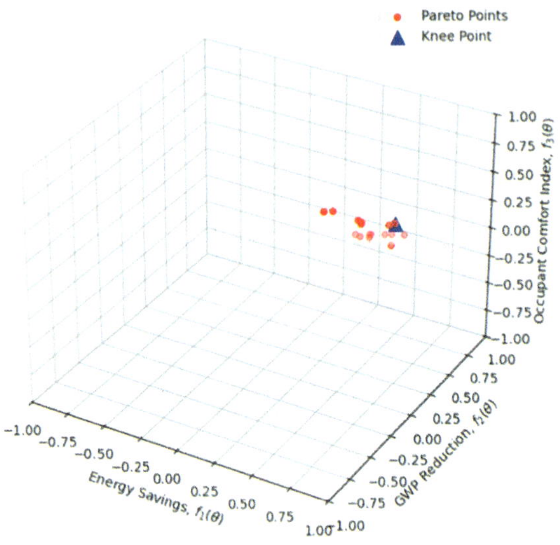

(a) Conventional NSGA-II, using 165 true simulations.

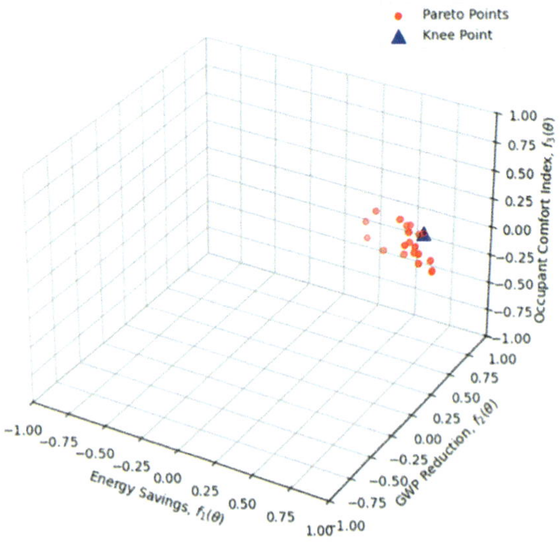

(b) Surrogate Assisted NSGA-II (SA-NSGA-II), using 10 true simulations and 140 surrogate evaluations.

Table 7.5 Knee-point solution comparison

Knee-point	NSGA-II	SA-NSGA-II
$[f_1^*(\theta), f_2^*(\theta), f_3^*(\theta)]$	[0.61, 0.47, 0.23]	[0.82, 0.43, 0.25]
θ_1^*	0.78	0.88
θ_2^*	0.64	0.70
θ_3^*	0.28	0.20
θ_4^*	2.46	2.02
θ_5^*	20.0	15.5
θ_6^*	24.8	22.58
θ_7^*	15.05	16.20
θ_8^*	24.55	21.35
θ_9^*	0.1	0.75
θ_{10}^*	0.01	0.05
θ_{11}^*	1.97	0.08
θ_{12}^*	1.32	0.17
θ_{13}^*	0.43	0.13
θ_{14}^*	0.21	0.04

the `SA-NSGA-II` algorithm used in `PyE+`, demonstrating the surrogate model's improving predictive accuracy over successive retraining steps (every 5 generations). This progressive refinement underscores the efficacy of integrating periodic surrogate retraining within optimization loops.

The results unequivocally demonstrate that the Surrogate-Assisted (SA) approach significantly accelerates building energy simulation and optimization workflows, providing rapid yet reliable solutions suitable for real-time decision-making contexts. This computational acceleration enables practical integration into operational Building Energy Management Systems (BEMSs), substantially enhancing the adaptability and responsiveness required by modern Digital Twin (DT) and cognitive city implementations.

$$\text{RMSE} = \sqrt{\frac{1}{N}\sum_{i=1}^{N}\left(\hat{f}(\theta_i) - f(\theta_i)\right)^2}$$

where:

- N is the number of data points (e.g., validation set),
- $\hat{f}(\theta_i)$ is the surrogate prediction for the ith candidate,
- $f(\theta_i)$ is the corresponding true value from EnergyPlus™ simulation.

Fig. 7.8 Surrogate model (SA-NSGA-II) performance in PyE+ for $G = 100$ and $ps = 30$

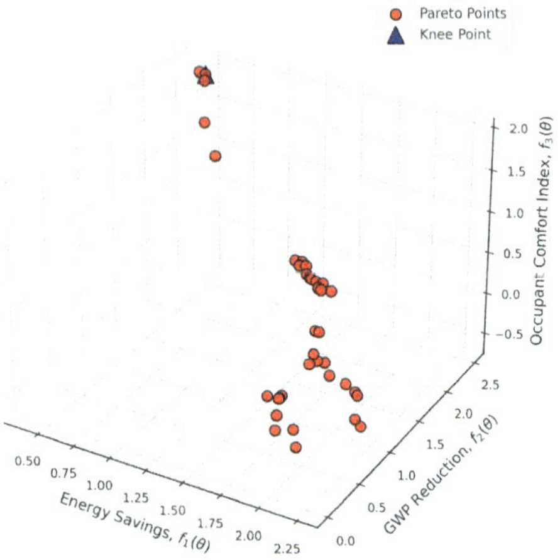

(a) Pareto Front with Knee-Point solution.

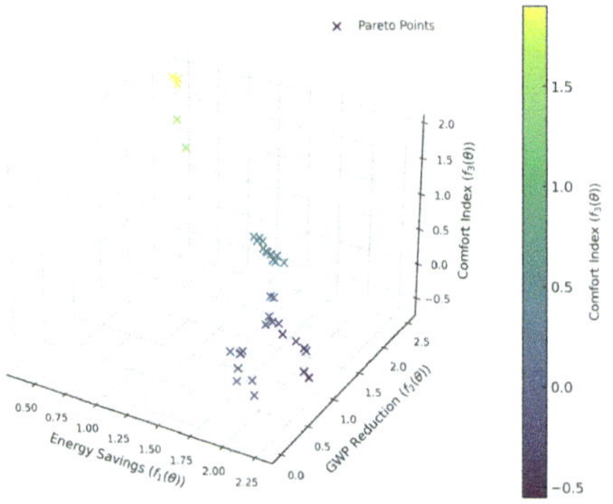

(b) Trade-off between the *Comfort Index*, $f_3(\theta)$, *GWP Reduction*, $f_2(\theta)$, and *Energy Savings*, $f_1(\theta)$.

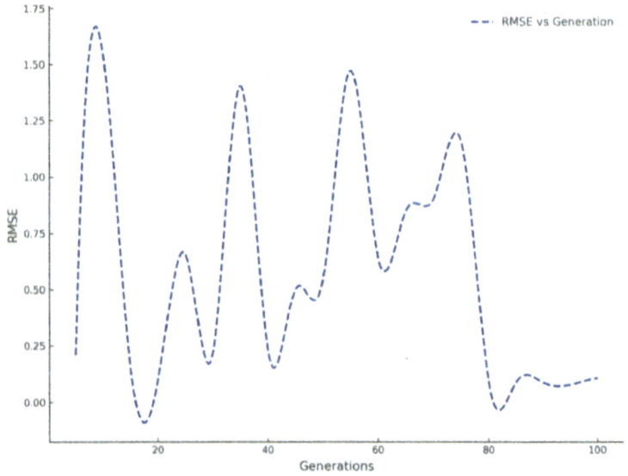

(c) RMSE accuracy evolution over generations.

7.4 Embedding Intelligent Digital Twins Into BEMS and Operational Workflows

This section shifts perspective from "*how the twin is built*" (Sect. 7.3) to "*how the twin is used*" in realistic building workflows. The same computational machinery (Surrogate-Assisted (SA) inference and Multi-Objective Optimization (MOO)) can support multiple modes of Digital Twin deployment, ranging from design-time decision support to operational tuning and grid-interactive strategies.

7.4.1 Offline Analysis Versus Online Decision Support

In building applications, a Digital Twin often operates across a spectrum between offline and online usage. Offline usage includes design exploration, retrofit evaluation, and commissioning studies where the key requirement is fidelity. Online or near-online usage emphasizes responsiveness: the Digital Twin (DT) must provide actionable guidance quickly enough to support operational decisions.

Since, building dynamics are comparatively slow, sub-minute inference is often sufficient to enable meaningful decision support, especially for supervisory setpoint adjustment, Demand Response (DR) evaluation, and operational planning.

In `PyE+`, this spectrum is addressed through the Surrogate-Assisted (SA) switching mechanism: EnergyPlus™ is invoked when retraining is needed or when model reliability is insufficient, while the surrogate provides rapid objective inference for the majority of candidate evaluations. This provides a practical pathway toward decision-capable twins which remain physically anchored while achieving operationally useful latency.

7.4.2 Applications in Energy Optimization and Demand Response

Buildings increasingly participate in Demand Response (DR) and flexibility markets, where operational decisions must balance energy, emissions, and comfort objectives under time and uncertainty constraints. *Surrogate-Assisted (SA) optimization can enable rapid evaluation of candidate setpoint strategies and operational adjustments which respect comfort while responding to grid signals.* In this context, the primary value of `PyE+` is its ability to generate high-quality trade-off solutions (Pareto-optimal sets) at a fraction of the computational cost of exhaustive simulation.

Scenario: Surrogate-Assisted DR Optimization

Consider a hotel building participating in a DR event where the grid operator requests a 15% load reduction for a 2-h window. Using conventional simulation-based optimization, evaluating 150 candidate thermostat and lighting setpoint strategies could take over 3 h. In contrast, the Surrogate-Assisted NSGA-II implemented in `PyE+` achieves comparable Pareto-optimal solutions in under 2 min by leveraging a trained surrogate model for 90% of evaluations. This acceleration enables real-time decision support, allowing the building to select a strategy which may, for example, reduce energy use by 18%, cuts GWP by 12%, and maintains an occupant comfort index above 0.7.[7]

7.4.3 Implications for Grid-Interactive and Smart Buildings

At a broader level, surrogate-enabled Digital Twins (DT) support the vision of grid-interactive and smart buildings by enabling fast scenario testing, preference-aware exploration of trade-offs, and iterative tuning of operational strategies. While the results presented in Sect. 7.3.11 demonstrate computational acceleration primarily in a simulation-based context, the framework is structured to support future coupling with real building data streams and BEMS platforms. This creates a credible pathway from simulation-driven optimization toward operational integration, consistent with the Digital Twin vision articulated earlier [10, 11] and the broader context of Cognitive Cities [13].

7.5 Limitations, Scalability, and Future Directions

While the Surrogate-Assisted (SA) Digital Twin (DT) framework presented in this chapter demonstrates substantial computational acceleration and decision-support capability, several limitations and practical considerations merit discussion.

First, the effectiveness of surrogate modeling is inherently dependent on the quality, coverage, and representativeness of the training data. Although Latin Hypercube (LHC) sampling and periodic retraining mitigate extrapolation errors, surrogate accuracy may degrade in regions of the design space which are sparsely sampled or under

[7] Note: The performance values in this scenario are illustrative and intended to demonstrate the concept rather than report empirical results.

rapidly changing operational conditions. *This highlights the importance of adaptive sampling strategies and uncertainty-aware surrogate selection, particularly when transitioning from simulation-based studies to real-world deployments.*

Second, the current implementation of `PyE+` primarily targets supervisory-level optimization and decision support, operating on time scales compatible with building thermal dynamics. Fast inner-loop control actions (e.g., actuator-level control with sub-second response requirements) remain outside the intended scope of the framework and are better addressed through dedicated control strategies embedded within Building Energy Management Systems (BEMSs).

Scalability also presents both challenges and opportunities. While Surrogate-Assisted (SA) optimization significantly reduces computational burden, extending the framework to large-scale multi-building or district-level Digital Twins (DTs) will require careful coordination of model fidelity, communication overhead, and computational resources. *Hierarchical or federated Digital Twin (DT) architectures, where local twins interact with higher-level supervisory twins, represent a promising direction for addressing such scalability concerns.*

Finally, the present framework assumes limited coupling with real-time operational data streams. Future extensions will focus on tighter integration with BEMS platforms, sensor networks, and field controllers, enabling continuous model calibration and closed-loop operation. Such integration is essential for realizing fully operational Digital Twins (DTs) which evolve alongside the physical building, rather than remaining static or periodically updated representations.

7.6 Summary and Outlook

This chapter examined how Artificial Intelligence (AI), and more specifically Computational Intelligence (CI) techniques, can augment Building Digital Twins (BDTs) to enable fast, interpretable, and decision-capable Building Energy Management (BEM). By framing intelligence as a computational and decision-enabling layer—rather than a replacement for physics-based simulation—the chapter established a pragmatic pathway toward operational Digital Twins (DTs) suitable for real-world building applications.

A surrogate-assisted Digital Twin architecture was introduced and instantiated through the `PyE+` framework, demonstrating how Machine Learning (ML) surrogates and evolutionary Multi-Objective Optimization (MOO) can dramatically reduce simulation latency while preserving physical fidelity. Through detailed formulation of objectives, penalty handling, learning mechanisms, and performance evaluation, the chapter illustrated how such architectures support energy, emissions, and comfort trade-offs within realistic building contexts.

Crucially, the discussion extended beyond algorithmic design to consider how intelligent Digital Twins (DTs) can be embedded into Building Energy Management Systems (BEMSs) and operational workflows, including Demand Response (DR) and grid-interactive scenarios. This positioning underscores the role of Digital Twins not merely as analytical tools, but as active participants within cyber-physical building ecosystems.

The subsequent chapters build directly upon this foundation. Chapter 8 provides a critical review of Integrated Building Energy Management Systems (iBEMSs), examining their architectures, capabilities, and limitations as the physical and cyber-physical substrate upon which intelligent Digital Twins (DTs) must operate. Chapter 9 then advances toward coordinated and distributed control, exploring how AI-augmented Digital Twins (DTs), iBEMS platforms, and consensus-based control strategies can be unified into a hierarchical, closed-loop framework for smart and technologically resilient buildings.

References

1. de Wilde P (2023) Building performance simulation in the brave new world of artificial intelligence and digital twins: a systematic review. Energy Build 292:113171. https://doi.org/10.1016/j.enbuild.2023.113171, https://www.sciencedirect.com/science/article/pii/S0378778823004012
2. Palley B, Martins JP, Bernardo H, Rossetti R (2025) Integrating machine learning and digital twins for enhanced smart building operation and energy management: a systematic review. Urban Sci 9(6):202. https://doi.org/10.3390/urbansci9060202
3. Sghiri A, El Bhiri B, Assoul S (2025) Digital twins and energy efficiency in buildings: a literature review. In: El Bhiri B, Assoul S, Essaaidi M (eds) Technology and the environment: implementing smart and sustainable solutions into our cities. Springer Nature Switzerland, Cham, pp 55–63
4. Hu W, Wang X, Tan K, Cai Y (2023) Digital twin-enhanced predictive maintenance for indoor climate: a parallel LSTM-autoencoder failure prediction approach. Energy Build 301:113738. https://doi.org/10.1016/j.enbuild.2023.113738, https://www.sciencedirect.com/science/article/pii/S0378778823009684
5. Liang J, Lou Y, Yu M, Bi Y, Yu K (2025) A survey of surrogate-assisted evolutionary algorithms for expensive optimization. J Membr Comput 7:108–127. https://doi.org/10.1007/s41965-024-00165-w
6. Barkanyi A, Chovan T, Nemeth S, Abonyi J (2021) Modelling for digital twins - potential role of surrogate models. Processes 9(3):476. https://doi.org/10.3390/pr9030476
7. Nuevo-Gallardo C, del Barrio IL, Iglesias MF, Trueba JBE, Bandera CF (2025) Real-Time digital twins for building energy optimization through blind control: functional mock-up units, docker container-based simulation, and surrogate models. Appl Sci 15(24):12888. https://doi.org/10.3390/app152412888
8. Wani M, Hafiz F, Swain A (2025) Accelerating building energy simulations with PyE+: a surrogate-driven AI framework for real-time decision-making. In: 2025 international conference on power engineering and electrical technology (PEET), pp 1–10. https://doi.org/10.1109/PEET65412.2025.11340811
9. Wani MH (2025) PyE+: a Python-EnergyPlus co-simulation and optimization framework with expedited computation using ML-based surrogate modeling. https://doi.org/10.5281/zenodo.15877753, version 1.0.0, Accessed: 2025-06-28
10. Ye F, Wan JD, Xie Y, X (2023) A digital twin framework of reflow soldering based on a novel high dimension surrogate model. In: Proceedings of the 2023 24th international conference on electronic packaging technology (ICEPT), pp 1–5. https://doi.org/10.1109/ICEPT59018.2023.10492063

11. Alden RE, Jones ES, Poore SB, Ionel DM (2025) Smart home HVAC digital twin ML metamodel for electric power distribution systems with solar PV and CTA-2045 controls. IEEE Trans Ind Appl 61(1):572–582. https://doi.org/10.1109/TIA.2024.3458941
12. Wani M, Hafiz F, Swain A, Ukil A (2021) Estimating thermal parameters of a commercial building: a meta-heuristic approach. Energy Build 231:110537. https://doi.org/10.1016/j.enbuild.2020.110537, https://www.sciencedirect.com/science/article/pii/S0378778820316832
13. Cui X, Liu B, Li Y, Wang Y (2025) A "Smart Model-Then-Control" strategy for the scheduling of thermostatically controlled loads. IEEE Trans Smart Grid 16(3):2246–2260. https://doi.org/10.1109/TSG.2025.3542544

Chapter 8
Integrated Building Energy Management Systems (iBEMS)

Abstract Integrated Building Energy Management Systems (iBEMS) constitute the operational backbone of modern intelligent buildings, providing sensing, actuation, communication, and supervisory control across heterogeneous subsystems. As buildings evolve from passive energy consumers into active cyber–physical entities interacting with grids, storage, and occupants, the role of iBEMS has expanded significantly. However, despite advances in integration and automation, most existing iBEMS implementations remain fundamentally execution-oriented, with limited predictive, learning, or system-level decision-making capability. This chapter presents a critical and pedagogical review of iBEMS architectures, examining their evolution from conventional Building Management Systems (BMS), their communication and interoperability foundations, functional decomposition, and energy-centric and multi-energy coordination strategies. Current intelligence paradigms—including rule-based control, optimization-driven approaches, and AI-enhanced modules—are systematically reviewed, highlighting both their practical benefits and their structural limitations. The chapter asserts that iBEMS should not be treated as standalone intelligent systems, but rather as the physical–cyber execution layer underpinning Building Digital Twins (BDTs). Within this layered interpretation, Digital Twins (DTs) provide continuously synchronized system representations, predictive modeling, and learning-enabled decision support, while iBEMS deliver reliable interaction with the physical building. This reframing clarifies the architectural boundary between execution and cognition and establishes the necessity of Digital Twin (DT) augmentation for autonomous and adaptive building operation. By synthesizing these insights, the chapter motivates a shift toward distributed and coordinated building intelligence, setting the stage for consensus-based control frameworks and multi-agent coordination strategies explored in the subsequent chapter.

Keywords Building control · Energy-centric BEMS · Integrated building energy management system · Multi-energy coordination

M. H. Wani et al., *Digital Twins for Building Optimization: Enhancing Energy and Comfort*, Studies in Infrastructure and Control,
https://doi.org/10.1007/978-3-032-23274-8_8

8.1 Introduction and Motivation

This chapter provides a critical and pedagogical review of Integrated Building Energy Management Systems (iBEMS),[1] positioning them as the *physical–cyber execution layer* underpinning intelligent Building Digital Twins (BDTs). While Chaps. 4–7 established the theoretical foundations of Computational Intelligence (CI), learning-enabled optimization, and Surrogate-Assisted (SA) Digital Twin (DT) architectures, the realization of these concepts in real buildings ultimately depends on the capabilities of the underlying Building Energy Management (BEM) infrastructure.

Conventional Building Management Systems (BMS) were historically designed to provide local monitoring and rule-based control of building subsystems such as HVAC, lighting, and electrical distribution. Although effective for supervisory control, such systems were not conceived to support predictive analytics, coordinated optimization, or system-level decision-making under uncertainty. As buildings evolved into active participants within energy systems—interacting with grids, storage, renewables, and occupants—the limitations of "siloed BMS" architectures became increasingly evident.

Integrated Building Energy Management Systems (iBEMSs) emerged as a response to these challenges. By incorporating energy-centric monitoring, communication interoperability, and coordinated control across multiple subsystems, iBEMSs aim to provide a unified operational view of building energy behavior. However, despite significant advances in sensing, communication, and control integration, most existing iBEMS implementations remain reactive in nature, with limited predictive or decision-making capability.

This chapter asserts that iBEMS should be understood not as an alternative to Digital Twins (DTs), but as their indispensable physical substrate. Intelligent Digital Twins (DTs) require reliable sensing, actuation, data aggregation, and supervisory control—all of which are delivered by iBEMS. At the same time, iBEMSs require augmentation through intelligence and predictive models to move beyond static or rule-based operation. Establishing this mutual dependency is essential for understanding how intelligent Digital Twins (DTs) can be deployed in practice.

---

[1] **Note on Terminology—"Intelligent" versus "Integrated" BEMS:** While the author's PhD thesis and the rest of this book refer to BEMS as "*intelligent*", it is important to note the subtle difference between *Intelligent BEMS* and *Integrated BEMS*. Clarification:

- **Integrated BEMS:** Primarily focuses on connecting multiple building systems (HVAC, lighting, security, etc.) into a single platform for centralized monitoring and control.
- **Intelligent BEMS:** Goes beyond integration by incorporating advanced analytics, AI, and adaptive algorithms for predictive and autonomous optimization.
- **Key Difference:** Integration = connectivity and rule-based control; Intelligence = learning, proactive decision-making, and optimization.
- **Overlap:** An integrated system with smart features may appear "intelligent", but true intelligence implies self-learning and adaptive control beyond mere integration.

The objective of this chapter is therefore threefold: (i) to trace the evolution from conventional BMS to modern iBEMS, (ii) to critically assess the architectural and functional limitations of current iBEMS implementations, and (iii) to position iBEMS as the execution layer within AI-augmented Digital Twin (DT) frameworks. This discussion provides the necessary groundwork for Chap. 9, which explores distributed coordination and consensus-based control mechanisms built upon this foundation.

8.2 From Conventional BMS to Integrated BEMS (iBEMS)

The transition from conventional Building Management Systems (BMS) to Integrated Building Energy Management Systems (iBEMS) reflects a broader shift in the role of buildings within modern energy systems. Early BMS architectures were primarily designed to provide localized monitoring and supervisory control of individual subsystems, whereas iBEMSs aim to coordinate energy-related functions across multiple domains, time scales, and stakeholders.

This evolution has been documented in both early integrated energy management frameworks and more recent studies emphasizing predictive and online control capabilities [1, 2]. Nonetheless, as this section critically examines, integration by itself does not ensure intelligence or effective decision-making capability.

8.2.1 Conventional BMS Architectures and Their Limitations

Traditional BMS architectures are typically hierarchical and vendor-centric, comprising field devices (sensors and actuators), local controllers, and supervisory control stations. Their primary functions include maintaining indoor environmental conditions within predefined setpoints, executing schedules, and responding to alarms. Control logic is predominantly rule-based, relying on static thresholds and deterministic control sequences.

While such systems provide robustness and reliability, they exhibit several fundamental limitations. First, subsystem-level optimization dominates, with minimal coordination between HVAC, lighting, plug loads, and on-site energy resources. Second, control actions are largely reactive, responding to measured deviations rather than anticipating future conditions. Third, conventional BMS architectures are poorly suited to accommodate uncertainty arising from weather variability, occupant behavior, or dynamic pricing signals.

As highlighted by Antunes et al. [1], early integrated management efforts were constrained by both computational and communication limitations, resulting in centralized supervisory control structures with limited adaptability. These constraints fundamentally limit the ability of traditional BMS to support advanced energy management objectives.

Scenario: Traditional BMS in Action

Building Context: A 20-story commercial office building in Dubai (UAE) uses a conventional BMS architecture. The system comprises temperature sensors, VAV controllers, and a supervisory workstation.
Situation: At 2 : 00 PM, outdoor temperature rises sharply due to an unexpected heatwave. Indoor zones begin to exceed the cooling setpoint of 24 °C.

System Response:

- The BMS detects deviations and commands chillers to ramp up.
- Lighting remains at full intensity despite abundant daylight.
- Plug loads and on-site PV generation are ignored in decision-making.

Outcome: Energy demand spikes, peak tariffs apply, and occupant comfort is restored only after a delay. The system reacts to conditions but fails to anticipate the heatwave or optimize across HVAC, lighting, and energy resources.
Key Limitation: Control remains reactive and siloed, illustrating the lack of predictive capability and cross-domain coordination in traditional BMS.

8.2.2 Emergence of Energy-Centric Integration

The emergence of energy-centric integration marked a decisive shift from comfort-driven supervisory control toward holistic energy management. This transition was motivated by rising energy costs, environmental concerns, and the increasing penetration of distributed energy resources. Integrated BEMS (iBEMS) architectures began to incorporate energy metering, Demand-Side Management (DSM), and coordination across multiple subsystems.

Antunes et al. proposed one of the early conceptualizations of an integrated energy management system, emphasizing the coordination of energy supply and demand within buildings through centralized optimization [1]. While this approach represented a significant step forward, it largely relied on static models and offline optimization, limiting its responsiveness in dynamic operational contexts.

More recent work by Zhu et al. represents a further step toward online control, in which fast prediction models are used to estimate indoor environmental conditions for HVAC regulation [2]. Crucially, though, these predictive models are integrated into particular control loops rather than incorporated into a comprehensive Digital Twin (DT) of the building. Consequently, their integration is limited to specific functions instead of encompassing the entire system.

8.2.3 Defining Characteristics of iBEMS

Based on the existing literature [1–3], iBEMS can be characterized by three defining features: (i) integration across multiple building subsystems and energy carriers, (ii) unified data acquisition and supervisory control infrastructure, and (iii) explicit consideration of energy performance alongside comfort and operational constraints.

Nevertheless, a critical examination reveals that most existing iBEMS implementations focus on *structural integration* rather than *cognitive integration* [4]. While data and control pathways are unified, decision-making remains largely localized, reactive, or rule-driven. Predictive modeling, uncertainty handling, and multi-objective trade-off analysis are typically addressed in isolation, if at all.

This observation is particularly evident in fast-prediction-based HVAC control frameworks such as that of Zhu et al., where Machine Learning (ML) is employed to accelerate control decisions but without explicit coupling to system-level optimization or Digital Twin feedback loops [2]. Consequently, while such approaches improve responsiveness, they do not fully realize the potential of intelligent, decision-capable iBEMS.

Conceptual Positioning of iBEMS within Intelligent Digital Twin Architectures

Figure 8.1 conceptually summarizes the role of iBEMS within intelligent building architectures. Rather than replacing iBEMS, Digital Twins build upon them by introducing predictive models, optimization, and learning-enabled decision support. This layered perspective clarifies why iBEMS are a necessary but insufficient condition for intelligent building operation, motivating the deeper architectural and coordination discussions developed in subsequent sections.

8.3 Communication and Interoperability Foundations

Communication infrastructure is a foundational enabler of Integrated Building Energy Management Systems (iBEMS). Figure 8.2 illustrates the notion of *communication layers* in iBEMS.[2] Without reliable, interoperable, and scalable communication across heterogeneous field devices, higher-level energy coordination, predictive analytics, and Digital Twin (DT) feedback loops cannot be realized. This section reviews field-level protocol heterogeneity and examines standardized and hybrid communication frameworks which have shaped modern iBEMS architectures.

---

[2] IoT in "Protocol & Interoperability Layer" of Fig. 8.2 indicates support for IoT protocols such as *MQTT, CoAP, AMQP*. Note that, IoT itself is **not** a protocol.

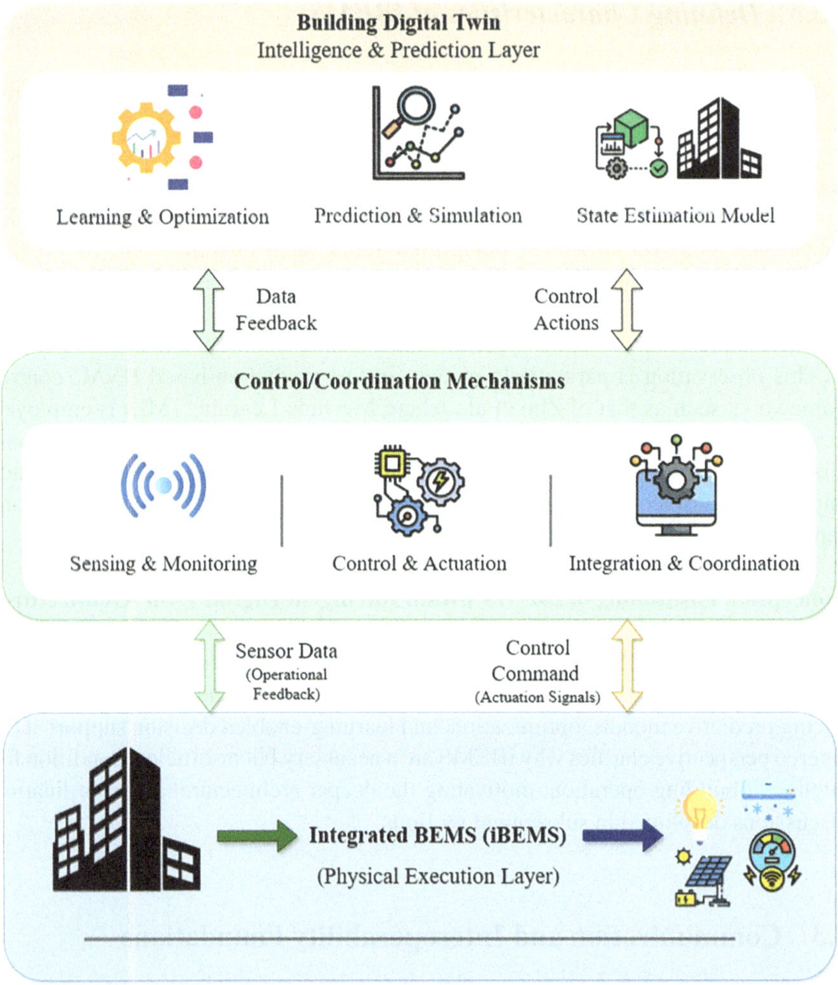

Fig. 8.1 Conceptual relationship between Integrated BEMS (iBEMS), Building Digital Twins (BDTs), and control/coordination mechanisms

8.3.1 Field-Level Integration and Protocol Heterogeneity

At the field level, iBEMS must integrate a diverse ecosystem of sensors, meters, actuators, and controllers deployed across HVAC, lighting, electrical, and auxiliary subsystems. Historically, these devices have relied on a variety of communication protocols—often proprietary or subsystem-specific—resulting in fragmented automation islands.

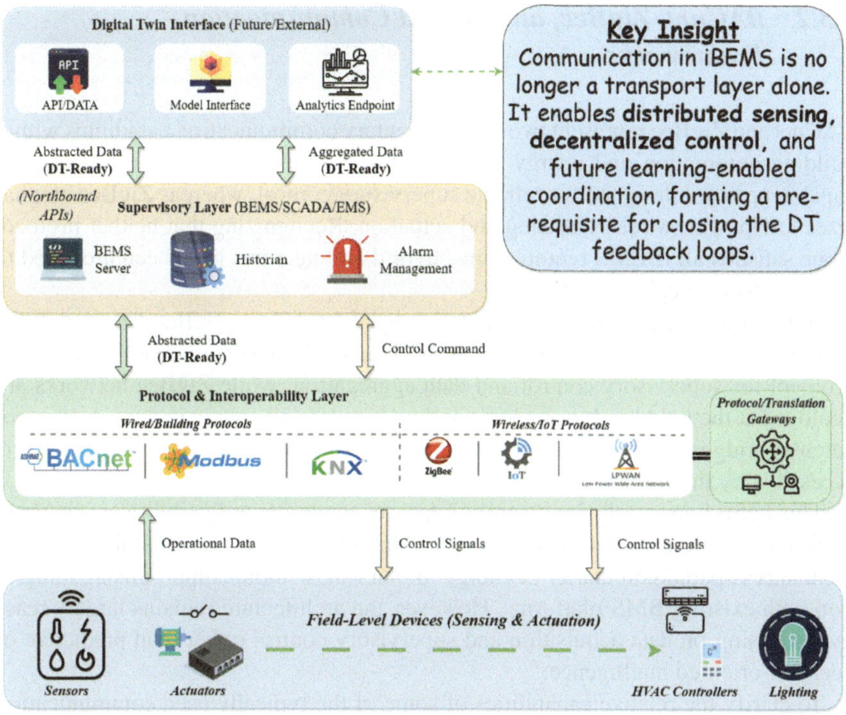

Fig. 8.2 Communication (Protocol) and interoperability layers in Integrated Building Energy Management Systems (iBEMS)

This heterogeneity poses *three* fundamental challenges:

- First, *interoperability* across vendors and subsystems becomes non-trivial, increasing integration cost and engineering effort.
- Second, *data consistency* and *semantic alignment* are difficult to ensure when devices expose heterogeneous data models.
- Third, *scalability* is constrained, as adding new devices or functionalities often requires protocol gateways or custom `middleware`.

Early Building Automation Systems (BAS) addressed this issue through standardized fieldbus protocols such as `BACnet`, `KNX`, and `LonWorks`, which enabled multi-vendor interoperability at the supervisory level. However, as energy monitoring requirements expanded and wireless sensing proliferated, traditional wired protocols alone proved insufficient—*particularly for retrofit scenarios and fine-grained energy metering*. This gap motivated the adoption of low-power wireless protocols and hybrid communication architectures in iBEMS.

8.3.2 *BACnet, ZigBee, and Hybrid Communication Frameworks*

BACnet and ZigBee represent two complementary communication paradigms within building automation and energy management systems. BACnet is a mature, IP-capable protocol designed for reliable supervisory control, whereas ZigBee emphasizes low-power, wireless sensing and actuation. Recognizing that neither protocol alone satisfies all iBEMS requirements, hybrid architectures have been proposed to combine their strengths.

Hong et al. [5] present a representative hybrid BACnet–ZigBee framework for building energy management. In their architecture, BACnet operates as the backbone protocol for supervisory control and data aggregation, while ZigBee networks are deployed at the field level to support wireless sensors and energy meters. A protocol gateway bridges the two domains, translating ZigBee device data into BACnet objects accessible by the BEMS.

This hybrid approach significantly improves deployment flexibility, particularly in existing buildings where installing wired infrastructure is impractical. Moreover, it enhances spatial granularity of energy monitoring without compromising integration with existing BMS platforms. However, the architecture remains largely reactive, focusing on data acquisition and supervisory control rather than predictive or decision-oriented intelligence.

To clarify the relative capabilities of some of the typically used communication paradigms (protocols) within industry, Table 8.1 summarizes their characteristics in the context of iBEMS and Digital Twin (DT) feasibility.[3]

The comparison highlights that hybrid communication frameworks offer the most practical pathway toward Digital Twin-enabled iBEMS, provided they are augmented with semantic data models, synchronization mechanisms, and higher-level intelligence layers.

8.3.3 *IoT-Enabled Architectures for iBEMS*

The integration of Internet of Things (IoT) technologies into building automation has further expanded the communication capabilities of iBEMS. Ożadowicz and Grela [7] propose an IoT-enabled energy metering architecture which embeds intel-

[3] While BACnet and KNX were traditionally designed as *wired* communication protocols for building automation, utilizing physical wiring like twisted pair cables—both standards have evolved to include options for wireless and IP-based communication as well, however, in limited capacities. KNX offers KNX RF (Radio Frequency) for wireless retrofits, but coverage and robustness are weaker compared to wired KNX TP, making it less suitable for large or interference-prone environments. BACnet/IP can leverage Wi-Fi infrastructure for wireless connectivity, yet this is not optimized for low-power IoT devices and lacks standardized support for dense sensor networks.

Table 8.1 Communication/platform frameworks for iBEMS in the context of digital twin (DT) feasibility

Framework	Developers/Bodies	Strengths	Limitations	DT feasibility
BACnet	ASHRAE	Standardized objects, robust supervisory control, vendor interoperability	Mostly wired at field level; limited dense low-power sensing; higher install cost	High for DT supervisory layers; limited for fine-grained real-time sensing
ZigBee	ZigBee Alliance	Low power, wireless, scalable sensing; cost-effective retrofits	Lower bandwidth and range; weak native system semantics	Moderate; solid for sensing layer, not sufficient alone for DT integration
Hybrid BACnet–ZigBee	Multi-vendor/Research [5]	Combines robust control with flexible sensing; retrofit-friendly	Gateway dependence; added complexity; integration latency	High when paired with common data models and intelligence layers
KNX	KNX Association	Open standard; strong room/zone control across HVAC, lighting, shading; multi-media (TP, IP, RF); broad vendor ecosystem [6]	Project-specific engineering (ETS); backbone integration to enterprise BMS required for portfolio-scale coordination	High at field/room layer; DT-ready when bridged to BACnet/IP or IP data hubs
EcoStruxure (EBO)	Schneider Electric	Open, IP-centric BMS; natively supports BACnet/Lon/Modbus + MQTT; unified data & control across HVAC, lighting, power, microgrids/EV; cloud/analytics	Vendor platform; integration quality depends on connectors/APIs and data modeling governance	High: portfolio-level DT via unified data backbone, APIs, and analytics-ready architecture
IPv6-based (e.g., 6LoWPAN, Thread)	IETF/Thread Group	Massive address space; native IP integration; scalable IoT; strong security features	Requires IPv6 infrastructure; limited legacy device support; higher initial complexity	Very high feasibility; ideal for DT-ready IoT layers and edge-cloud integration

ligence directly at the field level using `IPv6-based` communication, functional profiles, and interoperable network variables.

Their work demonstrates how IoT-enabled energy meters can function as autonomous nodes within an integrated BEMS (iBEMS), supporting energy measurement, historical logging, and demand monitoring. By adopting standardized functional blocks and peer-to-peer communication principles, the proposed architecture enhances interoperability and scalability across heterogeneous devices.

Importantly, this approach does not introduce a Digital Twin in the strict sense. Instead, it provides the foundational data acquisition and interoperability mechanisms upon which higher-level intelligence could be built. The contribution is therefore best understood as enabling *Digital Twin readiness*, rather than realizing a full Digital Twin.

From a pedagogical perspective, IoT-based iBEMS architectures illustrate a critical shift: communication is no longer merely a transport mechanism, but an active enabler of distributed sensing, decentralized control, and future learning-enabled coordination. These capabilities are essential prerequisites for closing the feedback loops required by intelligent Digital Twins.

8.4 Functional Decomposition of iBEMS

Integrated Building Energy Management Systems (iBEMS) are best understood not as monolithic platforms, but as layered cyber–physical architectures in which sensing, actuation, control, and data management functions are distributed across multiple abstraction levels. *This functional decomposition is essential for scalability, interoperability, and real-time operability, and forms the structural basis upon which Digital Twin (DT) intelligence can be deployed.*

Figure 8.3 illustrates a representative integrated BEMS (iBEMS) communication architecture commonly deployed in practice. The figure highlights the layered decomposition of sensing, control, integration, and management functions across HVAC, security, fire, lighting, and auxiliary subsystems.

It is important to note that this architecture, while highly integrated, remains fundamentally execution-oriented. Intelligence, prediction, and decision-making are largely external to this structure, motivating the need for Digital Twin (DT) augmentation.

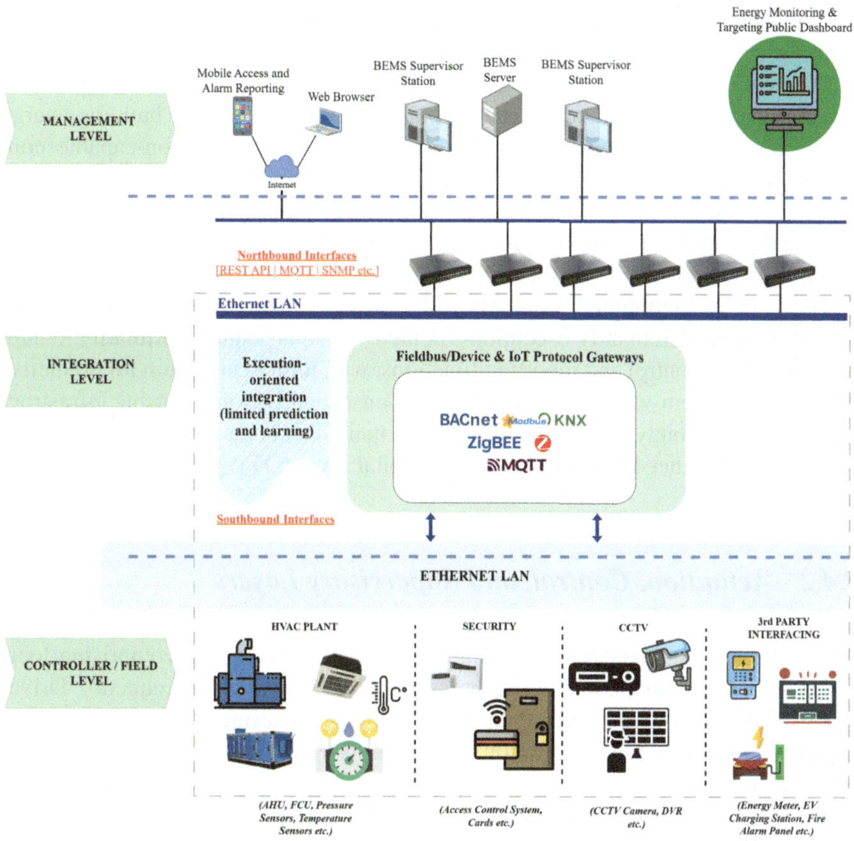

Fig. 8.3 Representative functional decomposition of an Integrated Building Energy Management System (iBEMS), illustrating sensing, control, integration, and management layers across multiple building subsystems. While execution-complete, such architectures lack intrinsic predictive, learning, and decision-making intelligence, motivating Digital Twin (DT) augmentation

The subsequent subsections examine these layers in detail, beginning with the sensing and metering layer which forms the foundation for actuation, supervisory control, and ultimately, data aggregation and abstraction.

8.4.1 Sensing, Metering, and Data Acquisition Layers

The sensing and data acquisition layer constitutes the *physical interface* between the built environment and the digital infrastructure of iBEMS. This layer encompasses environmental sensors (temperature, humidity, CO_2), electrical meters, flow meters, and status sensors embedded across building subsystems. Its primary function is the reliable and timely acquisition of raw operational data required for monitoring and control.

Ozadowicz and Grela [7] emphasize the critical role of IoT-enabled energy meters and distributed sensing nodes in enabling fine-grained visibility of building energy flows. By integrating smart metering devices with communication-capable controllers, the study demonstrates how energy consumption data can be captured at subsystem and component levels rather than solely at the building aggregate. This shift enables more granular diagnostics and localized control actions, particularly for energy-intensive systems such as HVAC.

However, as highlighted implicitly in [7], sensing layers in contemporary iBEMSs remain predominantly execution-oriented.[4] Data is acquired primarily to support immediate control and reporting functions, with limited provision for predictive inference or system-wide contextualization. Consequently, while sensing infrastructures are increasingly dense and connected, their data streams are not inherently structured for higher-level intelligence or Digital Twin (DT) synchronization.

8.4.2 Actuation, Control, and Supervisory Layers

Above the sensing layer lies the actuation and control hierarchy, comprising local controllers, Programmable Logic Controllers (PLCs), Variable Frequency Drives (VFDs), and supervisory control stations. These layers execute deterministic control logic to regulate equipment operation based on sensor feedback, predefined setpoints, and scheduling rules.

In most deployed iBEMS, control logic is hierarchically organized, with fast inner loops handling equipment-level dynamics and slower supervisory loops coordinating subsystem behavior. While this structure ensures stability and robustness, it inherently constrains adaptability. *Control decisions are typically reactive and rule-based, lacking foresight into future disturbances, occupant behavior, or energy market signals.*

> Importantly, the functional separation between control execution and decision intelligence becomes evident at this level. Control layers are optimized for real-time responsiveness and safety, whereas strategic decision-making—if present at all—is externalized to analytics platforms or human operators. This disconnect directly motivates the need for Digital Twin augmentation discussed in later sections.

---

[4] Meaning the sensing layer is not used for more advanced purposes like analysis, prediction, or high-level decision-making.

8.4.3 Data Management, Aggregation, and Abstraction

The data management and aggregation layer represents the highest functional abstraction within conventional iBEMS. Its role is to consolidate heterogeneous data streams from distributed sensing and control nodes into unified representations suitable for visualization, reporting, and high-level coordination.

Das et al. [8] provide a representative example of energy-centric aggregation strategies within building systems. Their work demonstrates how subsystem-level measurements—particularly from air-conditioning units—can be aggregated and abstracted to enable coordinated energy management decisions based on grid conditions, tariff structures, and power quality considerations. By operating at the dc bus level and exploiting aggregation of electrical variables, the proposed methodology reduces grid stress while maintaining uninterrupted HVAC operation.

From an iBEMS perspective, [8] illustrates both the strength and limitation of current aggregation approaches. While effective in consolidating control-relevant variables and enabling energy-aware operation, aggregation remains tightly coupled to specific subsystems and predefined control objectives. Data abstraction is functional rather than semantic; system states are summarized for control execution rather than interpreted within a broader building-wide or predictive context.

As a result, although modern iBEMS can aggregate and manage large volumes of operational data, they lack a unified, continuously calibrated system representation. This limitation fundamentally restricts their ability to support Digital Twin (DT) functionalities such as predictive simulation, adaptive optimization, and bidirectional learning.

Collectively, the functional decomposition presented in this section reveals that iBEMS architectures are highly capable execution platforms, yet structurally incomplete from an intelligence standpoint. The sensing, control, and aggregation layers provide the necessary physical and cyber foundations, but the absence of integrated predictive and decision-making intelligence necessitates the introduction of Digital Twin (DT) frameworks, as explored in subsequent sections.

8.5 Energy-Centric and Multi-energy iBEMS Architectures

As buildings transition from passive energy users to active cyber–physical entities embedded in wider energy networks, iBEMS architectures have gradually moved away from single-domain supervisory control toward explicitly *energy-centric* and *multi-energy coordination* frameworks (see Fig. 8.4); a high-level comparison between these two paradigms is provided in Table 8.2. This evolution reflects the growing penetration of Distributed Energy Resources (DERs), sector coupling technologies, and grid-interactive operation requirements, which collectively demand tighter coordination across electrical, thermal, and storage subsystems.

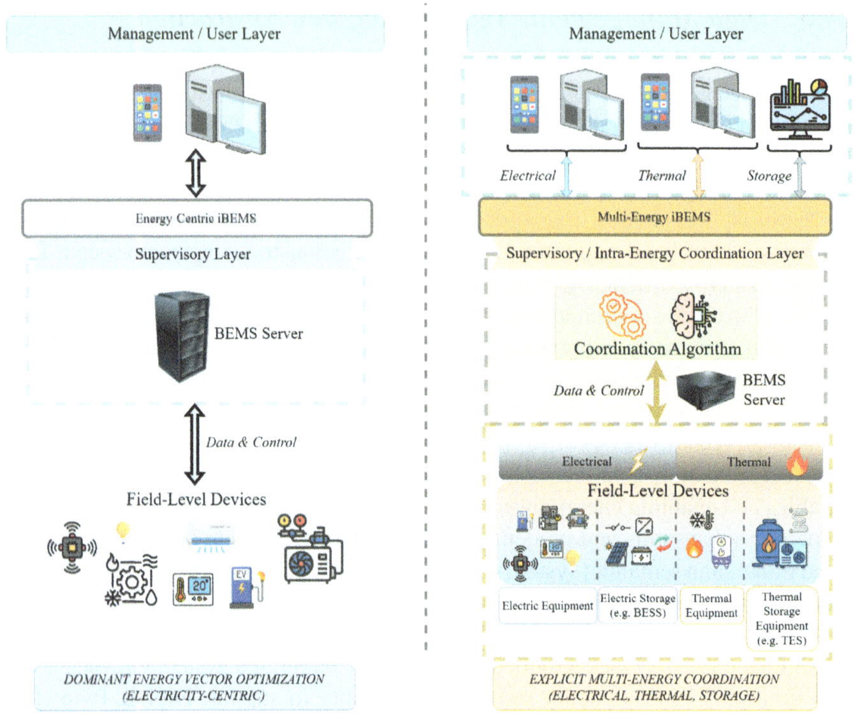

Fig. 8.4 Conceptual comparison between energy-centric and multi-energy iBEMS architectures. Energy-centric iBEMS primarily optimizes a dominant energy vector (typically electricity), whereas multi-energy iBEMS explicitly coordinates electrical, thermal, and storage subsystems across hierarchical control layers

Unlike conventional BMS designs which treat energy flows implicitly or in isolation, energy-centric iBEMS architectures explicitly model, monitor, and coordinate multiple energy vectors, often across heterogeneous temporal and spatial scales. This section reviews representative architectural patterns for multi-energy coupling and hierarchical energy management, highlighting both their capabilities and structural limitations in the absence of Digital Twin (DT) augmentation.

8.5.1 Electrical, Thermal, and Multi-energy Coupling

Early energy-centric iBEMS research focused on improving the coordinated operation of electrical and thermal subsystems within individual buildings, particularly under uncertainty and fluctuating demand conditions. In this context, [9, 10] illustrate the transition from rule-based coordination toward optimization-driven multi-energy management.

Table 8.2 Comparison between energy-centric and multi-energy iBEMS architectures

Aspect	Energy-centric iBEMS	Multi-energy iBEMS
Primary focus	Optimization of a dominant energy vector (typically electrical energy)	Coordinated management of multiple energy vectors (electrical, thermal, storage, gas)
Energy coupling	Implicit or loosely coupled subsystems	Explicit modeling and coordination of interdependent energy flows
Control objective	Energy efficiency, peak reduction, cost minimization	Multi-objective trade-offs including efficiency, flexibility, robustness, and resilience
Temporal scope	Single or limited time scale (often supervisory)	Multi-time-scale coordination (planning, supervisory, and fast control)
System awareness	Subsystem-level visibility	System-level energy interaction awareness
Scalability	Effective for single buildings or isolated subsystems	Better suited for complex buildings and grid-interactive operation
Adaptability	Limited adaptability to uncertainty and non-stationarity	Improved robustness under uncertainty, but still model-dependent
Digital twin readiness	Partial (execution-focused, limited prediction)	Higher structural compatibility, but lacking continuous synchronization
Representative studies	Energy-aware HVAC and electrical coordination	Bi-level optimization, hierarchical MPC, multi-energy co-simulation

For instance, the bi-level energy management framework presented in [10], formulates building energy coordination as a hierarchical optimization problem, accounting for electrical and thermal interactions *while explicitly handling uncertainties in demand, renewable generation, and market signals.* The proposed architecture demonstrates that coordinated multi-energy control can significantly improve operational robustness and cost efficiency when compared to decoupled strategies.

However, such formulations typically rely on static or slowly updated models and assume well-defined system boundaries. While effective for "day-ahead" or "intra-day" optimization, these approaches exhibit limited adaptability to real-time disturbances, occupant-driven variability, and evolving system configurations—constraints which become increasingly pronounced in large-scale or highly dynamic deployments.

Hybrid multi-model co-simulation frameworks, such as the one proposed by [11], further extend multi-energy coordination by enabling the joint simulation of electrical networks, building energy systems, and control layers within a unified environment.

By coupling software-based simulators with *real-time hardware-in-the-loop* components, this approach highlights the feasibility of integrating iBEMS operation with grid-level services, such as voltage regulation and flexibility provision.

From an architectural standpoint, these studies collectively demonstrate that multi-energy coupling is technically achievable within iBEMS. Nevertheless, the coupling logic remains largely external to the building control fabric itself, relying on pre-defined coordination schemes rather than continuously evolving system representations.

8.5.2 Hierarchical and Multi-time-Scale Energy Management

The increasing complexity of multi-energy buildings has motivated the adoption of hierarchical and multi-time-scale control architectures within iBEMS, see Fig. 8.5. These architectures decompose energy management tasks across supervisory, coordination, and local control layers, each operating at distinct temporal resolutions.

A representative example is the modular hierarchical Model Predictive Control (MPC) framework proposed by [11], which separates long-term planning, mid-term coordination, and fast local control. Such hierarchical decomposition improves scalability and computational tractability while enabling coordinated decision-making across multiple subsystems.

Despite these advantages, hierarchical iBEMS architectures often rely on simplified predictive models and assume reliable state estimation across layers. As a result, discrepancies between modeled and actual system behavior can propagate through the hierarchy, degrading control performance. Moreover, while multi-time-scale structures manage complexity effectively, they remain fundamentally execution-oriented, focusing on control action delivery rather than continuous system understanding.

Crucially, the above reviewed approaches treat intelligence as an add-on layer rather than as an intrinsic property of the energy management architecture. Predictive capabilities are typically embedded within optimization routines or MPC formulations, with limited feedback from real-world performance to model structure or parameters. This separation constrains the ability of iBEMS to adapt autonomously over time, particularly under non-stationary operating conditions.

Synthesis and Implications. The literature reviewed in Sect. 8.5 demonstrates that energy-centric and multi-energy iBEMS architectures represent a substantial advancement over conventional BMS designs. By enabling explicit coordination

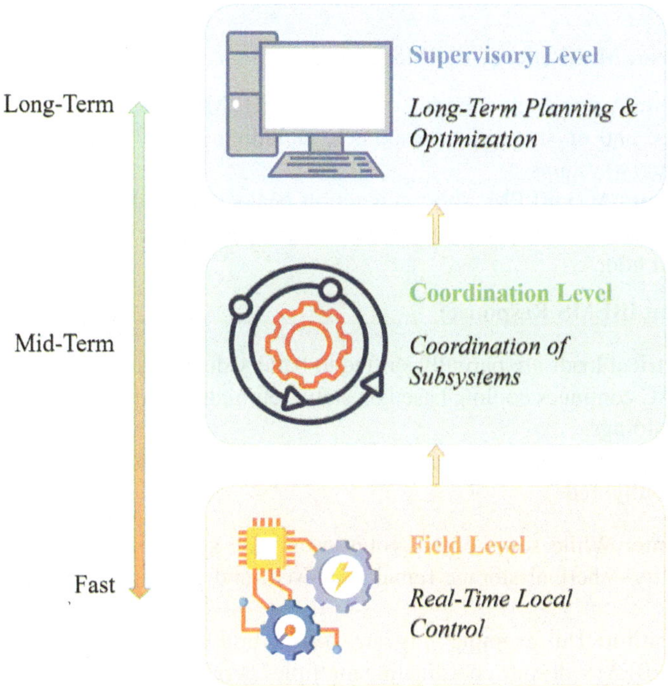

Fig. 8.5 Hierarchical and multi-time-scale energy management architecture in iBEMS. Long-term supervisory layers handle planning and optimization, mid-level coordination layers manage subsystem interactions, and fast local controllers execute real-time control actions. Intelligence is typically embedded within optimization routines rather than continuously synchronized system representations (like physics-based models)

across energy vectors and temporal scales, these systems lay the structural groundwork for grid-interactive and flexible building operation.

However, the absence of continuously synchronized system representations limits their ability to evolve beyond preconfigured coordination strategies. While optimization and hierarchical control improve performance within known operating envelopes, they do not inherently provide the persistent situational awareness, bidirectional data flow, and adaptive intelligence required for truly autonomous operation.

These limitations motivate the integration of Digital Twin (DT) frameworks, wherein iBEMS serve as the *physical execution layer*, and higher-level intelligence emerges from continuous model synchronization, learning, and predictive feedback. This transition is examined in the subsequent sections and forms the conceptual bridge toward the distributed intelligence and consensus-based control paradigms discussed in Chap. 9.

Scenario: Multi-energy iBEMS in Action

Context: A university campus operates an iBEMS which manages HVAC, lighting, and on-site energy resources, including solar PV, thermal storage, and battery systems.
Situation: At 3:00 PM, solar generation peaks while outdoor temperatures rise. The campus faces a dynamic tariff signal indicating high grid prices for the next hour.

Current iBEMS Response:

- Electrical loads are partially shifted to battery discharge.
- HVAC continues cooling based on static setpoints without considering thermal storage.
- Coordination between electrical and thermal subsystems is rule-based and preconfigured.

Outcome: While some cost savings occur, the system fails to exploit full flexibility—thermal storage remains unused, and predictive optimization is absent.
Implication: This example illustrates a structural limitation of current multi-energy iBEMS: despite coordinating multiple energy carriers, decision-making remains largely static and non-predictive, resembling traditional energy-centric approaches. A Digital Twin (DT) framework could continuously synchronize models, forecast conditions, and optimize across energy vectors, enabling proactive strategies such as pre-cooling using thermal storage before peak tariffs.

8.6 Intelligence in iBEMS: Current Practices

While modern iBEMS architectures exhibit increasing levels of automation and optimization, the notion of "intelligence" within these systems remains narrowly scoped. In most deployed systems, intelligence is embedded as task-specific logic layers designed to improve operational efficiency, rather than as a system-level capability enabling learning, adaptation, and anticipatory decision-making. This section critically examines prevailing intelligence paradigms in iBEMS, highlighting both their practical strengths and their structural limitations.

8.6.1 *Rule-Based and Optimization-Based iBEMS*

Historically, intelligence in iBEMS has been dominated by `rule-based` and `optimization-based` strategies. Table 8.3 provides a high-level comparison of both strategies. Rule-based systems encode expert knowledge in the form of deterministic "*if–then*" logic, schedules, and threshold-based responses. These approaches are widely deployed due to their transparency, predictability, and ease of commissioning.

Optimization-based iBEMS extend this paradigm by formulating energy management tasks as "*constrained optimization problems*". Typical objectives include energy–cost minimization, peak demand reduction, and comfort maintenance, often solved using linear, nonlinear, or Model Predictive Control (MPC) techniques. As reviewed in Sect. 8.5, hierarchical MPC and bi-level optimization frameworks enable coordinated control across energy vectors and time scales.

Despite their effectiveness within predefined operating envelopes, both rule-based and optimization-driven approaches exhibit fundamental limitations. Rule-based logic lacks adaptability and scales poorly as system complexity increases. Optimization-based methods, while more flexible, rely heavily on accurate models, fixed constraints, and assumed disturbance profiles. In practice, model mismatch, uncertainty, and evolving building usage patterns can significantly degrade performance over time.

Crucially, in both cases, intelligence is procedural rather than representational: decision-making logic is executed *on* the system, not *within* a continuously evolving system representation.

Table 8.3 Taxonomy of intelligence paradigms in iBEMS

Paradigm	Primary Function	Strengths	Limitations
Rule-Based Control	Equipment scheduling, setpoint enforcement	Transparent, easy to deploy, low computation	Rigid, non-adaptive, poor scalability
Optimization-Based Control	Peak shaving, tariff optimization, MPC	Performance guarantees, constraint handling	Model-dependent, execution-oriented
Data-Driven Prediction	Load, price, or demand forecasting	Improved short-term accuracy	Limited feedback integration
AI-Enhanced Modules	Demand Response (DR), storage dispatch	Improved efficiency in narrow tasks	Isolated intelligence, offline learning

8.6.2 AI-Enhanced iBEMS: Capabilities and Gaps

Recent advances in Artificial Intelligence (AI) and Machine Learning (ML) have led to growing interest in AI-enhanced iBEMS, with data-driven models applied to load forecasting, indoor environment prediction, Demand Response (DR), and energy pricing strategies. Representative implementations include ANN-based Demand Response (DR) models for optimizing retail electricity pricing and energy storage dispatch under time-varying tariffs [12], as well as integrated frameworks combining electrical and thermal energy management using forecasting-based operational control algorithms [13]. These approaches demonstrate tangible benefits in reducing peak demand and energy costs, particularly in data-rich scenarios where explicit physical modeling is challenging.

However, in most reported deployments, AI components are introduced as isolated functional modules rather than as integral parts of the iBEMS architecture. Learning models are typically trained offline, applied in fixed control pipelines, and operate in parallel with conventional supervisory logic—providing predictions or recommendations without tight coupling to actuation, system state evolution, or long-term learning. Feedback from real-world performance rarely informs model structure or decision logic in a systematic manner, limiting adaptability over time.

From a Digital Twin (DT) perspective, this represents a critical shortcoming. While AI, in the above context, improves local performance metrics, it does not, by itself, establish a persistent, synchronized, and interpretable system representation capable of supporting autonomous decision-making and continuous optimization under dynamic conditions.

Synthesis: What Is Missing in Current iBEMS Intelligence?

Existing iBEMS implementations primarily treat intelligence as an add-on capability rather than as an intrinsic system property. Learning models support isolated tasks such as forecasting or Demand Response (DR), but lack continuous feedback, self-calibration, and cross-layer coordination. As a result, most iBEMSs remain execution-oriented systems which optimize predefined objectives without developing situational awareness or long-term adaptability.

This structural limitation prevents iBEMS from functioning as autonomous cyber–physical systems. Bridging this gap requires Digital Twin (DT)–enabled architectures capable of unifying sensing, prediction, optimization, and control within a closed-loop learning framework.

8.6.3 Implications for Digital Twins and Coordinated Intelligence

The reviewed literature demonstrates that while AI-enhanced iBEMS can achieve incremental performance improvements, they fall short of enabling system-level intelligence. Without persistent model synchronization, bidirectional data exchange, and decision feedback, intelligence remains fragmented and fragile.

These limitations motivate the transition toward Digital Twin (DT)–enabled iBEMS, where predictive models, control strategies, and physical system states are continuously aligned. Such architectures provide the foundation for distributed and consensus-based control frameworks, which are explored in Chap. 9.

8.7 Critical Limitations of Existing iBEMS

The preceding sections have shown that modern iBEMS architectures are structurally sophisticated, energy-aware, and increasingly integrated across subsystems and energy carriers. However, these advances have not translated into truly intelligent, adaptive, or autonomous building operation. This section consolidates the critical limitations of existing iBEMS implementations at a system level, emphasizing why these limitations cannot be resolved through incremental improvements alone.

Rather than reiterating architectural or functional shortcomings already discussed, the focus here is on *emergent limitations*—those that arise from how sensing, control, communication, and intelligence are composed and operationalized in practice.

8.7.1 Scalability and Interoperability Challenges

While iBEMS frameworks are explicitly designed to integrate heterogeneous subsystems, their scalability remains constrained by protocol fragmentation, vendor-specific implementations, and brittle integration logic. As discussed in Sects. 8.3 and 8.4, interoperability is often achieved through gateways, middleware, or project-specific mappings rather than through unified semantic models.

Scenario: Portfolio-Scale Deployment

Consider a facilities operator managing a portfolio of commercial buildings across different regions. Each site employs an iBEMS composed of BACnet HVAC controllers, KNX lighting, proprietary fire systems, and IoT-based energy meters. Although each building functions adequately in isolation, scaling analytics or coordinated control across the portfolio requires extensive re-engineering of data models, APIs, and control logic.
Key limitation: Integration scales linearly with engineering effort rather than system intelligence. As a result, iBEMS architectures struggle to support portfolio-level optimization, cross-building learning, or city-scale Digital Twin (DT) deployment.

8.7.2 Limited Predictive and Decision-Making Capability

Most iBEMS implementations remain fundamentally reactive, even when equipped with optimization routines or predictive components. Forecasting models, where present, are typically embedded within isolated control loops (e.g., HVAC MPC or DR scheduling) and operate on *fixed horizons* with limited feedback from real-world performance.

This limitation was highlighted in Sects. 8.5 and 8.6: *although both hierarchical optimization and AI-augmented modules deliver superior performance in well-characterized (known) scenarios, they do not generalize reliably under non-stationary operating conditions.*

Scenario: Non-Stationary Operation

A mixed-use building undergoes gradual occupancy pattern changes due to hybrid work policies. Although the iBEMS continues to optimize HVAC and lighting based on historical schedules and static models, energy savings degrade over time, and comfort complaints increase.
Key limitation: Prediction does not translate into understanding. Without continuously synchronized system representations, iBEMSs lack the ability to reinterpret data, recalibrate models, or revise objectives autonomously.

8.7.3 Disconnect Between Control and System-Level Intelligence

Perhaps the most *fundamental limitation* of existing iBEMS lies in the structural separation between control execution and system-level intelligence. Control layers are optimized for stability, safety, and real-time responsiveness, whereas intelligence—if present—is externalized to analytics platforms, offline optimization tools, or human operators.

This separation leads to what may be described as *procedural intelligence*: decisions are executed *on* the system rather than emerging *within* a continuously evolving system representation.

Scenario: Control Without Context

During a grid contingency, an iBEMS sheds electrical load and discharges batteries based on predefined DR rules. However, it fails to recognize that thermal storage could have been pre-charged earlier, or that comfort constraints could be relaxed temporarily in unoccupied zones.
Key limitation: Control actions lack contextual awareness. Without a persistent Digital Twin (DT) layer, iBEMS cannot reason about trade-offs, anticipate consequences, or coordinate decisions across subsystems and time scales.

Synthesis. Taken together, these limitations reveal that existing iBEMS architectures are *execution-complete but cognition-incomplete*. They provide the physical and cyber infrastructure required for intelligent operation, yet lack the representational, predictive, and learning capabilities necessary for autonomous decision-making.

This observation motivates the conceptual repositioning explained in the next section: iBEMS must be understood as the *physical execution layer* of Building Digital Twins, rather than as standalone intelligent systems.

8.8 iBEMS as the Physical Layer of Building Digital Twins

The preceding sections have shown that Integrated Building Energy Management Systems (iBEMS) provide a robust and mature execution infrastructure for monitoring, control, and energy management across heterogeneous building subsystems. At the same time, they have revealed a persistent structural characteristic: existing iBEMS architectures are predominantly *execution-oriented*, with intelligence embedded locally, episodically, and largely within predefined control or optimization routines. System-level reasoning, learning, and predictive decision-making remain external to the core iBEMS structure.

This observation motivates a conceptual repositioning of iBEMS within a broader Digital Twin (DT) framework. Throughout this chapter, and in the book as a whole, iBEMSs are not regarded as standalone intelligent systems; rather, they are interpreted as the "*physical–cyber execution layer*" upon which Building Digital Twins (BDTs) are realized. This distinction is critical. While iBEMS excels at sensing, actuation, and supervisory control, Digital Twins (DTs) introduce the representational, predictive, and learning capabilities required for higher-level reasoning across subsystems and time horizons.

This layered interpretation is consistent with the CI-based formulations developed in the author's doctoral research [14], where the "twin" is not defined merely as a simulation model, but as a continuously synchronized, bidirectional construct linking physical assets, cyber representations, and decision logic. *Within that framework, intelligence emerges at the level of prediction, optimization, and learning, while physical execution remains explicitly external.*

8.8.1 Why Digital Twins Cannot Operate Without iBEMS?

A Building Digital Twin (BDT) is fundamentally dependent on reliable and timely interaction with the physical building. Without iBEMS, a DT lacks the mechanisms required to (i) observe system states, (ii) influence system behavior, and (iii) validate or recalibrate its internal models. In practical terms, iBEMS provides the only scalable and standardized interface through which Digital Twins (DTs) can interact with real buildings. Specifically, iBEMS supply:

- **Sensing and data acquisition**, enabling the DT to receive *real-time* or *near-real-time* measurements of energy flows, temperatures, equipment states, and environmental conditions;
- **Actuation pathways**, allowing DT-generated decisions to be translated into concrete control actions via setpoints, schedules, and supervisory commands;
- **Operational context**, including alarms, constraints, safety interlocks, and rule-based protections—*which bound feasible decision spaces.*

From this perspective, a Digital Twin (DT) without iBEMS is reduced to an offline simulation or analytical tool, disconnected from operational reality. Conversely, an iBEMS without a Digital Twin (DT) remains *reactive* and *myopic (short-sighted)*, executing predefined logic without the ability to anticipate consequences, learn from experience, or reason about long-term trade-offs. The two are therefore structurally complementary rather than interchangeable.

8.8.2 Mapping iBEMS Components to DT Feedback Loops

The integration of iBEMS and Digital Twins (DTs) can be understood through the lens of *nested feedback loops*, see Fig. 8.6, where physical processes, cyber representations, and decision mechanisms are continuously coupled. Building on the CI-based

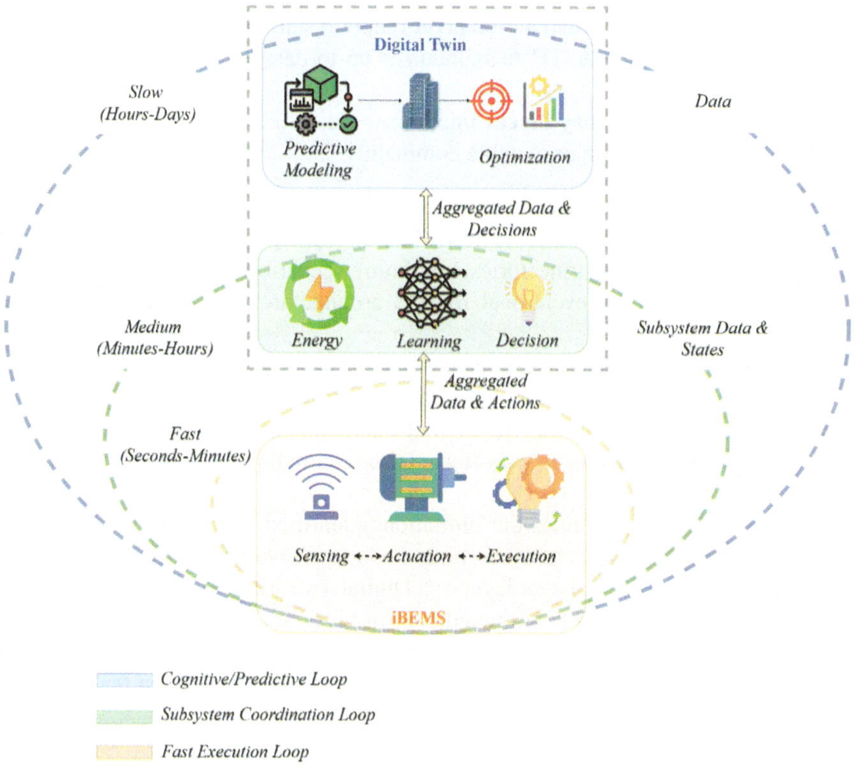

Fig. 8.6 Conceptual mapping between iBEMS components and Building Digital Twin (BDT) feedback loops. iBEMS provides sensing, actuation, and execution capabilities, while predictive modeling, learning, and decision-making reside within the Digital Twin layer

optimization frameworks for BEMSs articulated in [14],[5] iBEMS components map naturally onto the lower layers of these (feedback) loops. At the most fundamental level:

- **Sensors and meters** within iBEMS close the measurement loop, feeding physical state information into DT's cyber model;

[5] The author's doctoral research demonstrated that while Computational Intelligence (CI) can substantially enhance the predictive and optimization capabilities of building energy models, such intelligence remains inherently decoupled from physical execution. Throughout the thesis, sensing, actuation, and infrastructure-level control are abstracted away, revealing a fundamental "architectural gap" between cognition and execution. This gap motivates the interpretation of Integrated Building Energy Management Systems (iBEMS) not as intelligent entities in their own right, but as the indispensable physical–cyber execution layer upon which Building Digital Twins (DTs) must operate. From this perspective, Digital Twins (DTs) provide prediction, learning, and decision support, while iBEMS supplies the sensing, actuation, and interoperability required to realize cognitive decisions in practice.

- **Data aggregation and abstraction layers** support state estimation and model synchronization, enabling the DT to maintain an up-to-date representation of building behavior;
- **Control and supervisory layers** implement the actuation loop, translating DT-generated decisions into executable commands.

Crucially, the predictive, learning, and optimization loops reside above the iBEMS layer, within the Digital Twin (DT) itself. These loops operate on abstracted system representations, historical trajectories, and future scenarios—capabilities which are explicitly absent from conventional iBEMS architectures, as discussed earlier in Sects. 8.6 and 8.7.

Taken together, this section establishes that iBEMSs are *necessary but not sufficient* for intelligent building operation. They constitute the physical execution substrate of Building Digital Twins (BDTs), but cannot, on their own, support predictive reasoning, adaptive learning, or coordinated decision-making across subsystems and time horizons.

This interpretation reframes the limitations identified earlier not as implementation shortcomings, but as "*architectural boundary conditions*". Once iBEMS is explicitly positioned as the lower layer of a Digital Twin (DT) stack, the path toward distributed intelligence and coordinated control becomes both technically and conceptually clear. These issues are explored in the next section, where the focus shifts from individual buildings toward aggregation, coordination, and *consensus-based intelligence*.

8.9 Toward Distributed and Coordinated Building Intelligence

The preceding sections have established that while Integrated Building Energy Management Systems (iBEMS) provide a robust and indispensable execution infrastructure, they remain structurally constrained in their ability to deliver autonomous, adaptive, and system-level intelligence. Even when augmented with optimization routines or AI-based modules, decision-making in most iBEMS implementations remains localized, static, and weakly coordinated across subsystems and temporal horizons.

As buildings increasingly operate as active participants within interconnected energy ecosystems—interfacing with grids, districts, campuses, and cities—these limitations become more pronounced.

> *Intelligence can no longer be confined to isolated control loops or centralized supervisory layers; instead, it must emerge from coordinated decision-making across distributed entities operating under shared objectives and constraints.*

This transition motivates a shift from *integrated execution* toward *distributed and coordinated building intelligence*.

8.9.1 Aggregation, Coordination, and Control Boundaries

Modern iBEMS already perform limited forms of aggregation and coordination, such as consolidating subsystem-level measurements, enforcing building-wide power constraints, or executing Demand Response (DR) strategies. However, as demonstrated throughout Sects. 8.5–8.7, these capabilities are typically implemented through predefined hierarchies, static aggregation logic, or centralized optimization routines.

A critical limitation lies in how control boundaries are defined. Most iBEMS architectures implicitly assume that coordination occurs at a single supervisory level, with lower-level controllers executing commands and higher-level analytics operating offline or episodically. This rigid separation constrains scalability and adaptability. As the number of controllable assets grow—across zones, buildings, or portfolios—centralized coordination becomes computationally burdensome and increasingly brittle under uncertainty and non-stationary operating conditions.

From a system perspective, aggregation should not merely summarize data or enforce constraints; it should enable *coordinated decision-making* among interacting subsystems. Achieving this requires architectures in which control authority, information exchange, and decision responsibility are distributed across multiple "*agents*" while remaining aligned with global objectives such as comfort, energy efficiency, and grid compliance.

8.9.2 Motivation for Consensus-Based Control Architectures

Consensus-based control frameworks provide a principled mechanism for achieving coordinated behavior among distributed agents without relying on fully centralized decision-making. In such architectures, individual devices or subsystems—*such as HVAC zones, lighting groups, or storage units*—act as autonomous agents which iteratively exchange limited information with their neighbors to reach agreement on shared variables or objectives.

The motivation for consensus-based approaches in the context of iBEMS is twofold. First, they align naturally with the physical and organizational decentralization of building systems, where control authority is already distributed across multiple controllers and subsystems. Second, consensus mechanisms offer inherent scalability and robustness, enabling coordination to emerge even as system size and complexity increase.

Importantly, consensus-based control does not replace Digital Twin (DT) intelligence; rather, it complements it. Within the architectural interpretation developed in this chapter, iBEMS provides the `physical execution layer`, Digital Twins

(DTs) supply `predictive and learning-enabled intelligence`, and *consensus-based control acts as the* `coordination logic` *which reconciles local actions with global objectives*. This layered interaction enables closed-loop operation in which predictions inform coordination, coordination shapes control actions, and physical responses continuously update the system representation.

This perspective directly motivates the focus of Chap. 9, where consensus-based control strategies are demonstrated in context of the Digital Twin (DT) environments. By bridging local actuation, distributed coordination, and predictive intelligence, such frameworks can provide a viable pathway toward scalable, adaptive, and autonomous building operation across multiple spatial and temporal scales.

8.10 Summary and Outlook

This chapter has provided a critical and pedagogical examination of Integrated Building Energy Management Systems (iBEMS) from architectural, functional, and operational perspectives. By tracing the evolution from conventional Building Management Systems (BMS) to modern, energy-centric, and multi-energy iBEMS architectures, the discussion has highlighted both the substantial progress achieved in integration and the persistent limitations which constrain autonomous and adaptive operation.

Through detailed analysis of communication infrastructures, functional decomposition, hierarchical control structures, and prevailing intelligence paradigms, the chapter has shown that contemporary iBEMS are highly capable execution platforms. They enable reliable sensing, actuation, supervisory control, and energy-aware coordination across building subsystems. However, these capabilities remain predominantly execution-oriented, with intelligence embedded locally, episodically, and within predefined control or optimization routines.

By explicitly positioning iBEMS as the physical–cyber execution layer of Building Digital Twins (BDTs), the chapter reframes these limitations as architectural boundary conditions rather than implementation shortcomings. Digital Twins (DTs) introduce the representational, predictive, and learning-enabled capabilities required to complement iBEMS and support system-level reasoning, adaptive optimization, and long-term decision-making.

Finally, the chapter has motivated a transition beyond centralized and siloed intelligence toward distributed and coordinated building intelligence. As buildings increasingly operate as active participants within interconnected energy ecosystems, scalable coordination mechanisms become essential. This motivation leads naturally to the consensus-based and distributed control frameworks, discussed in Chap. 9, where Digital Twin (DT) intelligence and iBEMS execution are unified within coordinated, multi-agent architectures.

References

1. Antunes CH, Soares A, Gomes Á (2017) An integrated building energy management system. In: Sayigh A (ed) Mediterr green buildings renew energy. Springer, Cham, pp 191–199
2. Zhu HC, Ren C, Cao SJ (2021) Fast prediction for multi-parameters (concentration, temperature and humidity) of indoor environment towards the online control of HVAC system. In: Building simulation, vol 14, pp 649–665. Springer
3. Wang S, Xie J (2002) Integrating building management system and facilities management on the internet. Autom Constr 11(6):707–715
4. Continental Automated Buildings Association (CABA) (2020) Intelligent Building Energy Management Systems. Tech. rep., CABA, chrome-extension://efaidnbmnnnibpcajpcglclefindmkaj/ https://www.ashb.com/wp-content/uploads/2020/12/CABA-IBEMS-Report-2020-FULL-WEB.pdf, prepared by Harbor Research
5. Hong SH, Kim SH, Kim JH, Kim YG, Kim GM, Song WS (2013) Integrated BACnet-ZigBee communication for building energy management system. In: IECON 2013-39th annual conference of the IEEE industrial electronics society, IEEE, pp 5723–5728
6. Merz H, Hansemann T, Hübner C (2018) Building automation: communication systems with EIB/KNX, LON and BACnet, 2nd edn. Signals and communication technology. Springer, Cham. https://doi.org/10.1007/978-3-319-73223-7
7. Ozadowicz A, Grela J (2015) Control application for internet of things energy meter - a key part of integrated building energy management system. In: 2015 IEEE 20th conference on emerging technologies & factory automation (ETFA), IEEE, pp 1–4
8. Das D, Mishra S, Singh B (2021) An aggregated energy management methodology for air conditioning system with DAB converter. IEEE J Emerg Sel Top Ind Electron 3(1):124–132
9. Sharma S, Verma A, Xu Y, Panigrahi BK (2019) Robustly coordinated bi-level energy management of a multi-energy building under multiple uncertainties. IEEE Trans Sustain Energy 12(1):3–13
10. Vašak M, Banjac A, Hure N, Novak H, Marušić D, Lešić V (2021) Modular hierarchical model predictive control for coordinated and holistic energy management of buildings. IEEE Trans Energy Convers 36(4):2670–2682
11. Barbierato L, Schiera DS, Orlando M, Lanzini A, Pons E, Bottaccioli L, PE (2024) Facilitating smart grids integration through a hybrid multi-model co-simulation framework. IEEE Access
12. Hwang HK, Yoon AY, Kang HK, Moon SI (2020) Retail electricity pricing strategy via an artificial neural network-based demand response model of an energy storage system. IEEE Access 9:13440–13450
13. Jeong S, Wi YM (2024) Research on development and implementation of integrated energy management system for buildings. J Electr Eng Technol 19(8):5393–5405
14. Wani MH (2022) Intelligent control strategies for efficient building energy management systems (BEMS). Department of Electrical and Electronic Engineering, University of Auckland, PhD thesis

Chapter 9
Consensus-Based Control Strategies for Smart Appliances and Building Automation in Digital Twin Environments

Abstract This chapter discusses consensus-based control strategies for smart appliances and building automation within Digital Twin (DT)-enabled integrated Building Energy Management Systems (i-BEMS). Motivated by the growing scale, heterogeneity, and connectivity of modern buildings, the chapter reframes building control as a multi-agent coordination problem rather than a purely centralized optimization task. A hierarchical architecture is adopted in which physical controllers execute local actions, consensus mechanisms enable distributed coordination, and Digital Twins (DTs) provide predictive intelligence for validation and adaptation. A convex quadratic optimization framework is formulated to balance local comfort–energy objectives, inter-agent agreement, and global power constraints. Centralized and distributed primal–dual algorithms are introduced, illustrating how coordinated behavior can emerge through local interactions and limited centralized supervision. Python™-based examples are used to demonstrate convergence, consensus formation, and constraint handling under simplified assumptions, serving as pedagogical illustrations rather than deployable solutions. The chapter further examines how Digital Twins (DTs) can be integrated into consensus loops to support closed-loop prediction, validation, and adaptive constraint management. Practical implementation considerations, including controller mapping, timing, robustness, privacy, and partial observability, are discussed. Overall, the chapter positions consensus-based control as a scalable and flexible coordination paradigm which complements traditional building control architectures and supports the development of intelligent, distributed, and resilient Energy Management Systems (EMSs).

Keywords Consensus-based control · Centralized control · Communication graphs · Decentralized control · Distributed coordination · Multi-agent control

9.1 Introduction and Motivation

The increasing penetration of sensing, actuation, and connectivity in modern buildings has transformed Building Energy Management Systems (BEMS) from passive supervisory platforms into active cyber–physical infrastructures. Contemporary

M. H. Wani et al., *Digital Twins for Building Optimization: Enhancing Energy and Comfort*, Studies in Infrastructure and Control,
https://doi.org/10.1007/978-3-032-23274-8_9

BEMSs are expected not only to monitor energy consumption but also to coordinate heterogeneous devices—including HVAC systems, lighting, and auxiliary loads—under dynamic operational constraints, occupant comfort requirements, and grid-level objectives. This evolution has motivated the development of *integrated* BEMS (i-BEMS), where multiple subsystems are jointly optimized rather than controlled in isolation.

In parallel, the concept of the Digital Twin (DT) has emerged as a powerful paradigm for representing, predicting, and validating the behavior of complex physical systems. Within the building domain, Digital Twins (DTs) enable data-driven forecasting, scenario evaluation, and performance assessment by maintaining a continuously updated virtual counterpart of the physical building. However, while Digital Twins (DTs) provide predictive intelligence, they do not inherently specify how coordinated control actions should be executed across distributed devices in real time.

A key challenge therefore arises at the intersection of execution and intelligence: how should a large population of smart building devices be coordinated in a scalable, robust, and energy-aware manner, while remaining compatible with Digital Twin (DT)-based prediction and validation? Centralized optimization approaches, although conceptually appealing, often face scalability, communication, and robustness limitations when deployed across distributed field controllers. These challenges motivate the adoption of *consensus-based control strategies*, in which coordination emerges through local interactions among agents rather than through fully centralized decision-making.

This chapter positions consensus-based control as the *coordination logic* which bridges physical execution and predictive intelligence. Specifically, we adopt the following hierarchical interpretation:

- Integrated BEMS (i-BEMS) acts as the *physical execution layer*, interfacing directly with sensors and actuators.
- Consensus-based control provides a *distributed coordination layer*, enabling multiple agents to align their actions while respecting global constraints.
- Digital Model-Driven Digital Twins (DMDTs) form the *predictive intelligence layer*, supporting forecasting, validation, and adaptation.

Under this hierarchy, closed-loop optimization is achieved by allowing the physical layer to feed operational data to the consensus layer, which computes coordinated control actions which are subsequently evaluated and refined using Digital Twin (DT) predictions. The resulting architecture supports scalability, modularity, and resilience, while remaining compatible with existing BEMS deployments.

To concretize these ideas, this chapter formulates a simple consensus-based optimization problem for smart building devices subject to comfort objectives and global power constraints. A Python™-based reference implementation is used to demonstrate the approach, followed by centralized and distributed algorithmic formulations suitable for practical deployment in i-BEMS environments.

9.1.1 From Centralized i-BEMS Control to Distributed Coordination

Traditional Building Energy Management Systems (BEMSs) have largely relied on centralized supervisory control architectures, in which a single optimization or decision-making entity computes control actions for multiple subsystems. Within an i-BEMS framework, such centralized approaches enable global visibility of system states and constraints, facilitating coordinated scheduling of HVAC, lighting, and auxiliary loads. However, as the scale and heterogeneity of building systems increase, centralized control strategies encounter practical limitations related to scalability, communication overhead, robustness, and deployment complexity.

Modern buildings increasingly consist of geographically distributed zones, heterogeneous devices with different actuation characteristics, and field controllers operating at varying time scales. Collecting all state information at a central location and issuing control commands to each device introduces latency and *single-point-of-failure* risks, particularly when operating under fast-changing environmental conditions or grid-level signals. These challenges are further exacerbated when buildings participate in Demand–Response (DR) programs or interact with smart grids, where responsiveness and resilience are critical.

Distributed coordination frameworks offer an alternative paradigm in which *local agents—representing zones or devices*—compute control actions using locally available information while exchanging limited information with neighboring `agents`. Rather than relying on a single centralized controller, coordination emerges through structured interactions among agents. In this context, consensus-based control provides a principled mechanism for aligning local decisions with system-wide objectives, such as total power limits or aggregated comfort targets, without requiring full centralization. Table 9.1 compares some of the key aspects of Centralized i-BEMS and Distributed Consensus-Based Control.

Table 9.1 Comparison between centralized i-BEMS control and distributed consensus-based coordination

Aspect	Centralized i-BEMS control	Distributed consensus-based control
Decision architecture	Single supervisory optimizer	Multiple interacting agents
Scalability	Limited for large systems	Naturally scalable with system size
Communication load	High, system-wide data aggregation	Local neighbor-to-neighbor exchange
Robustness	Single point of failure	Graceful degradation under failures
Deployment complexity	Central serve-dependent	Compatible with field-level controllers
Adaptability	Reconfiguration-intensive	Parameter-level adaptation

Within an i-BEMS environment, this transition enables individual controllers to retain autonomy while contributing to a coordinated building-wide response. *Consensus mechanisms thus serve as a natural "intermediary" between purely centralized optimization and fully decentralized control, offering a balance between global coordination and local flexibility.*

9.1.2 Why Consensus-Based Control in Digital Twin (DT) Environments?

Digital Twins introduce predictive and analytical capabilities which extend beyond traditional control architectures by enabling real-time forecasting, performance evaluation, and scenario analysis. In building applications, DTs are commonly used to predict thermal behavior, estimate energy consumption, and assess occupant comfort under varying operating conditions. While these capabilities provide valuable intelligence, they do not inherently specify how distributed control actions should be coordinated across multiple devices in real time.

Consensus-based control can complement Digital Twin (DT) environments by providing a coordination mechanism which can exploit predictive insights without imposing centralized execution. By embedding consensus logic within the control layer, local agents can adjust their actions in response to both real-time measurements and DT-informed predictions, while collectively respecting system-wide constraints. This separation of roles allows the Digital Twin (DT) to function as a predictive and validation layer, rather than as a direct low-level controller.

From a system architecture perspective, consensus-based control aligns naturally with the layered Digital Twin (DT) paradigm adopted in this book. The physical execution layer (i-BEMS) interfaces with sensors and actuators, the consensus layer coordinates distributed decisions, and the Digital Model-Driven Digital Twin (DMDT) layer provides predictive intelligence, see Fig. 9.1. This structure supports closed-loop operation in which predicted outcomes can be used to assess, refine, or adapt coordination strategies over time.

Moreover, consensus-based strategies are well suited to the data-driven and adaptive nature of Digital Twins (DTs). As models are updated or operating conditions change, consensus gains, constraints, or coordination objectives can be modified without restructuring the underlying control architecture. This flexibility is particularly valuable in real buildings, where system configurations, occupancy patterns, and operational priorities evolve over time.

In the remainder of this chapter, consensus-based control is formulated explicitly for smart building applications within an i-BEMS framework. The proposed approach integrates comfort objectives, energy penalties, and global power constraints, and is demonstrated through both centralized and distributed algorithmic implementations compatible with Digital Twin (DT)-enabled operation.

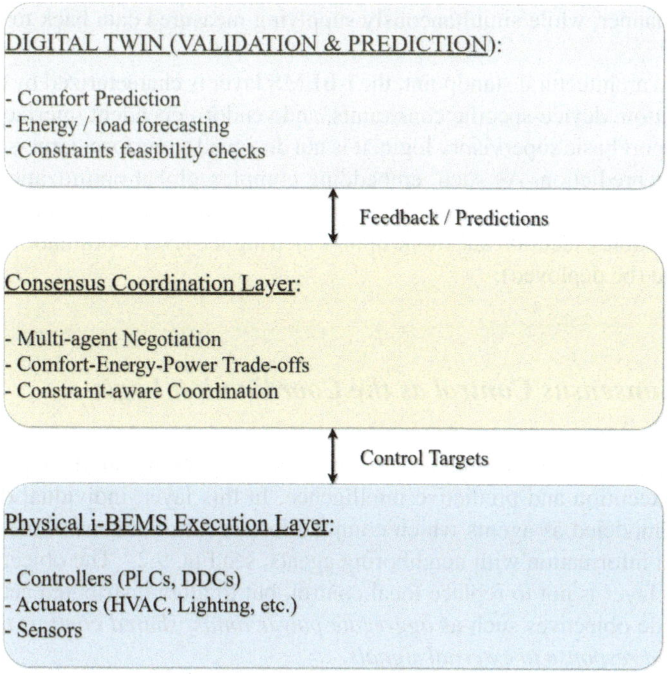

Fig. 9.1 Layered interaction between i-BEMS execution, consensus-based coordination, and Digital Twin (DT) prediction

9.2 Architectural Positioning of Consensus Control

This section situates consensus-based control within the overall architecture of an integrated Building Energy Management System (i-BEMS) operating alongside a Digital Twin (DT). Rather than viewing consensus as a standalone control technique, it is positioned here as an intermediate coordination layer which connects low-level physical execution with high-level predictive intelligence. This architectural separation clarifies the respective roles of execution, coordination, and prediction, and avoids overloading any single layer with conflicting responsibilities.

9.2.1 i-BEMS as the Physical Execution Layer

As outlined in Chap. 8, at its lowest level the i-BEMS functions as the physical execution layer, directly interfacing with sensors, actuators, and field controllers. This layer can include devices such as zone-level HVAC controllers, lighting controllers, variable-speed drives, and other building subsystems that must operate under real-time conditions. Its main function is to carry out control commands in a dependable and safe manner, while simultaneously supplying measured data back to the upper layers.

From an architectural standpoint, the i-BEMS layer is characterized by deterministic execution, device-specific constraints, and vendor-dependent interfaces. While it may support basic supervisory logic, it is not designed to perform large-scale coordination or prediction. As such, embedding complex global optimization directly within this layer can compromise robustness and maintainability. Instead, i-BEMS provides a stable execution backbone upon which higher-level coordination strategies can operate (be deployed).

9.2.2 Consensus Control as the Coordination Logic

Consensus-based control occupies the intermediate coordination layer, bridging physical execution and predictive intelligence. In this layer, individual devices or zones are modeled as agents which compute local control actions while exchanging limited information with neighboring agents, see Fig. 9.2.[1] The objective of the consensus layer is not to replace local control, but to align distributed actions with system-wide objectives such as *aggregate power limits, shared comfort targets, or coordinated response to external signals*.

By operating on locally available information and structured communication graphs, the consensus layer enables scalable coordination without requiring full centralization. This design naturally accommodates heterogeneous devices, asynchronous updates, and partial observability, making it well suited for real building deployments. Importantly, consensus control focuses on *how* distributed decisions are aligned, rather than *what* decisions should be made, leaving predictive reasoning to other layers—*virtual layer*.

9.2.3 Digital Model-Driven Digital Twins (DMDTs) as the Predictive Intelligence Layer

At the highest level of the proposed architecture, Digital Model-Driven Digital Twins (DMDTs) provide the predictive intelligence required to evaluate and anticipate the impact of coordinated control actions. DMDTs maintain a continuously updated virtual representation of the physical building by fusing sensor data, contextual information, and learned or physics-based models. Typical outputs include predictions of thermal comfort, energy consumption, and system response under candidate operating conditions.

[1] Adapted from [1].

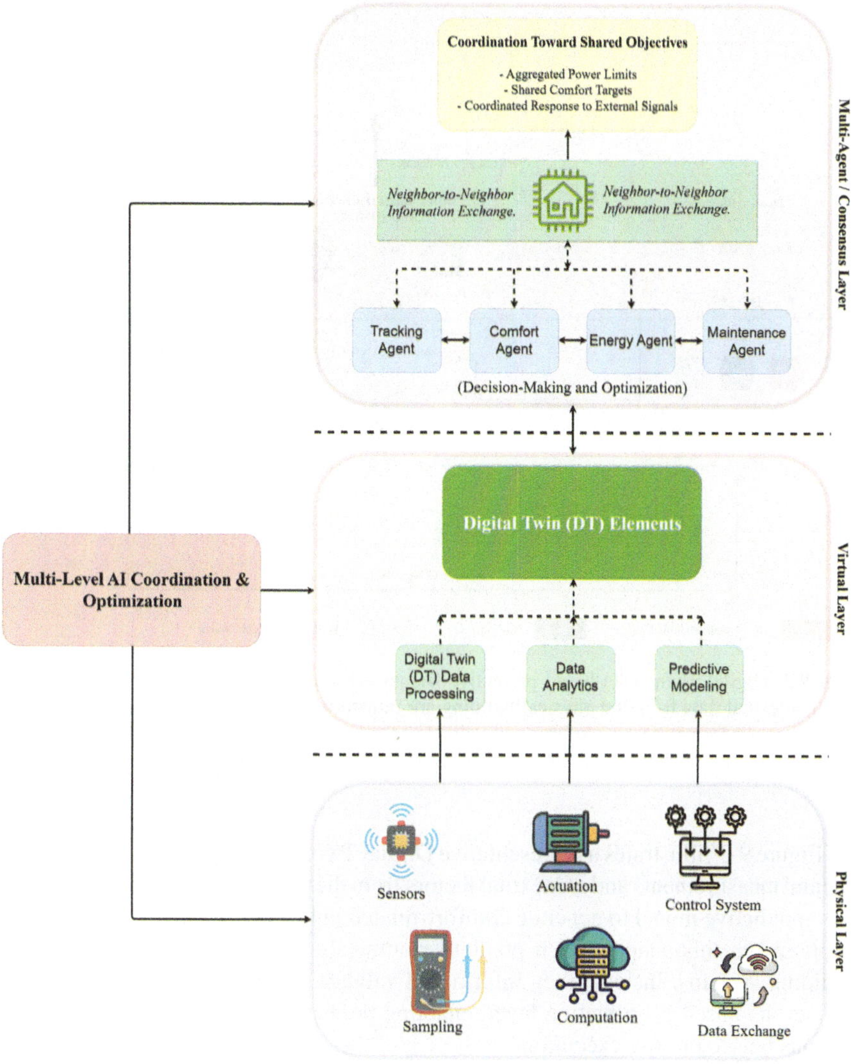

Fig. 9.2 Multi-layer consensus-based coordination framework for smart buildings, depicting how physical devices, Digital Twin (DT) analytics, and distributed agents exchange information to align local decisions with system-wide objectives

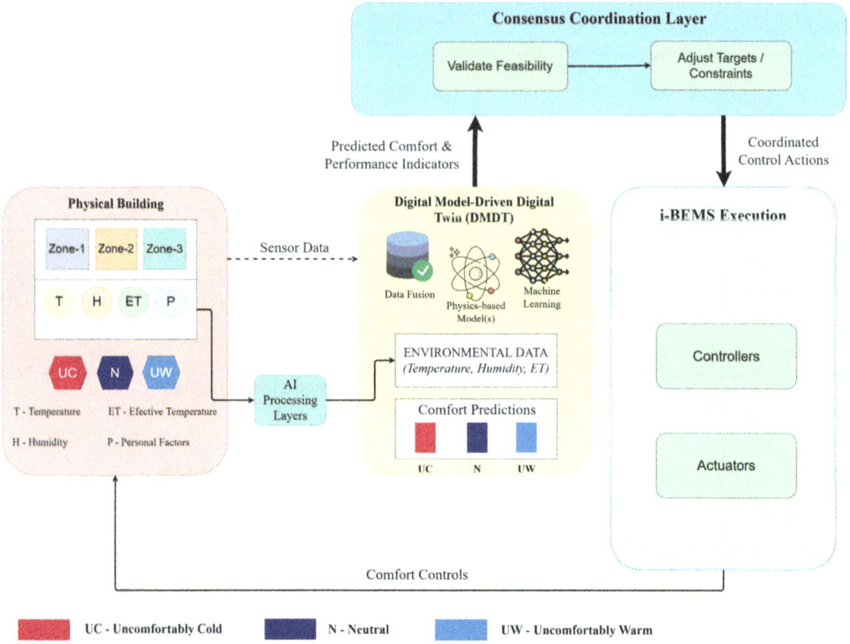

Fig. 9.3 Digital Twin (DT)-based predictive framework illustrating how sensed environmental and contextual data from the physical building are transformed into comfort-related performance indicators. These predictions inform and validate consensus-based coordination decisions, while control execution remains within the i-BEMS layer

Figure 9.3[2] illustrates a representative Digital Twin formulation in which environmental measurements and contextual factors from the physical building are processed by a predictive model to generate comfort-related indicators. Within the architecture proposed in this chapter, such predictive outputs are not used to directly actuate building systems. Instead, they inform and validate the decisions produced by the consensus-based coordination layer, enabling predictive assessment of coordinated actions before or after execution.

By decoupling prediction from actuation, the Digital Twin (DT) layer enhances transparency and adaptability. Models can be updated, retrained, or replaced without restructuring the underlying coordination or execution layers. This separation allows consensus-based control strategies to exploit predictive intelligence while preserving modularity, robustness, and deployability within existing i-BEMS infrastructures.

[2] Adapted from [2].

9.3 Multi-agent Modeling of Building Systems

Consensus-based control is naturally expressed in a *Multi-Agent System* (MAS) form: each controllable building component (or zone) is modeled as an agent, see Fig. 9.4,[3] with local objectives and limited communication, while coordination emerges through information exchange and agreement dynamics. This modeling view is widely used across building automation, residential Demand–Response (DR), and emerging Digital Twin (DT) settings, where device-/zone-level agents interact under global objectives such as comfort, energy, and power constraints [1, 5–9].

9.3.1 Smart Appliances and Zones as Agents

A recurring theme in the literature is that scalable coordination becomes difficult when a single supervisory optimizer must directly manage large populations of heterogeneous devices. For instance, consensus-based coordination has been investigated in building automation settings to enable distributed decision-making across building subsystems while retaining coordinated operation [5]. In residential Demand–Response (DR), a hierarchical distributed approach is proposed to coordinate large-scale participation through Home Energy Management Systems (HEMS), emphasizing that distributed architectures can address communication, privacy, and customer preference challenges [6]. Similarly, for smart homes with large numbers of appliances, distributed control using multi-agent consensus has been motivated as a means to avoid communication congestion and the "dimension curse" associated with centralized control [7].

Recent DT-oriented works extend this agent perspective further: device-/system-level agents can be paired with DT representations, and the MAS becomes a natural computational substrate for monitoring, optimization, and control of HVAC behavior within a DT-enabled framework [1]. In parallel, DT models are also being used to represent comfort-related dynamics (e.g., personalized thermal comfort prediction) and to support energy-efficient operation by coupling sensing, prediction, and control decisions [2, 10].

---

[3] A communication graph in a Multi-Agent System (MAS) defines who talks to whom, representing agents as nodes and communication links as edges [3, 4], crucial for coordination, consensus, and task performance, with designs ranging from fixed (chains, trees) to adaptive structures (learned graphs) for efficiency, especially under bandwidth constraints. These graphs dictate information flow, impacting system dynamics and enabling complex behaviors like distributed control and cooperative navigation.

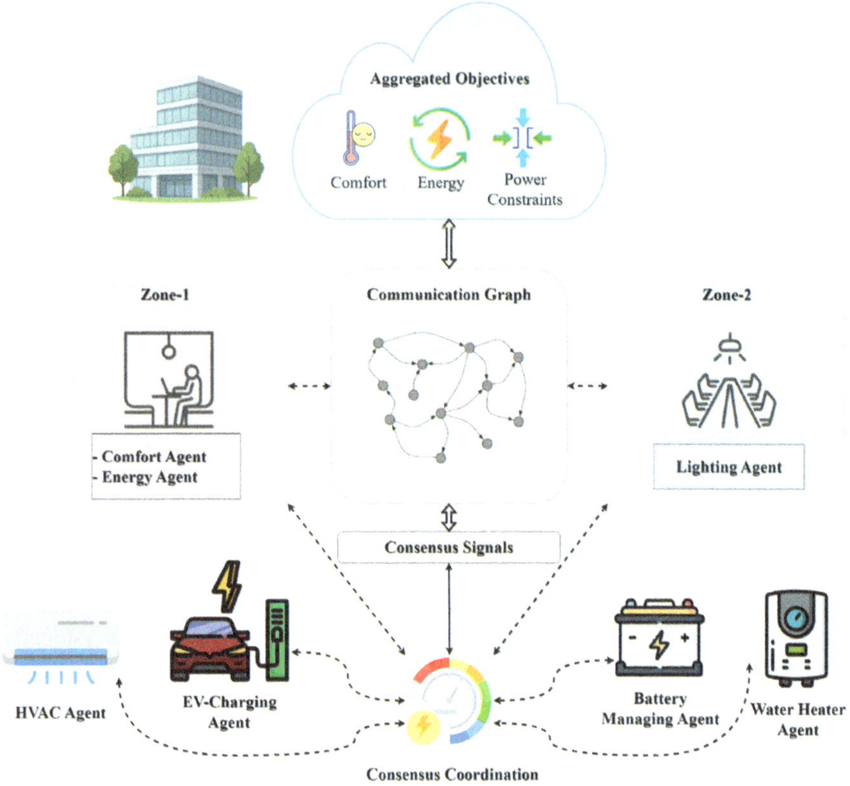

Fig. 9.4 Multi-agent representation of a smart building system with consensus-based coordination. Controllable zones and devices are modeled as agents connected through a communication graph, enabling local information exchange and coordination toward aggregated comfort, energy, and power objectives. The structure of the communication graph determines neighborhood interactions, consensus signal propagation, and the resulting collective behavior

9.3.2 Communication Graphs and Neighborhood Structures

Once devices/zones are cast as agents, coordination is defined by an underlying communication graph[4] (or neighborhood structure), where each agent exchanges information only with a subset of neighbors.

[4] A communication graph is a network representation where nodes correspond to agents (e.g., devices or zones), and edges represent direct communication links. The neighborhood of an agent is the set of nodes it can exchange information with.

Analogy: Communication Graphs

Imagine a neighborhood where each house can only talk to its immediate neighbors rather than the entire city. If one house learns about a new recycling rule, it shares it with its neighbors, who then pass it along. This local exchange forms a network of connections—similar to a communication graph—where information flows efficiently without overwhelming a central authority.

Distributed Demand–Response (DDR) schemes explicitly reduce communication requirements by limiting information exchange to aggregated signals (e.g., incentive/reward prices and aggregated load responses) rather than full device states [6]. In appliance-level Demand–Response (DR), consensus-based approaches are also framed as scalable alternatives to centralized scheduling, highlighting that exchanging information only between adjacent nodes can improve reliability and reduce central bottlenecks [7].

Beyond building-focused studies, advances in consensus theory provide additional perspectives relevant to practical deployments. For example, self-triggered output-feedback consensus approaches have been proposed for nonlinear Multi-Agent Systems (MASs), which is relevant to resource-constrained settings where continuous communication is undesirable and event/trigger-based coordination can reduce exchange rates [11]. In DT-enabled power tracking contexts, distributed coordination can also be framed at the device level, where DT agents support distributed tracking objectives through consensus-like mechanisms [8].

9.3.3 Consensus Variables and Coordination Objectives

Across the reviewed works, the choice of a "consensus variable" is dictated by the coordination objective.

Example: Consensus Variable in Action

Imagine a group of smart thermostats in different rooms trying to agree on a single indoor temperature setting:

- `Consensus Variable`: The target temperature (e.g., 22 °C).
- `Coordination Objective`: Minimize energy use while maintaining comfort.

Each thermostat adjusts its local setting based on neighbors' suggestions until all agree on a common value. This mirrors how consensus-based DR schemes work—agents iteratively align on an economically or comfort-driven variable without revealing full internal states.

In Demand–Response (DR) and load allocation, consensus can be formed over economically meaningful quantities. A typical instance is the incremental-cost, consensus-based allocation method for demand-side resources, which was introduced with explicit consideration of the discrete On/Off behavior of many household appliances and is supported by a priority-ranking mechanism and fitted cost functions to facilitate distributed optimization [7]. Related residential DR work also emphasizes coupling coordination with comfort-related constraints (e.g., indoor and water temperature constraints) and introduces discomfort considerations while preserving privacy by avoiding disclosure of detailed appliance operating states [6].

In DT-enabled contexts, coordination objectives broaden to include predictive intelligence and data-driven comfort/energy services. DT-based comfort modeling using BIM-extracted spatial–temporal information has been used to support preference prediction, illustrating how DT data streams and spatial context can shape coordination objectives beyond purely power-centric targets [10]. DT-based deep learning frameworks for personalized thermal comfort further reinforce the role of prediction as an enabling layer for energy-efficient decision-making [2]. Complementing these, data-driven thermal management studies have explored clustering-based ideas in smart building thermal contexts, including the use of consensus-oriented clustering constructs as part of a broader data-driven management approach [12].

Finally, some works discuss "consensus" in a broader systems sense, where agreement and trust mechanisms (e.g., blockchain-style consensus) are combined with DT concepts for smart-city applications. While such works are not specific to i-BEMS control loops, they highlight an adjacent motivation: trustworthy coordination of distributed agents and data streams in cyber–physical infrastructures [13].

9.4 Problem Formulation for Consensus-Based i-BEMS Control

The consensus-based control problem formulated in this section translates the multi-agent modeling concepts introduced in Sect. 9.3 into an explicit optimization framework suitable for implementation within an integrated Building Energy Management System (i-BEMS). Each controllable device or zone is modeled as an agent with a scalar actuation variable, while coordination is enforced through consensus penalties and global constraints. The formulation is intentionally kept *quadratic* and *convex* to facilitate scalable implementation and to align with the *centralized* and *distributed* algorithms presented in Sects. 9.5 and 9.6.

The optimization problem for consensus-based control in an i-BEMS context is defined as

$$\min_{\mathbf{x}} \sum_{i=1}^{N} \left[(x_i - s_i)^2 + w_i x_i^2 \right] + \beta \sum_{(i,j)\in\mathcal{E}} (x_i - x_j)^2 \tag{9.1}$$

subject to the global power constraint:

$$\sum_{i=1}^{N} c_i x_i \leq P_{\max}, \qquad x_i \in [0, 1] \tag{9.2}$$

Here, the decision vector $\mathbf{x} = [x_1, \ldots, x_N]^\top$ collects the normalized actuation levels of all participating agents, where:

- x_i denotes the normalized actuation level of agent "i" (HVAC, fan, or lighting)
- s_i is the comfort target of agent "i"
- w_i is an energy penalty weight
- β is the consensus gain
- c_i is the power coefficient associated with agent "i"
- $P_{\max}$ is a global power cap (e.g., imposed by a demand–response event)
- N is the total number of agents participating in the optimization problem[5]

The first term in Eq. (9.1) captures local comfort–energy trade-offs at the agent level, while the second term penalizes disagreement among neighboring agents connected by the communication graph "$\mathcal{E}$". The constraint Eq. (9.2) enforces compliance with building-wide or grid-imposed energy limits. Together, these elements reflect the layered architecture introduced earlier: local objectives are preserved at the agent level, coordination is enforced through consensus penalties, and system-level feasibility is ensured via a shared constraint.

To solve this constrained optimization problem in an online and distributed manner, a primal–dual gradient-based approach is adopted. The resulting update equations, which form the analytical basis of the centralized consensus algorithm presented in Algorithm 9.1, are given by

$$g_{\text{local},i} = 2(x_i - s_i) + 2w_i x_i \tag{9.3}$$

$$g_{\text{cons},i} = \begin{cases} 2\beta \sum_{j \in \mathcal{N}_i} (x_i - x_j), & i \in \mathcal{H} \\ 0, & \text{otherwise} \end{cases} \tag{9.4}$$

$$g_{\text{dual},i} = \lambda c_i \tag{9.5}$$

$$x_i^{(k+1)} = \text{clip}\Big(x_i^{(k)} - \eta\big(g_{\text{local},i} + g_{\text{cons},i} + g_{\text{dual},i}\big),\ 0,\ 1\Big) \tag{9.6}$$

$$\lambda^{(k+1)} = \max\left\{0,\ \lambda^{(k)} + \gamma\left(\sum_{i=1}^{N} c_i x_i^{(k+1)} - P_{\max}\right)\right\} \tag{9.7}$$

These update equations make explicit how local comfort objectives, consensus-based coordination, and global power constraints are jointly enforced. Their algorithmic realization and practical deployment within an i-BEMS are discussed in detail in the following sections.

---

[5] In the Python™ example considered later, $N = 7$ (3 HVAC zones, 2 fans, and 2 lighting groups).

9.4.1 Local Comfort–Energy Objectives

The local objective term $(x_i - s_i)^2 + w_i x_i^2$ represents a simplified but expressive model of comfort–energy trade-offs at the *agent level*. The `first component` penalizes deviation from a desired comfort target s_i, which may correspond to preferred thermal, lighting, or airflow conditions depending on the device type. The `second component` penalizes actuation magnitude through a weighting parameter w_i, reflecting the relative energy sensitivity of the device.

This quadratic structure is commonly adopted in distributed building control literature due to its convexity and interpretability. It allows each agent to balance comfort tracking against energy expenditure using a single tunable parameter, while remaining compatible with gradient-based optimization methods. Importantly, these local objectives are evaluated independently at each agent, preserving modularity and enabling heterogeneous device modeling within a unified framework.

9.4.2 Consensus Penalties and Coupling Constraints

Coordination among agents is enforced through the consensus penalty term in Eq. (9.1), which penalizes squared differences between neighboring agents connected by the communication graph "$\mathcal{E}$". This term encourages agreement among agents while allowing local autonomy, and its influence is regulated by the consensus gain "β".

In the present formulation, consensus penalties are selectively applied to a subset of agents, denoted by the index set "$\mathcal{H}$", corresponding to HVAC zones. This choice reflects the fact that thermal zones are often strongly coupled through shared building physics and comfort requirements, whereas other devices such as fans or lighting groups may be weakly coupled or independent. The resulting structure aligns with practical i-BEMS deployments, where coordination requirements differ across subsystems.

The neighborhood-based formulation allows coordination to emerge through local interactions rather than global synchronization, supporting scalability and robustness. From an algorithmic perspective, this structure leads directly to *Laplacian-based consensus terms* in the update equations, as seen in Eq. (9.4).

9.4.3 Global Power and Demand–Response (DR) Constraints

The global power constraint Eq. (9.2) represents building-level or grid-level limitations, such as Demand–Response (DR) commitments or feeder capacity limits. Rather than embedding this constraint directly into each agent's objective, it is enforced via a shared dual variable "λ", enabling coordinated compliance without centralized decision-making.

This dual formulation is particularly suitable for i-BEMS environments, where global constraints must be respected while maintaining distributed execution. Agents respond to the broadcast dual signal by adjusting their local decisions, effectively internalizing system-level scarcity or congestion effects. When the constraint is *inactive*, the dual variable remains zero and agents prioritize local objectives; when the constraint becomes *active*, the dual variable increases and coordinated curtailment emerges naturally.

The resulting primal–dual update structure, summarized in Eqs. (9.6)–(9.7), forms the basis for the centralized and distributed algorithms presented in the subsequent sections. This formulation ensures that comfort objectives, coordination requirements, and global feasibility are addressed within a single unified optimization framework.

9.5 Centralized Consensus Optimization with Dual Ascent

This section presents a centralized realization of the consensus-based i-BEMS control problem formulated in Sect. 9.4. The purpose of this section is twofold: first, to clarify how the primal–dual update equations are executed in practice, and second, to provide an executable reference implementation which serves as a baseline for the distributed control strategy introduced in Sect. 9.6.

9.5.1 Centralized View of Consensus-Based i-BEMS Control

In the centralized view, all agent actuation states are assumed to be available to a supervisory entity, such as a Building Energy Management System (BEMS) server. This entity evaluates local comfort–energy objectives, enforces consensus coupling among selected agents, and coordinates compliance with global power constraints.

Figure 9.5 illustrates the centralized execution perspective adopted in this section. While control decisions are computed centrally, the underlying optimization problem remains agent-based: each agent contributes a local objective, coordination is enforced through consensus penalties, and a shared dual variable regulates global feasibility.

Although full centralization simplifies analysis and validation, it may not be suitable for large-scale deployments due to communication and robustness limitations. Nevertheless, it provides a clear conceptual and computational reference for understanding the interaction between local objectives, consensus coordination, and global constraints.

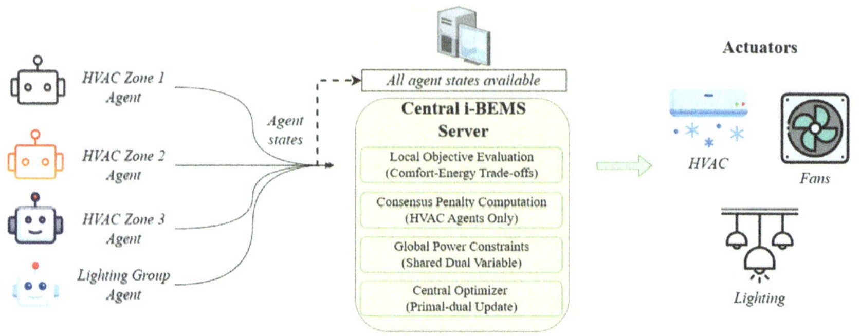

Fig. 9.5 Centralized execution view for consensus-based i-BEMS control with dual ascent. *All agent states are aggregated at a supervisory i-BEMS server, where local objectives, HVAC-only consensus penalties, and a global power constraint (via a shared dual variable) are jointly evaluated to compute coordinated control actions*

9.5.2 Primal–Dual Update Structure

The centralized controller implements the primal–dual update structure mathematically represented in Sect. 9.4. At each iteration, local gradients corresponding to Eq. (9.3), consensus gradients given by Eq. (9.4), and global constraint gradients from Eq. (9.5) are combined to update the agent actuation variables.

The primal update is performed first and projected onto admissible bounds, as specified in Eq. (9.6). The resulting actuation vector is then used to evaluate total power consumption, which drives the projected dual ascent step in Eq. (9.7). This update ordering ensures that the dual variable reflects the most recent system state and is consistent with the implementation logic used in the reference code (see Sect. 9.5.4).

When the global power constraint is inactive, the dual variable remains zero and agents prioritize local comfort objectives. As the constraint becomes active, the dual variable increases, and coordinated curtailment emerges naturally across agents, without requiring explicit centralized scheduling decisions.

9.5.3 Algorithmic Implementation: Pseudocode for Centralized Consensus Control

Algorithm 9.1 shows the pseudocode for *centralized consensus control* which implements the primal–dual update scheme described above. It structures the computation as a sequence of steps which clearly distinguishes local objective evaluation, consensus coordination, and enforcement of the *global constraints*.

This centralized routine forms the core of the reference implementation in Listing 9.1. The two views align line by line: the primal variables are updated before the dual variables, consensus coupling is applied exclusively to HVAC agents, and box constraints are imposed at every iteration via projection.

Algorithm 9.1 i-BEMS Consensus + Dual Ascent (Centralized View)

Inputs: Number of agents N, index sets $\mathcal{H}$ (HVAC), $\mathcal{F}$ (fans), $\mathcal{L}$ (lights)
Power coefficients $\mathbf{c} \in \mathbb{R}^N$, comfort targets $\mathbf{s} \in [0, 1]^N$, energy weights $\mathbf{w} \in \mathbb{R}_+^N$
HVAC Laplacian $L_{\text{hvac}} \in \mathbb{R}^{|\mathcal{H}|\times|\mathcal{H}|}$, power cap $P_{\max}$
Step sizes: primal $\eta > 0$, dual $\gamma > 0$, consensus gain $\beta > 0$
Initialize: $\mathbf{x}^{(0)} \leftarrow \text{clip}(\mathbf{s} + \text{small noise}, 0, 1)$; $\lambda^{(0)} \leftarrow 0$
for $k = 0, 1, 2, \ldots, T-1$ **do**
  Local gradient: $\nabla f(\mathbf{x}^{(k)}) \leftarrow 2(\mathbf{x}^{(k)} - \mathbf{s}) + 2\,\mathbf{w} \odot \mathbf{x}^{(k)}$
  Consensus term (HVAC only):
    Extract $\mathbf{x}_{\mathcal{H}}^{(k)}$ from $\mathbf{x}^{(k)}$
    $\mathbf{g}_{\text{cons},\mathcal{H}} \leftarrow 2\beta\, L_{\text{hvac}}\, \mathbf{x}_{\mathcal{H}}^{(k)}$
    Form full $\mathbf{g}_{\text{cons}}$ by placing $\mathbf{g}_{\text{cons},\mathcal{H}}$ on HVAC indices and zeros elsewhere
  Global-constraint term: $\mathbf{g}_{\text{dual}} \leftarrow \lambda^{(k)}\, \mathbf{c}$
  Primal update:
    $\mathbf{x}^{(k+1)} \leftarrow \text{clip}\Big(\mathbf{x}^{(k)} - \eta\big(\nabla f(\mathbf{x}^{(k)}) + \mathbf{g}_{\text{cons}} + \mathbf{g}_{\text{dual}}\big),\ 0,\ 1\Big)$
  Dual update (projected):
    $p^{(k+1)} \leftarrow \mathbf{c}^\top \mathbf{x}^{(k+1)}$
    $\lambda^{(k+1)} \leftarrow \max\{0,\ \lambda^{(k)} + \gamma\big(p^{(k+1)} - P_{\max}\big)\}$
end for
Output: final control $\mathbf{x}^{(T)}$, final dual $\lambda^{(T)}$, final total power $p^{(T)} = \mathbf{c}^\top \mathbf{x}^{(T)}$

9.5.4 Reference Python™ Implementation

Listing 9.1 presents a Python™ reference implementation of Algorithm 9.1. The implementation follows the same update ordering and variable definitions as the analytical formulation in Sect. 9.4, and is used to demonstrate convergence behavior, power constraint enforcement, and comfort trade-offs.

```
import numpy as np
import matplotlib.pyplot as plt

# ----------------------------------------
# i-BEMS multi-agent demo: consensus + dual ascent
# ----------------------------------------
# Agents: 3 HVAC zones (consensus-coupled), 2 fans, 2 lighting groups.
# Control variable x_i in [0,1]: normalized actuation level.
# Local objective per agent: (x_i - s_i)^2 + w_i x_i^2  (comfort vs energy)
# Global constraint: sum_i c_i * x_i <= Pmax  (power cap / DR event)
# Consensus applies ONLY to HVAC agents via Laplacian L_hvac.

np.random.seed(10)

```

```
# Indices
HVAC = [0, 1, 2]
FANS = [3, 4]
LIGHT = [5, 6]
N = 7

# Power coefficients c_i (kW per unit actuation)
c = np.array([
    4.5,  # HVAC-1
    4.0,  # HVAC-2
    4.2,  # HVAC-3
    1.5,  # Fan-1
    1.2,  # Fan-2
    0.6,  # Light-1
    0.5   # Light-2
], dtype=float)

# Comfort targets s_i (desired normalized actuation)
s = np.array([
    0.75,  # HVAC-1 wants strong cooling
    0.65,  # HVAC-2
    0.70,  # HVAC-3
    0.40,  # Fan-1 moderate ventilation
    0.50,  # Fan-2
    0.60,  # Light-1 brighter
    0.55   # Light-2
], dtype=float)

# Energy weights w_i (higher => bigger penalty on energy use)
w = np.array([0.25, 0.30, 0.25, 0.20, 0.18, 0.10, 0.10], dtype=float)

# Consensus coupling among HVAC agents (ring)
A_hvac = np.zeros((len(HVAC), len(HVAC)))
for i in range(len(HVAC)):
    A_hvac[i, (i - 1) % len(HVAC)] = 1
    A_hvac[i, (i + 1) % len(HVAC)] = 1
D_hvac = np.diag(A_hvac.sum(axis=1))
L_hvac = D_hvac - A_hvac

# Global power cap (e.g., DR event)
Pmax = 10.0  # kW

# Algorithm parameters
eta = 0.3          # primal step size
beta = 0.8         # consensus strength for HVAC
gamma = 0.6        # dual step size
T = 200            # iterations

# Initialize
x = np.clip(s + 0.05 * np.random.randn(N), 0.0, 1.0)
lam = 0.0  # Lagrange multiplier for global constraint

# Logs
x_hist = np.zeros((T, N))
p_hist = np.zeros(T)
lam_hist = np.zeros(T)
comfort_err_hist = np.zeros(T)

for k in range(T):
    # --- Store x before update for trajectory visualization ---
    x_hist[k] = x

    # Local gradients: d/dx_i [(x_i - s_i)^2 + w_i x_i^2] = 2(x_i - s_i) + 2
    w_i x_i
    grad_local = 2.0 * (x - s) + 2.0 * w * x

    # Consensus gradient for HVAC: 2*beta * L * x_hvac
```

```
    grad_cons = np.zeros(N)
    x_hvac = x[HVAC]
    grad_cons_hvac = 2.0 * beta * (L_hvac @ x_hvac)
    grad_cons[HVAC] = grad_cons_hvac

    # Global constraint gradient via dual ascent: + lam * c_i
    grad_dual = lam * c

    # --- Primal update ---
    x = x - eta * (grad_local + grad_cons + grad_dual)
    # Box constraints
    x = np.clip(x, 0.0, 1.0)

    # --- LOGGING LINES ---
    p_hist[k] = float(np.dot(c, x))                    # total power used for dual update
    lam_hist[k] = lam                                  # current lambda before update
    comfort_err_hist[k] = float(np.sum((x - s) ** 2))  # aggregate comfort error

    # --- Dual update: lam >= 0 ---
    lam = max(0.0, lam + gamma * (p_hist[k] - Pmax))

# Summaries
final_power = np.dot(c, x)

# Plot control variables
labels = ['HVAC-1', 'HVAC-2', 'HVAC-3', 'Fan-1', 'Fan-2', 'Light-1', 'Light-2']
plt.figure(figsize=(11, 6))
for i in range(N):
    plt.plot(x_hist[:, i], label=labels[i])
plt.title('Agent Control Variables x_i (consensus + global power constraint)')
plt.xlabel('Iteration')
plt.ylabel('Normalized actuation x_i [0..1]')
plt.legend(ncol=2)
plt.grid(True, alpha=0.3)
plt.tight_layout()
plt.savefig('ibems_controls.png')

# Plot total power vs cap & lambda
fig, ax = plt.subplots(1, 2, figsize=(12, 5))
ax[0].plot(p_hist, color='tab:blue')
ax[0].axhline(Pmax, color='k', linestyle='--', label='Pmax')
ax[0].set_title('Total Power vs Cap')
ax[0].set_xlabel('Iteration')
ax[0].set_ylabel('Power [kW]')
ax[0].legend()
ax[0].grid(True, alpha=0.3)

ax[1].plot(lam_hist, color='tab:red')
ax[1].set_title('Dual Variable (λ) Trajectory')
ax[1].set_xlabel('Iteration')
ax[1].set_ylabel('λ')
ax[1].grid(True, alpha=0.3)
plt.tight_layout()
plt.savefig('ibems_power_lambda.png')

# Comfort error
plt.figure(figsize=(8, 5))
plt.plot(comfort_err_hist, color='tab:green')
plt.title('Aggregate Comfort Error ∑(x_i - s_i)^2')
plt.xlabel('Iteration')
plt.ylabel('Error')
plt.grid(True, alpha=0.3)
```

```
plt.tight_layout()
plt.savefig('ibems_comfort_error.png')

print('Final x:', np.round(x, 3))
print('Final total power (kW):', round(final_power, 3))
print('Power cap Pmax (kW):', Pmax)
print('Final λ:', round(lam, 3))
```

Listing 9.1 Centralized Python™ reference implementation of consensus-based i-BEMS control with dual ascent.

The above Python™ implementation serves as an executable counterpart to Algorithm 9.1, enabling numerical experimentation and visualization while preserving full consistency with the mathematical and algorithmic descriptions.

Remark: The Python™ implementation presented in Listing 9.1 is intended as a compact reference realization of the centralized consensus algorithm rather than a deployment-ready controller. Its structure, however, directly maps to practical i-BEMS implementations when combined with distributed execution, real-time sensing, and communication layers, as discussed in the following sections.

Illustrative Results: Centralized Consensus-Based i-BEMS Control

This box summarizes representative outcomes obtained from the centralized Python™ reference implementation presented in Listing 9.1. The results are intended to *illustrate* the behavior of the consensus-based optimization framework introduced in this chapter and to complement the analytical and algorithmic descriptions provided in Sects. 9.4 and 9.5.

Evolution of Agent Control Variables. Figure 9.6 shows the trajectories of normalized actuation variables for all agents.[6] HVAC zones, which are coupled through consensus penalties, exhibit bounded oscillatory behavior as they coordinate toward a collective operating regime. In contrast, fan and lighting agents, which are not consensus-coupled, converge more smoothly toward their local optima. Such oscillatory patterns are characteristic of discrete-time consensus-based control with finite step sizes and are acceptable in building automation contexts, where actuation commands are naturally sampled and filtered.

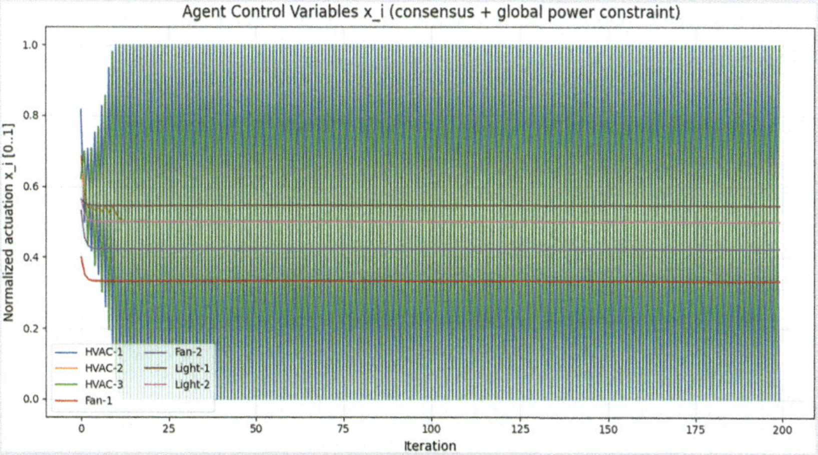

Fig. 9.6 Evolution of agent control variables under centralized consensus-based coordination and a global power constraint

Global Power Consumption and Dual Variable. Figure 9.7 illustrates the total building power consumption relative to the imposed *cap*, together with the trajectory of the dual variable. In this scenario, the power constraint remains inactive throughout the optimization, resulting in a zero-valued dual variable. This behavior indicates that the coordinated solution satisfies comfort objectives while naturally respecting system-level energy limits, without requiring explicit Demand–Response (DR) intervention.

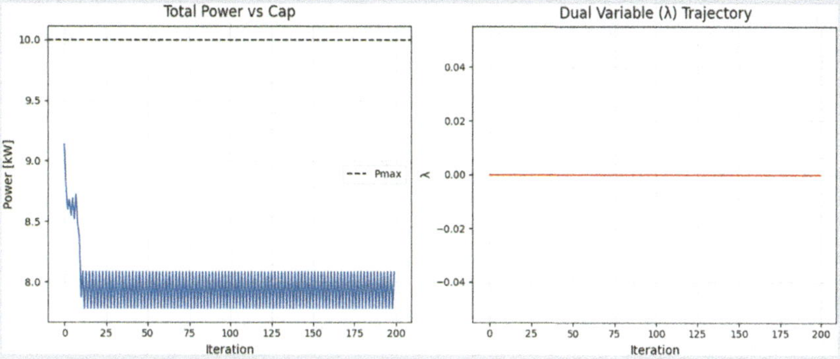

Fig. 9.7 Total power consumption relative to the global cap (left) and corresponding dual variable trajectory (right)

Aggregate Comfort Error. Figure 9.8 presents the aggregate comfort error over the optimization horizon. After an initial transient phase, the error converges to a bounded oscillatory regime, reflecting the trade-off between individual comfort tracking and consensus-based coordination among HVAC zones. In practical i-BEMS deployments, such oscillations are typically mitigated through actuator deadbands, temporal filtering, or slower supervisory update rates.

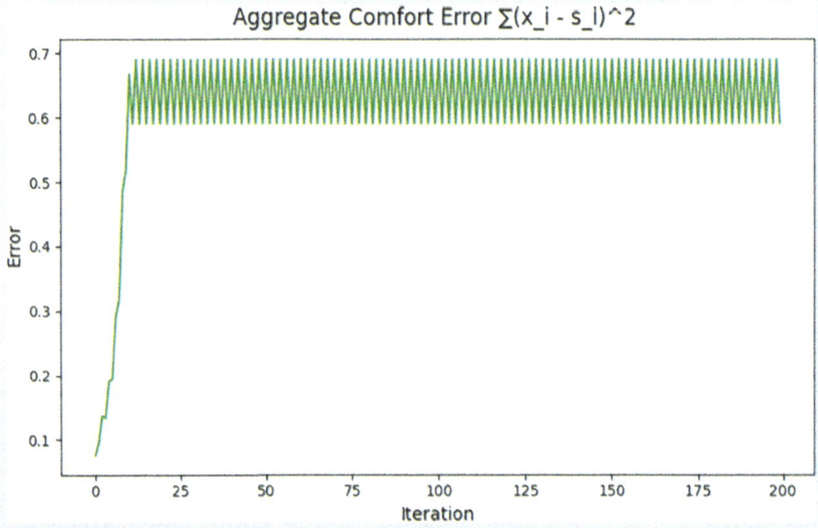

Fig. 9.8 Aggregate comfort error $\sum_i (x_i - s_i)^2$ over iterations

Interpretation. Overall, these results demonstrate that the centralized consensus-based formulation achieves coordinated operation under global constraints while preserving agent-level objectives. The example serves as a transparent validation of the proposed framework and provides intuition for the distributed implementations discussed in subsequent sections.

```
C:\Path\to\the\Python\Script.py

Final x: [1.    0.5   0.    0.333 0.424 0.545 0.5  ]
Final total power (kW): 8.086
Power cap Pmax (kW): 10.0
Final λ: 0.0

Process finished with exit code 0
```

Listing 9.2 Console output from the centralized consensus-based i-BEMS Python™ implementation.

[6] The apparent "bang–bang" oscillations in "x_i" trajectories are not inherent to consensus control. They arise from aggressive step-size and gain choices (e.g., η, β) combined with hard clipping. Real i-BEMS deployments mitigate such chatter using smaller steps, damping/ramping strategies, or actuator rate limits.

The console output in Listing 9.2 summarizes the final outcome of the centralized optimization process. The actuation vector, $\mathbf{x}^{(T)}$, represents the steady-state control decisions for all agents, subject to box constraints and consensus coordination. The resulting total power consumption remains below the imposed cap (viz. 10.0 kW), indicating that the global constraint is inactive at the optimum. Consequently, the dual variable converges to zero, which is consistent with complementary "slackness" conditions in primal–dual optimization.

9.6 Distributed Consensus Control for Field Deployment

While the centralized formulation in Sect. 9.5 provides a useful reference implementation, practical i-BEMS deployments typically rely on distributed execution across field controllers, supervisory servers, and communication networks. This section presents a distributed realization of the consensus-based control framework, in which each agent performs local computation and exchanges minimal information with neighboring agents and a central BEMS server.

The distributed structure preserves the mathematical formulation introduced in Sect. 9.4, while relaxing the assumption of centralized access to all agent states. Instead, coordination emerges through local information exchange and broadcast of a shared dual variable enforcing global constraints.

Algorithm 9.2 summarizes the distributed update logic used for deployment across field controllers and the BEMS server[7]—each agent computes locally and exchanges minimal information.

9.6.1 Agent-Level Updates and Local Computation

In the distributed setting, each controllable device or zone is implemented as an autonomous agent responsible for evaluating its own local objective and computing actuation updates. At each iteration, agent "i" evaluates the gradient of its local comfort–energy objective using only locally available parameters, as defined in Eq. (9.3). This computation does not require global system information and can be executed directly on embedded field controllers.

[7] **Notation clarification.** To avoid ambiguity, we note that the illustration for distributed consensus control in this section introduces a control variable "u_i" (denoted v_i in the accompanying Python™ code, Listing 9.3), which is *distinct* from the normalized actuation variable "x_i" used in the primal–dual formulation of Sects. 9.4 and 9.5. Specifically, $x_i \in [0, 1]$ represents an abstract, normalized decision variable used for optimization and power-constrained coordination, whereas u_i represents a virtual control signal or setpoint coordinated via consensus and subsequently tracked by a simplified thermal response model. This distinction reflects different abstraction layers rather than conflicting formulations.

Algorithm 9.2 i-BEMS Distributed Agent Updates with Central Dual

System-wide inputs: $P_{\max}$, step sizes η, γ, consensus gain β
Agent i **inputs:** c_i, s_i, w_i, bounds $x_i \in [0, 1]$, neighbor set $\mathcal{N}_i$ (HVAC only)
Initialize at each agent i**:** $x_i^{(0)} \leftarrow \text{clip}(s_i + \text{small noise}, 0, 1)$
Initialize centrally: $\lambda^{(0)} \leftarrow 0$
for $k = 0, 1, 2, \ldots, T - 1$ **do**
  (HVAC agents only) Exchange states: receive $\{x_j^{(k)}\}_{j \in \mathcal{N}_i}$
  Local gradient: $g_{\text{loc},i} \leftarrow 2\big(x_i^{(k)} - s_i\big) + 2\, w_i\, x_i^{(k)}$
  if i is HVAC **then**
    Consensus term:
    $g_{\text{cons},i} \leftarrow 2\beta \sum_{j \in \mathcal{N}_i} \big(x_i^{(k)} - x_j^{(k)}\big)$
  else
    $g_{\text{cons},i} \leftarrow 0$
  end if
  Global-constraint term (broadcast $\lambda^{(k)}$**):** $g_{\text{dual},i} \leftarrow \lambda^{(k)}\, c_i$
  Primal update: $x_i^{(k+1)} \leftarrow \text{clip}\Big(x_i^{(k)} - \eta\big(g_{\text{loc},i} + g_{\text{cons},i} + g_{\text{dual},i}\big), 0, 1\Big)$
  (Central aggregator / BEMS server):
    Collect $x_i^{(k+1)}$ from all agents and compute $p^{(k+1)} = \sum_i c_i x_i^{(k+1)}$
    $\lambda^{(k+1)} \leftarrow \max\big\{0, \lambda^{(k)} + \gamma\big(p^{(k+1)} - P_{\max}\big)\big\}$
    Broadcast $\lambda^{(k+1)}$ to all agents
end for
Outputs: $\{x_i^{(T)}\}_{i=1}^{N}, \lambda^{(T)}, p^{(T)}$

For agents participating in consensus coordination (e.g., HVAC zones), local state information is exchanged only with neighboring agents defined by the communication graph. The resulting consensus gradient term, consistent with Eq. (9.4), enforces coordination through local interactions rather than centralized synchronization. Devices not participating in consensus coordination simply omit this term, enabling heterogeneous device integration within the same framework.

The primal update is then computed locally using the projected gradient step in Eq. (9.6). Box constraints on actuation variables are enforced directly at the agent level, ensuring feasibility regardless of communication delays or partial observability.

9.6.2 Central Dual Variable Broadcasting Via BEMS

Global power and Demand–Response (DR) constraints are enforced through a shared dual variable, which is maintained by a central BEMS server or supervisory controller. After collecting updated actuation values from all agents, the BEMS computes the total power consumption and updates the dual variable according to Eq. (9.7).

The updated dual variable is then broadcast to all agents and incorporated into their subsequent local updates as a pricing or scarcity signal. This mechanism enables coordinated compliance with system-level constraints without centralized computation of individual control actions. When the global constraint is inactive, the dual

```
            +-----------------------------+
            |         BEMS Server         |
            |   - Digital Twin            |
            |   - Computes dual variable  |
            +--------------+--------------+
                           |
                           |  Broadcast 'lambda'
                           v
       +-------------------+-------------------+
       |                   |                   |
  +-------+          +-------+           +-------+
  |Agent A|<-------->|Agent B|<--------->|Agent C|
  +-------+          +-------+           +-------+
       ^                   ^                   ^
       |                   |                   |
       +-------- Local state / u(t) ---------+
```

Fig. 9.9 Distributed consensus-based i-BEMS execution with agent-level computation and central dual variable "λ" broadcasting. The BEMS server computes the shared dual variable based on aggregated system-wide information and broadcasts it to all agents, while agents exchange local information with neighbors and update their decisions accordingly

variable remains zero and agents prioritize local objectives; when the constraint becomes active, coordinated curtailment emerges naturally across the system.

Figure 9.9 offers a simple conceptual illustration of the manner in which distributed agents could engage with the central BEMS server in context of the execution model under discussion.

9.6.3 Communication and Scalability Considerations

The distributed consensus formulation significantly reduces communication requirements compared to centralized control architectures. Each agent exchanges information only with its immediate neighbors and receives a single scalar broadcast from the BEMS server. This structure scales linearly with the number of agents and is well suited for large buildings or campus-scale deployments.

From a robustness perspective, the reliance on local interactions improves resilience to communication failures and partial observability. Temporary loss of neighbor information affects only local coordination rather than global system operation. Moreover, the modular structure allows subsystems to be added or removed with minimal reconfiguration, supporting incremental deployment and system evolution.

9.6.4 *Illustrative Distributed Consensus Example*

To illustrate distributed consensus dynamics in a field-deployable setting, Listing 9.3 presents a Python™ example[8] in which multiple zones coordinate a virtual control variable using Laplacian-based consensus and local optimization. The example demonstrates agent-level updates and neighbor-based coordination consistent with Algorithm 9.2, while using a simplified thermal model for visualization.

```
import numpy as np import matplotlib.pyplot as plt

# ------------------------------
# Simple i-BEMS consensus demo
# ------------------------------
# Five zones (agents) connected in a ring graph represent rooms/subsystems.
# Each agent has: current temperature, comfort setpoint, energy price weight.
# Goal: reach consensus on a "virtual setpoint" u_i that balances
# comfort (close to s_i) and energy (lower u_i when price is high),
# subject to neighbor agreement through the Laplacian.

np.random.seed(42)

n = 5  # number of zones
# Ring adjacency
A = np.zeros((n, n))
for i in range(n):
    A[i, (i-1) % n] = 1
    A[i, (i+1) % n] = 1
D = np.diag(A.sum(axis=1))
L = D - A  # graph Laplacian

# Initial temperatures (Â,°C)
T0 = np.array([24.0, 23.0, 25.5, 22.5, 24.5])
# Comfort setpoints (Â,°C) -- may vary per zone/occupant
S = np.array([23.0, 23.5, 24.0, 23.0, 23.5])
# Energy price weights (higher means more costly to use energy in that zone)
alpha = np.array([1.0, 1.5, 2.0, 1.2, 0.8])

# Consensus + optimization dynamics
# Each agent updates a control variable u_i (HVAC actuation proxy)
# Minimize: sum_i [ (u_i - S_i)^2 + lambda * alpha_i * u_i^2 ]
# with consensus term beta * sum_{(i,j) in E} (u_i - u_j)^2
# Gradient descent with Laplacian coupling.

lambda_reg = 0.2   # energy penalty scaling
```

[8] **Clarification on the distributed illustration**. The Python™ example presented in Sect. 9.6.4 is intentionally framed as a *simplified distributed consensus illustration*, rather than a direct software realization of Algorithm 9.2. Algorithm 9.2 describes a primal–dual coordination scheme with a *central* dual variable "λ" which enforces a *global* power constraint through a broadcast–collect–update cycle (broadcast of λ, aggregation of total power, and dual update).

By contrast, the illustrative code omits both the global power constraint and the dual broadcast step. Instead, it performs local consensus and regularization on a virtual control variable ν, followed by a first-order thermal response:

$$T_{k+1} = T_k + \kappa\,(\nu_k - T_k)$$

so that zone temperatures exponentially track the locally agreed control signal. This abstraction is useful for visualizing agreement dynamics and thermal trajectories under distributed coordination, but it should not be interpreted as encapsulating the primal–dual λ-broadcast mechanism of Algorithm 9.2.

```
beta = 0.6          # consensus strength
eta = 0.15          # step size

steps = 120
U_hist = np.zeros((steps, n))
T_hist = np.zeros((steps, n))

# Initialize control variables near initial temperatures
nu = T0.copy()
T = T0.copy()

# Simple first-order room thermal model: T_{k+1} = T_k + kappa*(nu - T_k)
# nu acts like a target/setpoint influenced by consensus & energy
kappa = 0.12

for k in range(steps):
    U_hist[k] = nu
    T_hist[k] = T

    # Gradient of local objective: 2*(nu - S) + 2*lambda*alpha*nu
    grad_local = 2.0*(nu - S) + 2.0*lambda_reg*alpha*nu
    # Consensus coupling term: 2*beta * L * nu
    grad_cons = 2.0*beta * (L @ nu)
    # Update control
    nu = nu - eta * (grad_local + grad_cons)

    # Update temperature dynamics towards control variable
    T = T + kappa*(nu - T)

consensus_value = np.mean(U_hist[-1])

# Plots
plt.figure(figsize=(10,6))
for i in range(n):
    plt.plot(U_hist[:, i], label=f'u_zone_{i+1}')
plt.axhline(consensus_value, color='k', linestyle='--', alpha=0.6, label='final consensus ~')
plt.title('Consensus of Control Variables (u) across i-BEMS Zones')
plt.xlabel('Iteration')
plt.ylabel('Control variable u (proxy for setpoint) [Ã,°C]')
plt.legend(loc='best')
plt.grid(True, alpha=0.3)
plt.tight_layout()
plt.savefig('consensus_u.png')

plt.figure(figsize=(10,6))
for i in range(n):
    plt.plot(T_hist[:, i], label=f'T_zone_{i+1}')
for i in range(n):
    plt.axhline(S[i], color='gray', linestyle=':', alpha=0.3)
plt.title('Zone Temperatures Tracking the Consensus-Controlled Setpoints')
plt.xlabel('Iteration')
plt.ylabel('Temperature [Ã,°C]')
plt.legend(loc='best', ncol=2)
plt.grid(True, alpha=0.3)
plt.tight_layout()
plt.savefig('temperatures.png')

final_u = U_hist[-1]
final_T = T_hist[-1]
print('Final control variables u:', np.round(final_u, 3))
print('Approx. consensus value:', round(consensus_value, 3))
print('Final temperatures T:', np.round(final_T, 3))
```

Listing 9.3 Python™ demonstration of distributed consensus-based control with agent-level updates and local neighbor interactions.

```
C:\Path\to\the\Python\Script.py

Final control variables u: [19.017 18.218 17.723 18.584 19.517]
Approx. consensus value: 18.612
Final temperatures T: [19.017 18.218 17.723 18.584 19.517]

Process finished with exit code 0
```

Listing 9.4 Console output summarizing the final consensus control variables and resulting zone temperatures obtained from the distributed consensus-based i-BEMS example.

Listing 9.4 reports the final numerical outcome of the distributed consensus process. The control variables converge to closely clustered values, indicating agreement among agents despite heterogeneous comfort targets and energy weights. The resulting zone temperatures track these consensus-controlled setpoints through the simplified thermal dynamics, illustrating how distributed coordination can achieve coherent building-level behavior without centralized control of individual agents.

The thermal dynamics included in the example are used solely to visualize the effect of coordinated control on zone temperatures and are external to the coordination logic itself. Further, the consensus updates applied to the control variables directly correspond to the agent-level update steps in Algorithm 9.2.

Overall, the distributed formulation preserves the optimization objectives and coordination logic introduced earlier, while enabling scalable and deployable implementation within real-world i-BEMS infrastructures.

Illustrative Distributed Consensus Example with Control and Thermal Dynamics

This box illustrates the behavior of the distributed consensus-based control framework using the Python™ example presented in Listing 9.3. The example highlights how agent-level coordination emerges through local interactions and how this coordination propagates to physical zone behavior.

Consensus-Based Control Update. Each zone i maintains a local control variable u_i, which acts as a virtual setpoint or actuation proxy. At each iteration, agents update u_i using the following consensus-based gradient descent rule:

$$u_i^{(k+1)} = u_i^{(k)} - \eta \left[2\left(u_i^{(k)} - s_i\right) + 2\lambda_{\text{reg}}\, \alpha_i\, u_i^{(k)} + 2\beta \sum_{j \in \mathcal{N}_i} \left(u_i^{(k)} - u_j^{(k)}\right) \right]$$

where s_i is the comfort setpoint of zone i, α_i is an energy sensitivity weight, β is the consensus gain, and $\mathcal{N}_i$ denotes the neighboring zones. This update structure is consistent with the distributed coordination logic summarized in Algorithm 9.2, but without an explicit global power constraint.

Consensus of Control Variables. Figure 9.10 illustrates the evolution of the control variables $\{u_i\}$ across all zones. Despite heterogeneous initial conditions and differing comfort and energy weights, the control variables converge toward a common operating region. This behavior confirms that consensus is achieved through purely local information exchange, without centralized coordination.

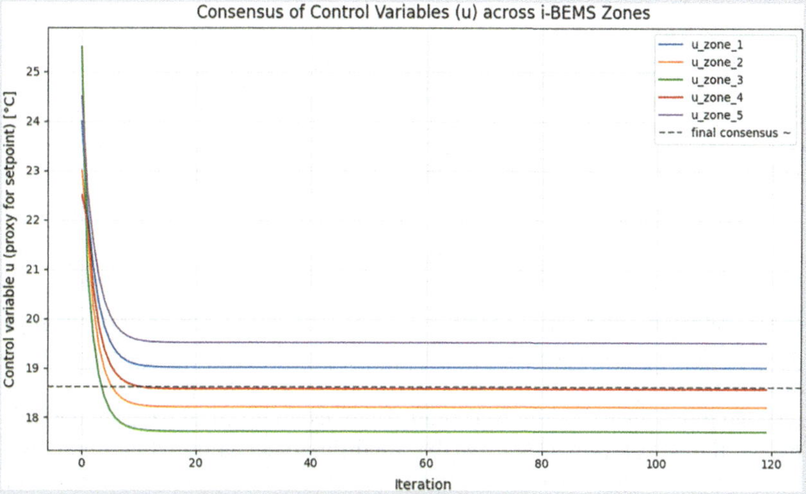

Fig. 9.10 Distributed consensus of control variables across zones

Thermal Response Model. To visualize the impact of consensus coordination on physical variables, a simplified first-order thermal response model is employed:

$$T_i^{(k+1)} = T_i^{(k)} + \kappa\left(u_i^{(k)} - T_i^{(k)}\right)$$

where T_i represents the zone temperature and $\kappa > 0$ is a thermal response coefficient. This model serves as an illustrative proxy rather than a calibrated building thermal model.

Zone Temperature Evolution. Figure 9.11 shows the resulting temperature trajectories. Starting from heterogeneous initial conditions, zone temperatures converge smoothly toward coordinated operating regimes driven by the consensus-controlled variables. While exact temperature agreement is not enforced, thermal behavior remains aligned across zones due to coordinated control actions.

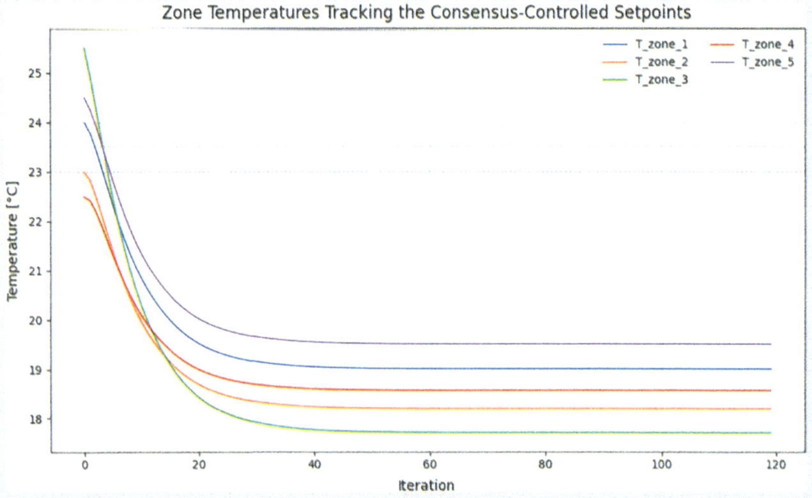

Fig. 9.11 Zone temperature trajectories under distributed consensus-based control

Interpretation and Scope. Together, Figs. 9.10 and 9.11 demonstrate the separation between coordination logic and physical response. Consensus emerges at the control layer through distributed interactions, while physical dynamics respond passively to coordinated control signals. In practical i-BEMS deployments, this structure enables scalable field implementation while remaining compatible with high-fidelity Digital Twin (DT) models, as discussed next in Sect. 9.7.

Illustrative Nature and Practical Deployment Considerations

The Python™-based implementations presented in Listings 9.1 and 9.3 are intended as *illustrative reference realizations* of the proposed centralized and distributed consensus-based control strategies. While they faithfully implement the mathematical formulations and algorithms introduced in Sects. 9.4–9.6, they are not intended to be directly deployable in production-grade i-BEMS environments.

Several simplifying assumptions are made for clarity and pedagogical purposes, including:

- idealized and synchronous agent updates with fixed step sizes;
- simplified device models and normalized actuation variables;
- static communication graphs without delays, packet loss, or failures;
- illustrative thermal dynamics rather than calibrated building models.

In real-world deployments, these elements must be extended to account for asynchronous execution, heterogeneous controller capabilities, communication latency, cybersecurity constraints, fault tolerance, and integration with existing Building Management Systems (BMS) and field protocols.

Nevertheless, the presented examples accurately capture the *core coordination logic* of consensus-based i-BEMS control. They serve as a transparent bridge between theory and practice, providing a concrete foundation upon which production-ready implementations can be built using industrial controllers, middleware platforms, and Digital Twin (DT)-enabled calibration and validation pipelines, as discussed in the following section.

9.7 Integration with Digital Twin (DT) Prediction and Validation

The consensus-based coordination strategies illustrated in Sects. 9.5 and 9.6 provide a scalable mechanism for aligning local device actions with building-level objectives. However, these algorithms operate on instantaneous measurements and static constraints unless augmented by predictive intelligence. The Digital Twin (DT) layer fulfills this role by enabling forward-looking assessment, validation, and adaptive refinement of consensus-based decisions.

Rather than replacing the coordination logic, the DT acts as an *intelligence overlay* which informs and modulates the parameters, targets, and constraints used by the consensus algorithms. This separation preserves the modularity of the i-BEMS architecture while enabling anticipatory control and improved robustness under uncertainty.

9.7.1 Closed-Loop Interaction Between Consensus and Digital Twin (DT)

In the architecture shown in Fig. 9.12, the DT layer operates in a closed-loop interaction with the consensus layer. Operational data collected from the physical i-BEMS layer—including zone temperatures, power consumption, occupancy indicators, and environmental measurements—are streamed to the DT for state estimation and pre-

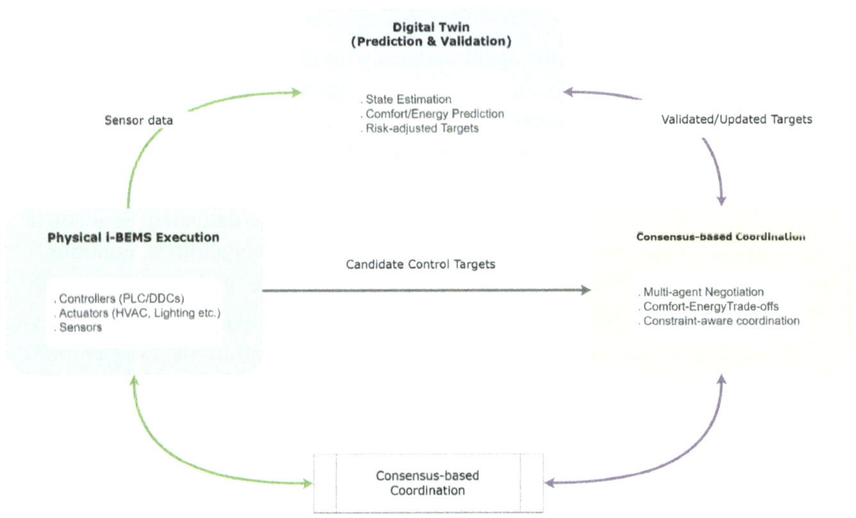

Fig. 9.12 Closed-loop interaction between consensus-based coordination and the Digital Twin layer

diction. Based on this information, the DT evaluates the expected impact of candidate control actions generated by the consensus algorithm.

The resulting predictions are then fed back to the coordination layer in the form of updated comfort targets, risk-adjusted constraints, or modified coordination weights. This feedback loop allows consensus-based decisions to be validated against predicted system behavior before execution, reducing performance degradation caused by model mismatch, disturbances, or delayed system response.

9.7.2 Predict–Coordinate–Execute Feedback Cycle

The integration of DTs enables a structured *predict–coordinate–execute* cycle which extends traditional feedback control paradigms. In this cycle, the DT first predicts near-future system trajectories under expected disturbances and operating conditions. These predictions inform the coordination stage, where the consensus algorithm computes control actions consistent with both local objectives and system-wide feasibility.

Once executed by the i-BEMS layer, the resulting system response is monitored and compared against DT predictions. Discrepancies between predicted and observed behavior trigger model recalibration or parameter updates within the DT, thereby closing the loop. Importantly, the core consensus algorithms described in Algorithms 9.1 and 9.2 remain unchanged; only their inputs are adaptively refined through DT feedback.

This layered feedback structure supports multi-time-scale operation (see this chapter, where fast consensus updates handle real-time coordination while slower DT updates capture *evolving building dynamics* and *occupant behavior.*

9.7.3 Digital Twin-Assisted Constraint Adaptation

A particularly impactful role of the DT lies in adaptive constraint management. Building-level constraints, such as Demand–Response (DR) limits or feeder capacity caps, are often specified conservatively to hedge against uncertainty. Using predictive simulation, the DT can estimate risk margins and dynamically adjust these constraints to balance safety and performance.

For example, a DT may tighten the global power cap during periods of high forecast uncertainty or relax it when confidence in system response is high. Formally, this can be represented as[9]

$$P_{\max}^{(k+1)} = P_{\max}^{(k)} - \Delta P_{\text{risk}}^{(k)} \tag{9.8}$$

where $\Delta P_{\text{risk}}^{(k)}$ is derived from DT-based uncertainty assessment. Similar adaptations can be applied to comfort targets or consensus gains, enabling context-aware coordination without altering the underlying optimization structure.

Through such mechanisms, the DT enhances adaptability and resilience while preserving the distributed and scalable nature of consensus-based i-BEMS control.

Example: Adaptive Power Cap Adjustment

Scenario: Consider a commercial building participating in a Demand–Response (DR) program with a nominal feeder capacity of $P_{\max} = 500$ kW. During a high-uncertainty forecast window (e.g., severe weather), the DT predicts a 15% risk of exceeding the feeder limit due to HVAC load variability.

DT-Based Adjustment:

$$\Delta P_{\text{risk}}^{(k)} = \alpha \cdot \sigma_{\text{forecast}} = 0.15 \times 500 = 75 \text{ kW}$$

where α is a risk sensitivity factor and σ_{forecast} represents normalized uncertainty.

Updated Constraint:

$$P_{\max}^{(k+1)} = 500 - 75 = 425 \text{ kW}$$

---

[9] This update rule follows a risk-aware adaptation principle commonly used in robust and predictive control frameworks. The term $\Delta P_{\text{risk}}^{(k)}$ is not empirical; it is derived from uncertainty metrics (e.g., forecast variance, confidence intervals) or scenario-based risk models, ensuring that constraint adjustments remain theoretically grounded.

Impact: The i-BEMS controller now optimizes loads under a tighter cap, prioritizing critical zones while deferring non-essential HVAC operations. When uncertainty drops (e.g., after weather stabilizes), the DT relaxes the cap back toward 500 kW, restoring comfort without compromising feeder safety.[10]

9.8 Implementation Considerations in Real i-BEMS

While the consensus-based control strategies developed in this chapter are formulated at an abstract algorithmic level, their practical realization within real integrated Building Energy Management Systems (i-BEMS) requires careful consideration of architectural mapping, time-scale separation, communication constraints, and robustness requirements. This section discusses key implementation aspects which bridge the gap between *algorithmic design* and *deployable system architectures*, without altering the underlying coordination logic.

9.8.1 Mapping Agents to Controllers and Zones

In practical i-BEMS deployments, the notion of an *agent* does not necessarily correspond one-to-one with a physical device. Instead, agents are logical entities which encapsulate control objectives, constraints, and communication responsibilities for one or more physical components. Typical mappings, see Fig. 9.13, include individual thermal zones, aggregated lighting groups, air-handling units, or virtual groupings spanning multiple subsystems.

Field controllers (e.g., PLCs, DDCs, or embedded controllers) often manage multiple zones or devices, executing low-level control loops at high frequency. In such cases, multiple consensus agents may be instantiated within a single controller, sharing local measurements and actuation interfaces while participating independently in the coordination process. Conversely, a single agent may abstract a collection of devices when coordinated actuation is desired.

This level of abstraction supports flexible deployment on diverse hardware platforms while maintaining a consistent coordination framework. Notably, the instructional examples of consensus algorithms discussed in Sects. 9.5 and 9.6 are formulated purely at the agent level, without dependence on specific control hardware or communication protocols.

[10] Importantly, these adaptations shall occur at the supervisory layer and do not alter the *consensus algorithm* itself; *agents* continue to negotiate under the updated constraints, preserving distributed scalability.

```
                             +---------------------------------+
                             |          BEMS Server            |
                             |  - Digital Twin                 |
                             |  - Dual Variable 'lambda'       |
                             |  - (Supervisory Agents)         |
                             |     > Aggregated Lighting Agent |
                             |     > Aggregated HVAC Agent     |
                             +---------^------------^----------+
                                       |            |
 'lambda' Broadcast / State Reports |            | State Reports / 'lambda' Broadcast
                                       |            |
-----------------------------------+------------+---------------------------------
                                       |            |
                                  (Logical Agents Layer)
                  +------------+  +------------+  +------------+  +------------+
                  | Agent A    |  | Agent B    |  | Agent C    |  | Agent D    |
                  +------^-----+  +------^-----+  +------^-----+  +------^-----+
                         |               |               |               |
-------------------+---------------+---------------+---------------+-------------
                         |               |               |               |
                  Floor 1         Floor 1         Floor 2         Floor 2
                  Zone 1          Zone 2          Zone 3          AHU
                         |               |               |               |
                         v               v               v               v
                  +--------------+ +--------------+ +--------------+ +--------------+
                  | Controller C1| | Controller C1| | Controller C2| |     PLC      |
                  +-------+------+ +------+-------+ +------+-------+ +-------+------+
                          |               |                |                 |
                  +-------v------+ +------v-------+ +------v------+ +---------v------+
                  |   BACnet     | |   BACnet     | |   Modbus    | |  EtherNet/IP   |
                  +--------------+ +--------------+ +-------------+ +----------------+
```

Fig. 9.13 Conceptual mapping between physical zones, logical consensus agents, field controllers, and the BEMS supervisory layer. Agents represent logical coordination entities that may abstract individual zones, aggregated subsystems, or shared equipment. Multiple agents may execute on a single controller, while a single agent may span multiple physical devices. Consensus coordination and global signals (e.g., dual variable λ) operate above the field protocol layer, enabling hardware agnostic deployment

9.8.2 Consensus Timing and Multi-time-Scale Operation

As outlined in Chap. 8, real building systems function over several distinct time scales, see Fig. 8.6. Fast inner-loop controllers regulate actuator dynamics and ensure safety, while higher-level supervisory control operates on slower intervals to coordinate energy usage and comfort objectives. Consensus-based coordination naturally fits within this supervisory layer, executing at time scales ranging from tens of seconds to several minutes.

In a typical deployment, local measurements are aggregated and preprocessed at the field-controller level before being exchanged with neighboring agents or the BEMS server. Consensus updates are then computed using the most recent available information, while low-level controllers track the resulting setpoints or actuation commands asynchronously.

This separation of time scales improves stability and robustness by preventing high-frequency communication or excessive computational burden at the coordination layer. It also allows Digital Twin (DT)-based prediction and validation, discussed in Sect. 9.7, to operate on even slower horizons without disrupting real-time control.

9.8.3 Robustness, Privacy, and Partial Observability

Consensus-based coordination offers inherent advantages in terms of robustness and scalability; however, real-world deployments must contend with imperfect information, communication constraints, and privacy considerations. Agents may operate under *partial observability*, relying only on local measurements and limited neighbor information, while communication links may be delayed, lossy, or intermittent.

Example: Consensus Under Partial Observability and Imperfect Communication

Consider a group of zone-level HVAC controllers attempting to reach consensus on a shared cooling-load allocation during peak hours. Each agent "i" can measure only its local temperature, occupancy estimate, and actuator state, but has no access to global building information or the internal states of distant zones *(partial observability)*. Furthermore, communication occurs only with immediate neighbors in the building's network topology and is subject to occasional packet loss due to Wi-Fi interference.

In this setting, each agent updates its coordination variable (let's designate it with "θ_i") based on neighbor exchanges[11]:

$$\theta_i(k+1) = \theta_i(k) + \alpha \sum_{j \in N_i} \big(\theta_j(k) - \theta_i(k)\big)$$

but missing messages are simply treated as absent neighbors for that iteration. Despite these limitations, the system converges reliably because consensus relies only on local interactions, is tolerant to asynchronous updates, and gracefully degrades when links fail temporarily. Sensitive data—such as real-time occupant counts or zone-level comfort profiles—never leave the local controller; only the abstract coordination variable θ_i is shared, preserving privacy while enabling global coordination.

From a robustness perspective, the local nature of consensus interactions limits the propagation of faults and allows graceful degradation in the presence of agent or communication failures. Privacy is preserved by design, as agents exchange only minimal coordination variables rather than raw sensor data or detailed occupant information. Nonetheless, practical systems must still address cybersecurity, authentication, and data integrity concerns at the communication layer.

Example: Cybersecurity Risks in Consensus-Based i-BEMS Coordination

Consider a cluster of Air-Handling-Unit (AHU) controllers participating in a building-wide consensus process to coordinate static pressure setpoints. Each agent communicates its local coordination variable θ_i over the building's IP network, assuming messages are authentic and unaltered. Although the exchanged variables contain no raw sensor data, they still form a control-relevant signal which influences global HVAC performance.

Suppose an attacker gains access to a poorly secured Wireless Access Point (WAP) and injects spoofed messages for a subset of agents. Rather than disabling communication entirely, the attacker subtly modifies transmitted coordination variables by adding small biases, e.g.,

$$\theta_j^{\mathrm{mal}}(k) = \theta_j(k) + \delta$$

where "δ" is designed to remain within typical operating ranges. These malicious updates cause neighboring agents to adjust their internal states toward incorrect values, gradually skewing the collective consensus point. *The impact*

[11] **Explanation of Symbols**. The update equation used in this example, is a standard discrete-time consensus rule. The symbols have the following interpretations:

- $\theta_i(k)$: The *coordination variable* of agent i at iteration k. It represents the quantity agents are trying to agree on (e.g., cooling-load share).
- $\theta_i(k+1)$: The updated value of the coordination variable at the next iteration. Each agent computes this using only locally available information.
- $\mathcal{N}_i$: The set of *neighboring agents* from which agent i receives messages. Communication is local; distant zones are never contacted.
- $\theta_j(k)$: The coordination variable of neighbor j at iteration k. These are the values agent i receives over the communication network.
- $(\theta_j(k) - \theta_i(k))$: The *consensus error* between agent i and neighbor j. If this difference is large, the update step pushes both agents toward agreement.
- $\sum_{j \in \mathcal{N}_i}(\cdot)$: A sum of all pairwise consensus errors with the neighbors of agent i.
- α: A *step size* or *learning rate* that controls how aggressively agent i adjusts θ_i. A small α improves stability under delays and packet loss.
- k: The discrete iteration index (not physical time). Agents may update at different rates, making the method compatible with asynchronous communication.

Overall, this update rule ensures that each agent moves its estimate toward the average of its neighbors' values. Even with missing messages or local-only observability, the network collectively converges to a shared value.

Table 9.2 Robustness, privacy, and observability considerations in consensus-based i-BEMS control

Challenge	Typical mitigation strategy
Partial observability	Local state estimation; reliance on neighborhood-level information rather than global states
Communication delays	Asynchronous or event-triggered consensus updates; conservative step-size selection
Packet loss or link failure	Redundant communication paths; graceful degradation via local autonomy
Privacy of occupant data	Exchange of abstract coordination variables instead of raw measurements
Cybersecurity risks	Secure communication channels; authentication and access control at the BEMS layer
Model mismatch and uncertainty	Digital Twin (DT)-assisted validation and adaptive parameter tuning

may manifest as persistent inefficiencies—such as unnecessary static pressure increases—or comfort drift in specific zones.

Since, consensus algorithms inherently trust neighbor messages, such integrity attacks can propagate slowly and remain undetected unless monitoring mechanisms are in place. Protection requires secure communication layers, message authentication tags (MACs), and network-level intrusion detection to ensure that exchanged variables originate from legitimate agents. Even though the consensus logic itself is lightweight and robust to random disturbances, intentional adversarial manipulation underscores the need for cybersecurity-aware i-BEMS deployments.

Table 9.2 summarizes key implementation challenges and typical mitigation strategies in consensus-based i-BEMS deployments.

These considerations reinforce the importance of viewing consensus algorithms as part of a broader Cyber–Physical System (CPS) architecture. While the algorithms themselves are lightweight and scalable, their successful deployment depends on careful integration with sensing, communication, and supervisory control infrastructure.

Overall, the implementation aspects discussed in this section highlight how consensus-based coordination can be embedded within real i-BEMS deployments in a robust, privacy-aware, and scalable manner, while preserving the modularity and adaptability emphasized throughout this chapter.

9.9 Discussion: Benefits and Limitations of Consensus-Based Control

The consensus-based control strategies described in this chapter provide a principled mechanism for coordinating large numbers of heterogeneous building subsystems while preserving modularity and scalability. However, as with any distributed control paradigm, their effectiveness depends on architectural context, operating conditions, and design choices. This section discusses the key benefits and limitations of consensus-based control in i-BEMS applications, with an emphasis on practical implications rather than algorithmic details.

9.9.1 Comparison with Centralized and Hierarchical Control

A primary advantage of consensus-based control lies in its contrast to fully centralized or rigidly hierarchical architectures. Centralized optimization approaches offer global optimality under complete information, but they scale poorly as system size and complexity increase. They also introduce *single points of failure* and significant communication overhead, particularly in large or geographically distributed buildings.

Hierarchical control structures mitigate some of these issues by decomposing the system into layers; however, they often rely on fixed aggregation schemes and predefined coordination paths. By contrast, consensus-based control distributes decision-making across agents, allowing coordination to emerge through local interactions rather than explicit top-down commands.

Importantly, consensus-based approaches occupy a middle ground between full centralization and complete decentralization. Global objectives are respected through shared coordination variables or constraints, while local autonomy is retained at the agent level. This balance is particularly well suited to i-BEMS environments, where subsystems differ in dynamics, ownership, and operational priorities. Consensus can be implemented with lightweight message exchanges (e.g., marginal costs or provisional setpoints), preserving privacy and limiting bandwidth usage, while maintaining robustness to intermittent links and delayed updates.

Example: Digital Twin (DT)-Enhanced Consensus via Parameter Refinement

Consider a simple i-BEMS scenario with three HVAC zones which jointly need to supply a total cooling load of 30 kW. Each zone i has a quadratic local cost $C_i(u_i) = a_i u_i^2$ for delivering cooling effort u_i, with nominal parameters[12,13]:

$$(a_1, a_2, a_3) = (1.0,\ 2.0,\ 3.0), \qquad \text{subject to } u_1 + u_2 + u_3 = 30.$$

Centralized solution (baseline). A central optimizer solves

$$\min_{u_1,u_2,u_3} \sum_{i=1}^{3} a_i u_i^2 \quad \text{s.t.} \quad u_1 + u_2 + u_3 = 30$$

Using *Lagrange multipliers*, the *KKT conditions* yield $2a_i u_i^\star = \lambda \Rightarrow u_i^\star = \lambda/(2a_i)$; enforcing the equality constraint gives

$$u^\star \approx (16.4,\ 8.2,\ 5.5)\ \text{kW}$$

which is globally optimal but requires full knowledge of all a_i and centralized coordination of all u_i.

Digital Twin intervention. A physics-based Digital Twin (DT) runs in parallel, continuously calibrating parameters using sensor data (e.g., supply air temperatures, fan power, damper positions, and zone heat gains). After detecting persistent deviations between predicted and actual energy consumption, the DT revises the cost coefficients to reflect actual equipment conditions:

$$(a_1, a_2, a_3)_{\text{DT}} = (1.2,\ 1.8,\ 2.5)$$

A centralized recomputation with these corrected parameters yields

$$u_{\text{DT}}^\star \approx (14.0,\ 9.33,\ 6.71)\ \text{kW}$$

steering effort toward the more efficient zones (lower a_i) and reducing operating cost.

Consensus-based reaction to DT updates. Under distributed consensus, agents exchange only coordination variables, such as marginal costs or provisional efforts, with their neighbors. When the DT publishes the revised a_i values *locally* to each zone controller, the marginal-cost expression:

$$\lambda_i(k) = 2a_i\, u_i(k)$$

updates accordingly. Starting from a shared initial allocation:

$$u(0) = (10,\ 10,\ 10)\ \text{kW}$$

agents iteratively adjust loads via neighbor exchanges with a small step size (e.g., $\alpha = 0.2$).[14] The distributed updates converge to a compromise close to the DT-refined optimum:

$$u(k) \to (14.0,\ 9.3,\ 6.7)\ \text{kW}$$

No agent exposes raw sensor streams or private model details; only abstract coordination signals are shared. The result adapts in real time to the DT's parameter corrections without requiring a central coordinator or rigid hierarchies.

Interpretation. This example illustrates how consensus-based coordination and Digital Twins (DTs) form a complementary pair: the DT provides high-fidelity, continuously updated parameters, while the consensus layer re-balances loads in a scalable and privacy-preserving manner. Compared to centralized control, the approach avoids single points of failure and heavy data aggregation; compared to hierarchical control, it reduces reliance on fixed aggregation rules and top-down orchestration, while still aligning with global objectives through shared coordination variables.

In summary, centralized optimization can produce globally optimal solutions when complete, high-quality information is available and reliable, but it incurs computational and communication burdens and creates architectural fragility. Hierarchical approaches offer structure and manageability but depend on predetermined aggregation and supervisory links. Consensus-based control, especially when enhanced by Digital Twin parameter corrections, achieves near-optimal behavior with local communication, preserves privacy by exchanging abstract variables, and scales gracefully across heterogeneous i-BEMS subsystems.

9.9.2 Applicability and Design Trade-Offs

Despite its advantages, consensus-based control is not universally optimal. Its performance and suitability depend on communication quality, system coupling strength, and the availability of accurate local models or measurements. Poorly chosen consensus gains, communication delays, or excessive heterogeneity can slow convergence or lead to oscillatory behavior, as illustrated in the examples earlier in this chapter, see Sect. 9.5.

Moreover, consensus algorithms typically converge to compromise solutions which balance local objectives rather than strict global optima. *In contexts where*

---

[14] In this context, "nominal parameters" refers to the baseline or default values of the model coefficients (a_1, a_2, a_3) used prior to any calibration by the Digital Twin (DT). These values typically come from design specifications, manufacturer data, or initial engineering assumptions, and represent the expected equipment behavior before sensor-driven refinement.

[14] For this formulation to be well-posed, the following assumptions are standard: (1) Each $a_i > 0$, ensuring $C_i(u_i) = a_i u_i^2$ is strictly convex and differentiable, which guarantees a unique global optimum under the load constraint. (2) The global constraint $u_1 + u_2 + u_3 = 30$ is feasible and known to all agents. (3) Each agent knows only its own parameter a_i and effort u_i but not those of other agents. (4) The optimality condition $2a_i u_i^\star = \lambda$ is valid due to convexity, and forms the basis for distributed marginal-cost consensus.

[14] The convergence of this distributed update relies on additional assumptions: (1) The inter-zone communication graph is connected and undirected (or weight-balanced), ensuring standard consensus convergence. (2) Agents exchange only marginal-cost or dual-like coordination variables rather than raw measurements, preserving privacy. (3) The Digital Twin updates parameters slowly enough that consensus iterations converge between updates. (4) Marginal-cost equalization across neighbors approximates the centralized KKT condition $2a_i u_i = \lambda$, enabling convergence toward the DT-refined optimum.

Table 9.3 Benefits and limitations of consensus-based control in i-BEMS applications

Aspect	Type	Discussion
Scalability	Benefit	Scales favorably with system size due to local interactions and limited information exchange
Robustness	Benefit	No single point of failure; local autonomy enables graceful degradation under faults
Optimality	Limitation	Typically converges to compromise solutions rather than strict global optima
Communication dependence	Limitation	Performance depends on network quality, latency, and synchronization assumptions
Design complexity	Limitation	Requires careful tuning of gains and update rates to ensure stability and convergence
Privacy	Benefit	Preserves privacy by exchanging abstract coordination variables instead of raw data
Integration with DTs	Benefit	Naturally complements Digital Twin (DT) layers which refine objectives and constraints

hard performance guarantees or strict constraint enforcement are required at all times, additional supervisory mechanisms—such as Digital Twin (DT)-assisted constraint adaptation—become essential.

Table 9.3 summarizes key trade-offs associated with consensus-based control in comparison to alternative control paradigms.

From a design perspective, consensus-based control is most effective when coordination requirements are moderate, subsystem interactions are localized, and adaptability is valued over centralized optimality. In such cases, it offers a compelling balance between performance, robustness, and deployability.

Overall, the discussion in this section highlights that consensus-based control should be viewed not as a replacement for centralized optimization, but as a complementary paradigm within a layered i-BEMS architecture. When combined with predictive intelligence from Digital Twins (DTs) and appropriate supervisory oversight, consensus-based coordination provides a flexible and resilient foundation for next-generation smart building control systems.

9.10 Summary and Outlook

This chapter presented a systematic treatment of *consensus-based control strategies* for smart appliances and building automation within Digital Twin (DT)-enabled environments. Starting from the motivation for moving beyond centralized and hierarchical control paradigms, the chapter framed integrated Building Energy Management Systems (i-BEMS) as inherently *multi-agent systems (MASs)*, where coordination, rather than centralized optimization, becomes the dominant control challenge.

A hierarchical interpretation was adopted in which physical i-BEMS controllers operate at the execution layer, consensus-based coordination provides distributed decision-making across agents, and Digital Twin (DT) models act as a predictive intelligence layer. Within this framework, a convex quadratic optimization problem was formulated to balance local comfort–energy objectives, inter-agent coordination, and global power constraints. Both centralized and distributed primal–dual algorithms were introduced, highlighting how coordination can be achieved using only local information exchange and minimal centralized supervision.

To support intuition, Python™-based examples were introduced for both centralized and distributed settings. These examples were deliberately framed as *illustrative implementations* rather than deployable control systems, demonstrating convergence behavior, consensus formation, and constraint handling under simplified assumptions. Their role is pedagogical: to bridge abstract formulations with observable system behavior, while preserving a clear distinction from real-world deployment requirements.

The chapter further explored how Digital Twins (DTs) can be integrated into the consensus loop—not as control replacements, but as supervisory elements which enable prediction, validation, and adaptive constraint management. This integration supports closed-loop operation without compromising distributed scalability. Practical implementation considerations were discussed, including agent-to-controller mapping, multi-time-scale operation, robustness to communication imperfections, and privacy under partial observability.

Outlook. Consensus-based control provides a powerful coordination mechanism for future smart buildings, particularly as systems grow in scale, heterogeneity, and autonomy. However, its effectiveness depends critically on architectural design, parameter tuning, communication reliability, and integration with higher-level intelligence such as Digital Twins (DTs). Future research directions include:

- hybrid hierarchical-consensus architectures for large portfolios of buildings,
- adaptive consensus gains informed by Digital Twin (DT) predictions,
- learning-assisted consensus under uncertainty and non-stationary occupancy patterns,
- and rigorous validation of consensus-based schemes in real-world BEMS deployments.

These directions position consensus control not as a replacement for existing methods, but as a complementary coordination layer within next-generation intelligent building systems.

References

1. Anik MA, Wasi AT, Rahman A, Ahsan MM (2025) A Digital Twin-based multi-agent framework for understanding and optimizing smart building HVAC systems. SSRN 5212671
2. Almadhor A, Ghazouani N, Bouallegue B et al (2025) Digital Twin based deep learning framework for personalized thermal comfort prediction and energy efficient operation in smart buildings. Sci Rep 15:24654. https://doi.org/10.1038/s41598-025-10086-y https://doi.org/10.1038/s41598-025-10086-y
3. Li Z, Duan Z (2017) Cooperative control of multi-agent systems: a consensus region approach. CRC Press
4. Zhang J, Zhang H, Yang D, Luo Y, Zhou B (2025) Consensus control of linear multiagent systems under switching networks and communication link faults. In: Adaptive collaborative control of multiagent systems by event-triggered mechanisms, intelligent control and learning systems, vol 14. Springer, Singapore. https://doi.org/10.1007/978-981-97-7866-9_2
5. Kim BY, Ahn HS (2015) Consensus-based coordination and control for building automation systems. IEEE Trans Control Syst Technol 23(1):364–371. https://doi.org/10.1109/TCST.2014.2317793
6. Kou X, Li F, Dong J, Starke M, Munk J, Kuruganti T, Zandi H (2019) A distributed energy management approach for residential demand response. In: 2019 3rd International conference on smart grid and smart cities (ICSGSC), pp 170–175. https://doi.org/10.1109/ICSGSC.2019.00004
7. Kong DY, Bao YQ, Hong YY, Wang BB, Huang HB, Liu L, Jiang HH (2019) Distributed control strategy for smart home appliances considering the discrete response characteristics of the on/off loads. Appl Sci 9(3):457. https://doi.org/10.3390/app9030457 https://doi.org/10.3390/app9030457
8. Wang Y, Guo Z, Pang Y, Gao K, Zhao J (2023) Consensus control for distributed power tracking by device-level digital twin agents. Front Energy Res 11–2023. https://doi.org/10.3389/fenrg.2023.1133516, https://www.frontiersin.org/journals/energy-research/articles/10.3389/fenrg.2023.1133516
9. Zhukovskii K (2025) Multi-agent energy-adaptive automation systems: architecture and implementation. Aalto University
10. Abdelrahman MM, Chong A, Miller C (2022) Personal thermal comfort models using Digital Twins: preference prediction with BIM-extracted spatial-temporal proximity data from Build2Vec. Build Environ 207:108532. https://doi.org/10.1016/j.buildenv.2021.108532 https://www.sciencedirect.com/science/article/pii/S0360132321009240
11. Pei W, Hua C, Cui H et al (2024) A hybrid systems approach to consensus of nonlinear multi-agent systems with self-triggered output feedback control. Int J Control Autom Syst 22:2074–2084. https://doi.org/10.1007/s12555-023-0409-4 https://doi.org/10.1007/s12555-023-0409-4
12. Chen H, Dai S, Meng F (2023) Smart building thermal management: a data-driven approach based on dynamic and consensus clustering. Sustainability 15(21):15489. https://doi.org/10.3390/su152115489 https://doi.org/10.3390/su152115489
13. Savaridassan P, Sahasra K, Kalaiselvi B, Nagoormeeran AR, Krishnaraj N (2024) Digital Twin consensus for blockchain-enabled intelligent transportation systems in smart cities, chap 14. John Wiley & Sons, Ltd., pp 297–322. https://doi.org/10.1002/9781394303564.ch14, https://onlinelibrary.wiley.com/doi/abs/10.1002/9781394303564.ch14

Glossary

Agent An autonomous decision-making unit in a distributed control or optimization setting, representing a zone, subsystem, or controllable load which updates its local control variable and exchanges limited information with neighbors or a supervisory layer.

Air Handling Unit (AHU) A centralized HVAC subsystem which conditions and circulates air (heating/cooling/ventilation). In this book, the AHU is modeled to study energy–comfort trade-offs and decision-aware control strategies.

Baseline (in optimization) A reference operating point or parameter set against which improvement is evaluated (e.g., energy, comfort, emissions). Baselines also support normalization and fair comparison across candidate solutions.

Black-box model A model that maps inputs to outputs without explicit physical interpretability (often purely data-driven). Used when first-principles modeling is infeasible, but may reduce transparency in control/decision-making.

Building Energy Management System (BEMS) A supervisory layer (software + hardware) that monitors, optimizes, and controls building energy use while maintaining comfort, coordinating HVAC, lighting, storage, and other flexible loads.

Building Management System (BMS) The operational automation platform used in buildings to monitor and control services (HVAC, alarms, power, etc.). In this book, BMS interfaces form a practical pathway to implement BEMS intelligence.

Building Energy Modeling (BEM) Physics-based or hybrid modeling of building energy behavior for simulation, analysis, and control design.

Building Digital Twin (BDT) A digital representation of a building that goes beyond offline simulation by integrating models with operational data streams for monitoring, prediction, calibration, and control-oriented services.

M. H. Wani et al., *Digital Twins for Building Optimization: Enhancing Energy and Comfort*, Studies in Infrastructure and Control,
https://doi.org/10.1007/978-3-032-23274-8

Calibration The process of adjusting model parameters (thermal, equipment, schedules, etc.) so that simulated behavior matches measured building data, improving fidelity for control and decision support.

Candidate control targets Proposed control setpoints or actions (e.g., temperature setpoints, power limits) generated by supervisory logic/optimization and assessed through simulation/twin-based evaluation before deployment.

Chiller A cooling plant component producing chilled water for HVAC; a major contributor to building electricity demand and a common target for optimization and demand flexibility.

Clip operator A projection/saturation operation that enforces bounds on a decision variable (e.g., mapping a control variable into an admissible interval such as [0, 1]). Used in constrained iterative updates.

Co-simulation Coupled execution of two or more simulators/tools (e.g., EnergyPlus™ and Python™) with data exchange for control, optimization, or learning workflows.

Consensus control A distributed coordination mechanism where agents adjust local variables to reach agreement (consensus) through neighbor interactions, often captured using graph/Laplacian structures and minimal communication.

Constraint handling Methods that enforce feasibility in optimization and control (bounds, power limits, comfort limits), using penalties, repair, projection, or primal–dual updates.

Control horizon The finite future window over which an optimization-based controller (e.g., MPC) plans control actions.

Cost function A scalar objective to be minimized/maximized (e.g., energy cost, discomfort). In multi-objective settings, multiple costs are optimized simultaneously.

Critical Peak Pricing (CPP) A tariff scheme where electricity price spikes during critical events; used as an external signal for demand response actions in DSM/DR programs.

Cyber-Physical Energy System (CPES) An integrated system combining physical energy infrastructure (buildings, grids, devices) with sensing, communication, and computation for intelligent operation.

Decision-Maker (DM) A human stakeholder (facility manager, occupant representative, operator) whose preferences may influence objective weighting, trade-off selection, or constraint priorities in decision-aware optimization.

Decision analytics Methods that structure and quantify choices under multiple criteria (energy, comfort, cost, emissions), enabling transparent selection among Pareto-optimal solutions.

Demand Response (DR) The operational arm of DSM where loads are adjusted in response to external signals (prices, incentives, grid constraints), implemented via flexible control of HVAC, lighting, EV charging, and storage.

Demand-Side Management (DSM) Coordinated strategies to shape, shift, reduce, or schedule end-use demand to improve efficiency, reliability, and economics; includes efficiency measures and flexibility orchestration across assets.

Digital Twin (DT) A digitally maintained counterpart of a physical system that is kept relevant through data integration, continuous (or periodic) updating, and service delivery such as prediction, diagnostics, and control.

Distributed control Control where computation and decision-making are spread across multiple agents/controllers rather than centralized in a single optimizer/server.

Dual variable (λ) A multiplier used in constrained optimization (primal–dual methods) to enforce global constraints (e.g., power limits). In this book, a central broadcast of λ coordinates distributed agents in i-BEMS consensus execution.

Electric Vehicle (EV) A flexible electrical load (and potentially storage) that can participate in DSM/DR through coordinated charging/discharging schedules.

EnergyPlus™ A building energy simulation engine used for physics-based modeling and evaluation. In this book, EnergyPlus™ is automated via Python™ for parametric studies and optimization workflows.

EnergyPlus™ EPW file Weather input file format used by EnergyPlus™ to drive simulations with location-specific climate data.

EnergyPlus™ IDF file Text-based input data file used by EnergyPlus™ that encodes building geometry, constructions, schedules, systems, and simulation settings.

EnergyPlus™ IDD file Input Data Dictionary used by EnergyPlus™ to define and validate IDF object structures; extending certain objects can resolve editor/parsing limitations in large models.

Energy efficiency Permanent reduction in energy use via improved envelopes, equipment, and control strategies; a foundational tier within DSM.

Energy flexibility The ability of building loads to shift or modulate consumption without unacceptable comfort/service degradation; central to DSM/DR participation.

Evolutionary Algorithm (EA) A population-based metaheuristic inspired by natural evolution, used to solve complex optimization problems with nonlinearities and multiple objectives.

Exploration–exploitation trade-off The balance between searching broadly (exploration) and refining known good regions (exploitation) in heuristic optimization and learning.

Fitness (in EA) A measure of solution quality (single or multiple objectives) used to compare and select candidate solutions during evolutionary search.

Flexible load A controllable demand component (HVAC, lighting, EV charging, storage) that can be shifted or modulated to support DSM/DR objectives.

Forecasting Prediction of exogenous signals (weather, occupancy, prices) to enable proactive control and scheduling, particularly within MPC and DT-enabled BEMS.

First-principles model A physics-based model derived from conservation laws and thermodynamics. Used to retain interpretability and support control design (e.g., AHU thermal dynamics).

Global constraint A system-wide condition (e.g., total power cap) that couples multiple agents and motivates coordination via primal–dual or consensus-based schemes.

Gradient-based update An iterative adjustment rule that moves a decision variable in the direction of decreasing cost (or increasing utility), often combined with projection to maintain feasibility.

Grey-box model A hybrid model combining simplified physics with data-driven parameter estimation, offering a practical balance between accuracy and interpretability.

Grid stability The ability of the power system to maintain reliable operation under changing demand/supply. DSM/DR participation of buildings supports stability under renewable variability.

Heating Coil Unit (HCU) A heating subsystem within an AHU plant; modeled in this book to capture heat exchange paths between water, coil, and air for control-oriented analysis.

Heuristic optimization Optimization using rule-based or metaheuristic search rather than guaranteed globally optimal mathematical programming; favored for nonlinear, multi-objective building problems.

High-Performance Computing (HPC) Parallel computing infrastructure used to accelerate large simulation workloads (e.g., repeated EnergyPlus™ runs in optimization and calibration).

i-BEMS Intelligent Building Energy Management System. In this book, i-BEMS denotes a DT-enabled and optimization-driven BEMS architecture that supports distributed, scalable execution across agents and field controllers.

Inference (ML) Producing predictions from a trained machine learning model (e.g., predicting objectives or states), often used to reduce simulation calls in surrogate-assisted optimization.

Key Performance Indicator (KPI) A measurable performance metric used to assess building operation (energy, comfort, cost, emissions, peak demand, etc.).

Laplacian (graph) A matrix representation of network connectivity used in consensus algorithms to model neighbor-based agreement dynamics.

Learning-based control Control approaches that use data-driven models or reinforcement learning to adapt policies under uncertainty and changing conditions.

Load shifting Changing the timing of energy consumption (e.g., pre-cooling, battery charging) to reduce peaks or align with renewable availability; a core DSM mechanism.

Local information exchange Communication limited to neighboring agents rather than global broadcast of full states; supports scalability and robustness in distributed i-BEMS deployment.

Machine Learning (ML) Data-driven modeling methods (e.g., neural networks, tree ensembles) used for prediction, surrogate modeling, and pattern extraction in energy systems.

Metaheuristic A high-level heuristic strategy (EA, swarm intelligence) designed to explore complex search spaces efficiently without strict assumptions on convexity or differentiability.

Model Predictive Control (MPC) An optimization-based control strategy that repeatedly solves a finite-horizon problem using a model to predict system response and apply the first control action.

Model drift Degradation of model accuracy over time due to changing equipment performance, occupancy, schedules, or sensor biases; motivates periodic recalibration in DT/BEMS.

Multi-Criteria Decision Making (MCDM) Methods that formalize and support decisions involving multiple criteria, enabling transparent selection among competing objectives in building operations.

Multi-Objective Optimization (MOO) Optimization involving two or more conflicting objectives (e.g., energy and comfort), producing a set of trade-off solutions rather than a single optimum.

Neighbor set ($\mathcal{N}_i$) The set of agents directly connected to agent i for communication/coordination in distributed consensus schemes.

Normalization (objectives) Scaling objectives into comparable ranges to improve numerical conditioning and fairness in optimization, especially when combining heterogeneous metrics (kWh, temperature deviations, cost).

Non-dominated solution A solution that is not worse in all objectives compared to any other solution; forms the basis of Pareto optimality in MOO.

Occupant comfort The perceived thermal (and broader) satisfaction of occupants; typically managed via setpoints, air quality, and controlled variability while pursuing energy efficiency.

Online optimization Optimization performed during operation (real-time or near real-time), using current measurements and forecasts to compute control decisions.

Pareto front The set (curve/surface) of non-dominated solutions representing best-achievable trade-offs among objectives in a multi-objective problem.

Peak shaving Reducing peak demand by shifting or curtailing loads; an important DSM/DR objective.

Penalty function A method to discourage infeasible solutions by adding a cost proportional to constraint violation in heuristic optimization.

Preference integration The process of incorporating DM priorities into optimization/decision-making, e.g., by interactive weighting, ranking, or MCDM-based selection.

Primal–dual method A constrained optimization approach that updates primal decision variables and dual multipliers iteratively to satisfy constraints while improving objective value.

Prosumer An entity (building or microgrid) that can both consume and produce energy (e.g., via PV) and participate in decentralized DSM/DR frameworks.

Python™–EnergyPlus™ workflow A scripting-driven automation pipeline where Python™ launches EnergyPlus™ simulations, parses outputs, and loops over scenarios for analysis, optimization, and learning.

Renewable integration The challenge of operating power systems reliably with variable generation (PV/wind); DSM/DR and building flexibility help align demand with supply variability.

Reinforcement Learning (RL) A learning framework where an agent learns a control policy by interacting with an environment to maximize long-term reward; relevant to adaptive building control.

Robustness (control) Ability to maintain acceptable performance under uncertainties (weather variability, occupancy changes, modeling errors, communication failures).

Scalability The capacity of an architecture/algorithm to maintain performance as building size, number of zones, or number of controllable assets increases; a key motivation for distributed i-BEMS designs.

Setpoint A desired operating target (e.g., temperature setpoint) that the controller attempts to track while balancing energy and comfort constraints.

Smart Grid A power grid enhanced with sensing, communication, and intelligent control, enabling advanced services such as DR, distributed energy coordination, and transactive energy.

Solar Photovoltaics (PV) A distributed generation technology frequently coordinated with storage and flexible loads for DSM (load shifting) and DR (peak support).

State variable A variable describing the system condition (temperatures, stored energy, etc.) used in modeling and control design.

Surrogate model A computationally cheap approximation (often ML-based) of an expensive simulator or evaluation function, used to accelerate optimization by reducing the number of full simulations.

Thermal dynamics Time-dependent evolution of temperatures and heat flows in building zones and HVAC components; central to comfort-aware control and DT modeling.

Time-of-Use (ToU) Electricity tariff structure with time-varying prices; provides an implicit DR signal for load shifting and cost-aware control.

Transactive energy Market-inspired coordination where energy prices and bids drive distributed scheduling and flexibility provision across prosumers and loads.

Valley filling Strategic increase/shift of consumption during low-demand periods to smooth demand profiles and better utilize generation; part of DSM load-shaping concepts.

Weather-driven uncertainty Uncertainty in building loads and comfort dynamics caused by stochastic weather variations, motivating predictive and robust control strategies.

White-box model A fully physics-based, interpretable model derived from first principles; generally more transparent but can be expensive to develop/calibrate for complex buildings.

Zone A thermal region in a building represented as a control volume in modeling/simulation (e.g., EnergyPlus™ zone), often used as the basic unit for control and optimization.

Index

M. H. Wani et al., *Digital Twins for Building Optimization: Enhancing Energy and Comfort*, Studies in Infrastructure and Control,
https://doi.org/10.1007/978-3-032-23274-8

M

N

O

P

R

S

T

W

The manufacturer's authorised representative in the EU is Springer Nature Customer Service Centre GmbH, Europaplatz 3, 69115 Heidelberg, Germany. If you have any concerns regarding our products, please contact ProductSafety@springernature.com

Printed and bound by CPI Group (UK) Ltd, Croydon, CR0 4YY
05/10/2026
02228057-0008